PROGRAMMING WITH DATA

Springer
New York
Berlin
Heidelberg
Barcelona
Budapest
Hong Kong
London
Milan
Paris
Singapore
Tokyo

John M. Chambers

PROGRAMMING WITH DATA

A Guide to the S Language

John M. Chambers
Bell Laboratories, Lucent Technologies
700 Mountain Avenue
Murray Hill, NJ 07974-0636
USA

With 4 figures.

Library of Congress Cataloging-in-Publication Data
Chambers, John M.
Programming with data: a guide to the S language / John M. Chambers.
p. cm.
Includes bibliographical references.
ISBN 0-387-98503-4 (pbk. : alk. paper)
1. S (Computer program language) 2. Statistics—Data processing.
I. Title.
QA76.9.A25F77 1998
519.5′0285′5133—dc21 98-13049

Printed on acid-free paper.

Production managed by Victoria Evarretta; manufacturing supervised by Thomas King.
Photocomposed pages prepared from the author's Postscript files.
Printed and bound by Hamilton Printing Co., Rensselaer, NY.
Printed in the United States of America.

9 8 7 6 5 4 3 2 1

ISBN 0-387-98503-4 Springer-Verlag New York Berlin Heidelberg SPIN 10673083

Preface

S is a programming language and environment for all kinds of computing involving data. It has a simple goal:

To turn ideas into software, quickly and faithfully

Any application involving data is suitable for S, particularly if some of your ideas involve the structure and meaning of the data.

You use S interactively, giving it tasks, looking at data, and creating objects that describe your projects. S can be, and is, used in a "non-programming" style, exploiting quick interaction and graphics to look at data. This use often leads to a desire to customize what you are doing, and S encourages you to slide into programming, perhaps without noticing.

This book is a guide to what happens next: to programming with S, starting from that first simple step and carrying on through increasingly ambitious stages, as your programming becomes more useful to you and to others.

The first three chapters of this book survey programming in S, emphasizing three different aspects:

1. *highlights* of the main techniques with examples, including an extended example in section 1.7.

2. the essential *concepts* underlying all S programming;

3. a *quick reference* to many of the tools and techniques in S.

The first chapter is for those who want to see typical examples of how to use the language. The second chapter is for those who like to understand the main ideas before getting into details. The third chapter is mostly tables; it and the index of the book should work together as a resource to learn more about what's in S as you go along. You can read all three chapters, or switch back and forth.

The remaining chapters get into more detail on various aspects of programming in S. Chapters 4 and 5 discuss computations and the objects in S, describing what's already available to use as building blocks in programming. Starting with Chapter 6, we discuss programming itself, first in general and then in terms of mechanisms for dealing with classes (Chapter 7), methods (Chapter 8), documentation (Chapter 9), connections (Chapter 10), interfaces (Chapter 11), and other tools. Two appendices deal with more specialized topics: programming in C with S; and compatibility with earlier versions of S.

Besides the book itself, there is also a large amount of online documentation. You should get quickly into the habit of typing `?` followed by a topic name, or using the help facility in a graphical interface, whenever you want to know more about a function or some other topic. The web site

```
http://cm.bell-labs.com/stat/Sbook
```

acts as an extension to the book. Look there for tools related to S, developments more recent than the printed version of the book, and pointers to other sources of information.

This book does not assume you have used S before, but it does assume you are interested in programming, in creating new software with S. If you have never used S before, I would recommend familiarizing yourself with the basics of computing with S in an interactive, non-programming mode. If you've purchased a copy of S-Plus, it will include documentation on the use of the system. You can learn the basics from the S-Plus manuals, plus the online documentation facilities. There are a number of books on S and S-Plus, some listed in the references (including books by Krause and Olson [6], Spector [8], and Venables and Ripley [10]). Some time spent with this material would make good preparation for the programming described in this book, if you can be patient enough to postpone those first programming steps for a while.

S-Plus has many sets of functions and optional extensions for various applications, beyond what is described in this book. To find more about S-Plus, look at the web site, currently

```
http://www.mathsoft.com/splus.html
```

In particular, if your system does not seem to have a copy of S or S-Plus, this is where to find out more. The version of S described in this book underlies S-Plus versions 5.0 and higher. The S-Plus libraries may extend or redefine some of the functions described in this book, as well as providing many other facilities. Make use of the online documentation to check, whenever you're unsure.

There are a wide variety of libraries and functions developed by users, datasets of general interest and other useful adjuncts to programming in S. The `statlib` archive is the most comprehensive, as well as having many links to other sources of software: look at the web site

```
http://lib.stat.cmu.edu/S/
```

This book discusses what underlies and is common to *all* this software: the S language and programming environment. If you have learned to program in earlier versions of S or S-Plus, you can use what you already know. You will find many extensions and new features, described throughout this book. These include new approaches to classes and methods, to documentation, to handling large objects, and to scheduling events.

There is a great deal of detailed information to dig into. Through it all, though, keep foremost in mind that fundamental goal: to turn ideas into software.

About the Cover

The images on the cover of this book were produced by the S-Wafers software, written by Mark Hansen and David James, and described in reference [4]. S-Wafers is a powerful and successful system for visualizing and modeling wafer test data (data from testing electronic chips and other devices). This application, used in Chapter 1 of the book, illustrates the style and power of programming with data. The front cover shows (top to bottom): the actual wafer, a plot of test results, and a mathematical model of spatial effects. The three images symbolize the path we often want to follow in programming with data. Starting from the original application and our ideas about it, programming with data enables us to organize and visualize the data in a direct way. Often, and certainly in this application, the ability to summarize and display the data conveniently proves to be of great value in practice. As our understanding grows, new and perhaps deeper ideas will be implemented, using quantitative models and other advanced techniques. Growing understanding leads to new questions and new challenges to computing with the data. The back cover of the book shows results from an experiment that varied two parameters in the manufacturing; the display shows a two-way table of the results, but a table not of numbers but of the results on whole wafers, condensing many numbers into each visual symbol.

Acknowledgments

It would be impossible to cite more than a small fraction of the contributions to S over the years. Starting with groups of people, the first debt is to colleagues past and present at Bell Labs, not only for so many specific ideas, but even more for a continually stimulating environment, probably never more so than now. At perhaps the opposite scale, from local to worldwide, the community of S users are the real owners of the language, and they have shaped it from the start by choosing how to use it. There is no substitute for a demanding user community. Books and other writing about S, in general and for particular applications, have helped shape the language; special mention here for the contributions of Bill Venables and Brian Ripley. Many colleagues contributed to earlier versions: Rick Becker, Allan Wilks, and Trevor Hastie all own important parts of S. In the evolution of the current version of S, the community of beta test users both within Lucent Technologies and outside have been of great help. The continuing productive relationship with MathSoft has contributed much to the joint growth of S and S-Plus.

The contributions from my colleagues at Bell Labs have been essential for the book and for S: Duncan Temple Lang's ideas have renewed the research and pointed to the future; Mark Hansen and David James, with S-Wafers, have provided a paradigm for S software, and have each added many important insights as well. The current approach to documentation needs special mention: Duncan Temple Lang's ideas and software made possible the SGML-based documentation, one of the most exciting recent directions in S; David James also provided essential contributions to this work. Major contributions as well have come from: Stephen Pope, an ideal beta tester; Kishore Singhal, with many years of pioneering use of S; Rafal Kustra, for work on the internals of S; Bill Cleveland and Stu Blank, for making the practical side of S a success. The editorial partnership with John Kimmel, over many years and many books, continues to be a pleasure. Valuable suggestions and comments came from Doug Bates, Linda Clark, Lorraine Denby, Bill Dunlap, Steve Golowich, Richard Heiberger, Sylvia Isler, Diane Lambert, Clive Loader, Maria Peman, José Pinheiro, Don Sun, Terry Therneau, Luke Tierney, and Scott Vander Wiel.

John Chambers
Bell Labs, April 1998

Contents

Chapter 1

Highlights

This chapter introduces the S language and programming environment, by presenting some of the most common and useful features. The chapter begins and ends with an example, presented in some detail. The sections in between introduce expressions, data, functions, classes and methods: the essential ingredients to programming with data in S. You can read the chapter straight through, but since it is fairly long, you might prefer to skip the example, by starting with section 1.2 on page 6. On the other hand, you can follow the example alone by reading section 1.1 and then section 1.7.

1.1 Computing with S

S specializes in *computing with data*: any application with interesting requirements for organizing, analyzing, or presenting data is a candidate. Much useful computation can be done with data using graphical or menu interfaces or other non-programming approaches; S can be used in this way and often is. The current S-Plus system provides a graphical interface to the S language. The user can select an object (a dataset) of interest and also select one or more tools to display the object, summarize it, or perform more elaborate computations such as fitting some form of model.

This book is about *programming with data*, meaning that we want to extend the tools available in some way, to *program* the system to implement some ideas we have. This is where the S language comes in: S aims to make the transition into programming easy, but to allow you to do as serious a job of programming as the application requires or as your time permits. This transition is easier because in fact the graphical interface really is an interface—behind it is the S language and programming environment. To start programming we just examine the expressions in the language that correspond to typical tools, and go on to modify these.

Because our focus is on programming, we present the tools in their language form, which also allows the discussion to be independent of different possible user interfaces. Every application will use S and the tools implemented in it differently, of course. However, what S does for you is likely to be similar. You can expect to use S in three main ways. It provides *organization* for your data, particularly by providing useful classes of objects and techniques for storing the objects in S databases. The existing S *tools* provide a wide range of visualization, computation, and modeling; in many applications these will include software specialized to the data you will encounter. Finally, S provides *programming*, a language and environment to turn ideas into new tools.

To make our discussion more real, let's introduce an actual application. The application is to data produced from modern manufacturing, specifically the manufacture of integrated circuits. All the processor and memory chips in the computer you are using, as well as countless similar devices hidden in the telephone system, automobiles, household appliances and elsewhere, represent a remarkable evolution of design and manufacture. Great reductions in size and power consumption of electronic devices, with equally significant increases in speed and capacity, require a manufacturing process of correspondingly increased complexity and precision.

Electronic devices (chips) are manufactured on *wafers*; as the name suggests, these are flat discs, though distinctly inedible. On the disc, a number of devices will be manufactured simultaneously, anywhere from a few devices to hundreds. Manufacturing here means many steps of depositing and etching layers of semiconductor material. The end result looks like Figure 1.1: rectangular chips laid out on the wafer.

To monitor and guide the manufacturing process, large quantities of data are generated during the many stages of manufacture. At the end of the process, the devices are subjected to a variety of tests to measure their performance before they can be shipped to users. Techniques for computing with such data, for visualizing, summarizing, and modeling the data, have scored

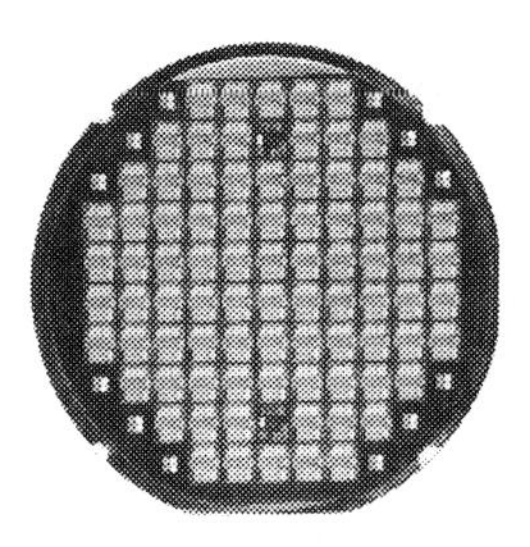

Figure 1.1: *What your computer's components look like at birth: a wafer containing devices (e.g, memory or processors). These are the larger light rectangles arranged in rows and columns. Smaller rectangles are other, special devices.*

some notable successes in improving our understanding of the underlying process. Software programmed in S has played an important role; we will use some of the ideas incorporated in that software to illustrate programming with data in S. In section 1.7 we will construct some actual software for this application. Here, we're considering the general ideas and how they relate to programming with data.

Let's start with one class of data. When the wafer comes off its "assembly line", the manufacturer needs to test whether the chips perform up to specifications. This is the data we want to use as an example. A complex automated machine, called the probe tester, attaches itself to the terminals on the chips and performs a programmed sequence of tests. In the specific kind of data we're considering, the result of the tests is a single character, representing the state of the chip. There will be one or more *ok* states and usually several failure states, indicating the stage at which the device failed. The *ok* devices will go on to be removed from the wafer and shipped (or at least put through further tests); the failed chips will be discarded.

Needless to say, the results of the test data get lots of attention from the people responsible for manufacturing the devices. The overall *yield*, the fraction of *ok* devices, is a critical quantity in measuring how well the manufacturing process is working. A long-running or relatively simple process is likely to produce high yields and (depending on your viewpoint) boring test data. But the constant pressure to improve and modify the devices means that new designs and manufacturing techniques are continually being introduced. Understanding the test data for these situations is critical. Remember that the process is extremely complex, and the data is often voluminous. Each wafer can have many devices; wafers are processed in batches, or *lots*, of as many as fifty wafers, and large manufacturing lines will be processing many lots simultaneously.

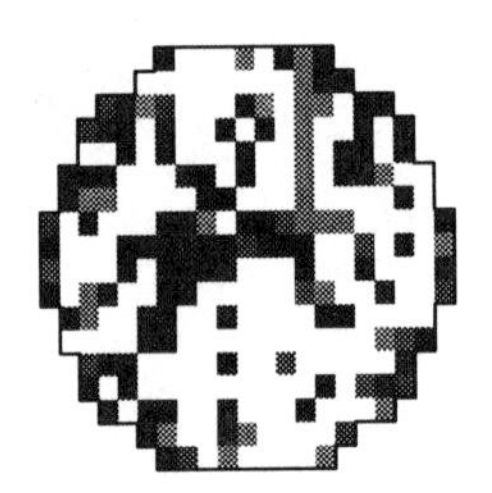

Figure 1.2: *Graphical display of the* `probeTest` *data. White (i.e., invisible) means no problem at that site; the various colors (or grey levels here) indicate the different failure modes.*

One of the first, and perhaps the most important, of the contributions of computing with such data was just to provide a plot, such as Figure 1.2, where the test result on each device is color coded. The importance of this plot is that it can present an enormous amount of information compactly, in a form that the viewer can understand easily. Many wafers can be presented simultaneously, perhaps in terms of other information (e.g., Figure 1.3). There may be some 10,000 test results in such a figure, but the eye has no trouble detecting patterns in the results.

The software implementing such plots allows the user to select interactively a variety of viewing modes: to view all or only one failure mode; to view the individual wafers or a composite representing the whole lot. With a web browser, you can look at this kind of display yourself, on our web site at

```
http://cm.bell-labs.com/stat/project/icmanuf
```

So where did programming with data, and in particular with S, come into this? S was used to organize the test data into objects representing the various data collected for wafers as they were manufactured. An S object named `probe48998`, for example, might contain the probe test data for the lot tagged as `48998`. The class of `probe48998` would automatically identify it as test data and the object would contain the information necessary to make the plots shown above. The plots themselves exist as tools in the form of S functions; more specifically, as methods for the S function `plot`. A plot like Figure 1.2 for each wafer could be produced by the S expression

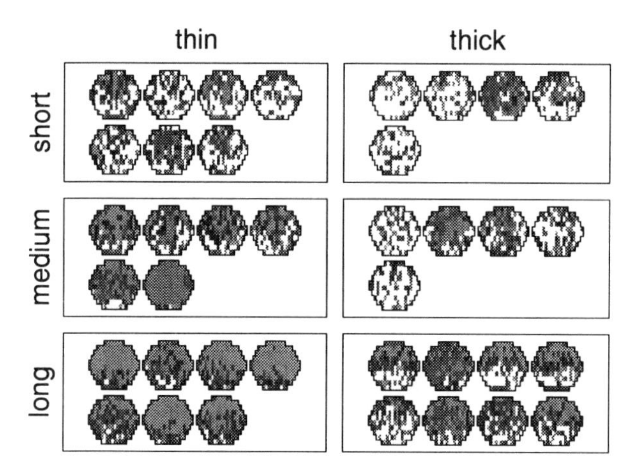

Figure 1.3: *Plots of* `probeTest` *objects from wafers manufactured under different experimental conditions. Each plot represents the "response" as measured by probe testing.*

```
plot(probe48998)
```

The plotting tools (and many others) were implemented by Mark Hansen and David James as a package called *S-Wafers*; Lucent Technologies engineers and others needing the tools could use them through S directly or through a graphical user interface. A paper by Hansen and James (reference [4]) describes S-Wafers; in section 1.7 we will recreate similar tools, to show how S programming with data can build up such a facility. For your own applications, the specific classes of data and the functions and methods needed to work with the data will differ, but the style and approach described in this example usually carry over.

To get a general introduction to programming with data before getting back to the example, continue to section 1.2 on the next page. If you want to continue the example immediately, jump ahead to page 44.

1.2 Getting Started

If you have never used S or S-Plus before, here are a few steps that should get you going. You should pick a directory in which you are going to play with the language. Since S creates files and directories to store objects, life will likely be simpler if you create a directory that is *just* for use with S, at least to start with. The recommended approach is to create a subdirectory, `S`, in your login directory. In a shell, execute the commands:

```
mkdir $HOME/S
cd $HOME/S
Splus CHAPTER
```

The last command sets the directory up for use with S. Now you can just type to the shell the command that invokes the version of S you are using. With S-Plus for example, you would type

```
Splus
```

This should start a session; S will type a banner with some information about the version and that sort of stuff, and then prompt you with an angle-bracket, `"> "`, to type expressions. With other user interfaces, you may start the session some other way, such as clicking on an icon.

By the way, you don't have to create the `S` directory, if you would prefer to use some other directory. You can run S in any directory that has been set up as an S chapter, and S will use the current directory to work in. There are several ways to tell S where to work, and if they all fail, S will create a temporary chapter for you. See section 4.1.2 or the online documentation `?Session`.

The main glitch in the prescription for making a chapter will be if your attempt to use the `CHAPTER` command gets the response that the command is not found. If so, the command itself is not in your path; check it out with friends, system administrators, or the shell command `locate`, if your system has that. If the command isn't in any reasonable place, your computer doesn't likely have S or S-Plus installed. Back to the preface of the book for getting a copy of the system.

Assuming you do get to run the `CHAPTER` command in `$HOME/S`, you can then run S from anywhere, though it's probably less confusing to run in `$HOME/S` in case you create ordinary files and hope to find them again later. There are many other ways to run S, often more elegant and user-friendly than the standard approach described above. I'm a fan of the `emacs` system, so I often run S through an interface in `emacs`. Other graphical user interfaces exist and more are being developed all the time. Use one that you find convenient or that works well in your environment. The descriptions in this book should apply to most any interface; there may be additional tools in a particular interface as well.

Once the session starts, S expects any sort of legal expression as a task: the S evaluator will parse and evaluate what you typed and, usually, print or plot some information in response. The S language should look familiar, if you have used languages in the style of C, C++, or Java. The following are legal S expressions:

```
x - avg
x[i] = -y
if(any(data < 0)) data = exp(data)
while(!converged)
 { model = refit(model); converged = state(model)}
```

The rest of this chapter provides a start on programming with S expressions and functions.

1.3 Using S Functions

S is a *functional* language. S expressions contain function calls, such as

```
summary(budget)
```

This expression gives S a task, to call the function named `summary`. S will construct a call to that function, giving it as an argument the name `budget`. An S function call returns an S object, the value of the call. Usually that's all it does: function calls are evaluated to get the value. When the function call itself makes up the user's whole task, the evaluation of the call completes the task. For the standard user interfaces, S then automatically shows the value of the call.

Time for an example. For the expression `summary(budget)`, suppose we have an object `budget` in our database, containing some typical information from household accounts: checks paid out with their check numbers, plus deposits and other withdrawals. When the S evaluator encounters the name of an object, it takes that as a shorthand request to get the object from the database. And if nothing else is done with the result, S will show the object to the user, by printing or plotting it. So about the simplest useful S expression is just the name of an object, say `budget`.

```
> budget
   Transaction    Category  Amount Date
1         1439 Credit Card  -50.56    1
2         1440     Service  -22.49    1
3         1441     Service  -13.42    1
4           NA  Withdrawal -100.00    2
5           NA     Deposit  652.50    2
6         1442   Telephone   55.13    2
7         1443     Service  -18.00    2
8         1444        Misc  -45.00    3
9           NA    Interest   33.50    3
10        1445         Tax -368.00   14
```

(In our examples, the lines typed by the user begin with the prompt `"> "`. Other lines are output, printed by S.)

Suppose we want to examine this data. We will use functions `summary` and `plot`.

```
> summary(budget)
  Transaction         Category          Amount              Date
 Min.    :1439   Service    :3   Min.    :-368.00   Min.    : 1.0
 1st Qu.:1440   Withdrawal:1   1st Qu.: -50.56   1st Qu.: 1.0
 Median :1442   Telephone :1   Median : -20.24   Median : 2.0
 Mean    :1442   Tax        :1   Mean    :  12.37   Mean    : 3.1
 3rd Qu.:1444   Misc       :1   3rd Qu.:  33.50   3rd Qu.: 3.0
 Max.    :1445   Interest   :1   Max.    : 652.50   Max.    :14.0
 NA's    :   3   (Other)    :2
> plot(budget$Date, budget$Amount)
```

The call to `summary` in the first expression returns a table that has the same column names as `budget`, with each column summarizing the corresponding data in `budget`. We didn't do anything with the result, so it was automatically printed. The second expression produces a scatter plot (not shown here) of the `Date` and `Amount` components of the data.

1.3.1 Arguments to Functions

Calls to functions can have any number of arguments, and the arguments can be any S expression.

```
today = date()
fit2 = update(fit, new[date == today, ] )
```

Each function is defined with some set of *formal arguments.* The arguments supplied in the call are matched to those in the definition. In the call to `update`, `fit` would be matched to the first argument and the expression `new[date == today, ]` to the second.

Not all the arguments in the definition need to be supplied in the call. The function can detect missing arguments. There may be a default expression included in the definition, or the function may do something else, or it may be that for this particular call the argument wasn't needed anyway. The user and the programmer have essentially complete flexibility in treating arguments. Trailing arguments can simply be left out; for example, `update` might have more than two arguments, but in this call they are all missing.

Optional arguments can conveniently be included by *name*: in the call we set the name of the argument to an S expression. For example:

```
history("budget", max = 2, function = "bshow")
```

If you don't remember the names of the formal arguments, the online documentation function `args` tells you:

```
> args(history)
history(pattern=".", max=10,
    evaluate, call, menu, graphics=TRUE,
    where=audit.file(), editor,
    function, file)
```

The documentation also gives the default values for arguments that have them. In this example the formal arguments `pattern`, `max`, and `function` match the expressions given and all other arguments are missing.

Named arguments are particularly helpful for functions that have many optional arguments used to fine-tune the way the function works: parameters for numeric computations, options for how the function should use data, what to do in special circumstances, and the like. As you can see, `history` is one of these functions. It's a common tendency for software in any language to accumulate these little features, even though the practice is sometimes deplored. S makes the habit a bit more bearable by allowing the user to omit irrelevant arguments and by allowing those arguments supplied to be given by name. S will complete argument names, so long as you give enough of the name to make it unique (and special user interfaces for S often provide additional help); in this example, two characters would have been enough for each argument.

The `history` function is a typical multipurpose function with many arguments. It goes through the expressions for recent tasks, and either returns some of these expressions (unevaluated) or does something specific with them. There are two frequently useful applications: to re-evaluate one of the tasks or to turn some expressions into a function definition. The optional arguments allow users to filter the expressions for particular strings, say how far back the search should go, and control what is done with the expressions found. The example above uses the argument `function` to create an S function from, in this case, the last 2 expressions containing the pattern `"budget"`.

```
> history("budget", max = 2, function = "bshow")
Function object bshow defined
and saved on file bshow.S
```

(Turn to page 21 to see what to do next.)

1.3.2 Arithmetic and Other Operators

S has the usual operators for arithmetic, comparisons, and logical operations (plus some other operators special to S).

```
residuals(fit) + offset
counts/(1 + rowMean(counts))
amount < 0
```

Operators look different from other functions because they appear in expressions in infix (scientific) notation. When we come to program with such operators, though, we will treat them just like ordinary functions, which is what they are. We put the operator `"<"` between two arguments instead in front of its arguments, but this is just because the interface we are using expects to parse such expressions. Most users prefer to see arithmetic expressions written this way, but in fact the S expression

```
"<"(amount, 0)
```

evaluates exactly the same as `amount < 0`.

To see all the standard S operators, use the online documentation; for example:

```
?Arithmetic; ?Logic
```

for the arithmetic and logical operators. S has a few other operators not found in most languages. The operator ":" creates a sequence in steps of ± 1 from its left operand to its right operand.

```
> 1:3
[1] 1 2 3
```

Operator "%%" computes modulus. Some special matrix operators do computations in numerical linear algebra, such as matrix multiplication.

Additional operators in S extract pieces of objects. Square brackets can be used to extract the portion of the object before "[" that corresponds to the expression between the square brackets:

```
amount[1:3]
amount[ amount < 0 ]
```

The first expression produces the first 3 elements of `amount`, the second the subset of elements corresponding to true values in the logical expression. Square brackets can be extended to apply to matrices, multiway arrays, and other similar objects. Multiple arguments between the square brackets refer to rows and columns, etc:

```
budget[amount < 0, "Date"]
```

This selects rows satisfying the logical expression and columns corresponding to the character string name given.

Other operators extract single elements, components of lists, or the slots of an object; see Chapter 3 or the online documentation, by typing `?"[["`, `?"$"`, or `?"@"`.

Assignment, written "=" in our examples, is another important S operator. S uses assignment in a very general way, both to assign an object corresponding to a name and also to modify an object in an open-ended variety of ways. If we give S an assignment task, such as:

```
budget = read.table("budget.data")
```

S will evaluate the right side of the assignment expression and save the result in a database, called in S the *working data*, under the name on the left side. Then the object `budget` is available for future tasks, whenever we are working with this database. Assignment operators are used to modify existing objects, when the expression on the left of the assignment is a function call instead of a name. We call these expressions *replacements*.

```
budget[8, 3] = -45.00
```

As you would expect, this expression replaces the `[8, 3]` entry of `budget` with the value on the right side. Replacements in this style are familiar from most languages, but in S replacements are much more general. Any function can appear on the left of the assignment if there is some interpretation of how the object should be modified, using the value on the right side of the assignment. For example, evaluating the expression

```
length(x1) = 14
```

means that the length of the object is set to the value `14` and the result re-assigned in place of the previous `x1`. We will take up assignments again on page 14.

1.4 Data, Objects, and Databases

S focuses on programming with data, emphasizing the ability to customize your view of data to match your programming needs. Data in S comes in the form of *objects*. *Everything is an object*: this is a fundamental concept in S. In many ways all objects are treated equally. All objects have a *class*, allowing computations on the objects to be customized. There are dozens of classes provided with S and defining new ones is very much part of the programming style. The behavior of classes can be customized by defining *methods* for S functions when they encounter arguments from particular classes.

S maintains *databases* for objects, allowing users to store and retrieve objects (any objects) by name. All the computations you do in S, from interactive analysis and visualization to the full extent of programming, use and create objects. By attaching databases, users get access to S objects and to libraries of S functions.

1.4.1 Objects and Classes

All S objects have a class, a character string that defines what the object is.

```
> class(x1); class(xm); class(class)
[1] "numeric"
[1] "matrix"
[1] "function"
```

An object's class defines the object in several ways.

First, it can define the *method* used by a function when the object appears as an argument. Whether we're plotting the object, using it in arithmetic, or extracting data from it, the computations can be tailored to the particular object by a method defined for that class.

Second, the *representation* of the object is defined by its class. A method using the class can count on finding certain information in the object, because the class representation is defined and accessible. The simplest classes contain *atomic* data—numbers of various kinds, logical (T or F) values, or character strings.

```
> class(x1)
[1] "numeric"
> x1 > 0
 [1] T T T T T T F T T T F T T T
> class(x1 > 0)
[1] "logical"
```

New classes are built up from existing ones, most often by including several *slots* in the representation, each slot containing some simpler class that defines part of the information in the new class. Slots have names, and programmers can extract or replace information by referring to the slots, using the operator "@". When we come to discuss defining new classes in section 1.6, we'll use slots to do so.

The third major contribution of classes comes from relations between classes. When one class includes all the behavior of another we say it *extends* the other class. Class extension helps greatly in defining new classes: all the methods defined for the earlier class are taken over by the new class, with no reprogramming.

One kind of extension is especially basic to the language. A *virtual* class exists only so other classes can extend it; it groups together classes that share some important behavior. For example, all the atomic classes mentioned above share the essential notion of having some data values, their elements, which can be extracted and replaced by referring to their *index*, their position in the object. S calls these objects *vectors*, and the virtual class `vector` exists so all vector classes can extend it. The numeric object `x1` is an example: because it's a vector we know that the expression `x1[1:5]` will return a vector of the first 5 elements, and `length(x1)` will return the number of elements.

When we come to define methods, we will see that `vector` and other virtual classes are enormously helpful: we can write methods for the virtual class that exploit the common behavior, and have the method apply to all the actual classes automatically. New classes can be defined extending the virtual class; they too automatically inherit all the methods.

1.4.2 Assignments to Databases

S maintains databases for users. An S database contains S objects, each associated with a name. The objects can be anything at all: datasets, functions, whatever information we need to keep and reuse (everything is an object, remember?). To give S a task that says "put this object in the database, with this name", we supply an assignment expression, with the name on the left and the object to assign on the right. This looks very familiar from most any programming language.

```
lx1 = log(x1 - min(x1) + 1)
```

As in other languages, the value of the expression on the right can now be retrieved and used by supplying the name `lx1`.

What happens to implement the assignment, however, is simpler from the user's point of view and more general than in most languages. The programmer does not need to make the assignment legal, by declarations or by making sure that the data being assigned is consistent with previous use of the name. When the assignment is supplied as a task for S to do, the object is saved on the *working database*, typically either the current directory for this S session or the user's home directory. This database is maintained throughout the S session and from one session to the next. Any object can be assigned to any name.

The expression on the left of the assignment operator can be more than just a name:

```
x[i] = 0
length(x) = max(10, length(y))
```

S assignment expressions with function calls on the left are called *replacements*. The model for them is simple but powerful. If the function on the left has a name as its first argument, S replaces the named object by something new. The new object is the value of a call to the *replacement function* corresponding to the name of the function on the left. The first argument to the replacement function is the original object and the last argument is the expression on the right side of the assignment. The name of the replacement function by convention is the name of the function on the left, concatenated with "<-" (another assignment operator). The second replacement expression above is equivalent to:

```
x = "length<-"(x, max(10, length(y)))
```

The generality of replacement expressions may take some getting used to, but it is a powerful programming technique.

1.4.3 Computing with Databases

Databases are collections of objects, each associated with a name. You always have a database, the *working* data, available when you're computing with S. The `CHAPTER` command created an empty working data, and assignment expressions you typed to S will have created some objects in the database. Other than assigning and maybe modifying the objects, you are not expected to manage them: S manages arbitrary objects in the database.

There are other databases always available; in particular one or more libraries of S functions, supplied with the language. These are not different in any essential way from your own database, except that you won't normally be assigning into them. When you refer to an object by name, whether it's a function or an object containing data, S *searches* for the object in the databases currently attached.

The function `search` returns the names of these databases:

```
> search()
[1] "."        "models" "main"
```

The first name on the list is always the working database: the name "." means the current directory. The other two names are, in this case, two S libraries.

You can get the names of all the objects on a database from the function `objects`. Give it the database you're interested in, either by its name or by its position in the `search()` output. By default, `objects` takes the working data.

```
> objects()
[1] ".Last.value" "budget"      "budget2"     "fit"
[5] "today"
```

When we typed the expression `summary(budget)`, S had to find both the function `summary` and the object `budget`. The rule is always that S looks in order in the databases in the search list. From the output of `objects()`, we can see that S will find budget on the working data, but not summary; that must be on one of the S libraries.

```
> find("summary")
[1] "models"
```

There are many tools for examining S databases; see the tables in Chapter 3. For most computations, however, it's enough to know that the data we assign is put into the working data.

If other databases have objects we need (either functions or data), we can add these to the search list by calling library or attach.

```
> library(data)
> search()
[1] "."       "models" "data"   "main"
```

Now the objects in the S library "data" are also available.

The function objects is a useful way to search in the very large collections of functions found on the libraries. Several optional arguments to objects let you restrict your search in any way you can imagine. The argument pattern returns only objects whose names match the argument. So, if we wondered about functions dealing with matrix objects, say, one step would be to look for objects with "matrix" in their name:

```
> objects("main", pattern = "matrix")
[1] "as.matrix"          "as.matrix.default"
[3] "is.matrix"          "matrix"
[5] "print.matrix"       "prmatrix"
[7] "smatrix"            "tsmatrix"
```

Other arguments restrict the class of object or apply an arbitrary test to each.

1.4.4 Getting Data into S; Connections

Most interesting programming will deal with data that originated in some "outside" application or process. The application will have left some data around somewhere, say on a file. We need to make a connection from the file to S in order to read the data in and work with it. S provides an unlimited variety of ways to do this, both through existing functions and through tools that let you design your own functions.

The function scan reads data items, as ordinary text, and interprets them as data for S.

```
> trx = scan()
1: 348 325 333 345 325 334 334 332
9: 347 335 340 337 323 327
15:
> trx
 [1] 348 325 333 345 325 334 334 332 347 335 340 337
[13] 323 327
```

When called with no arguments, `scan` reads from standard input (keyboard input), prompting with the index of the next item to be read. An empty line, or an `EOF` signal, terminates input. The `scan` function has a number of arguments, but the first two are the most important ones. The first argument tells `scan` to read from a file, rather than the keyboard. Most often, the argument is the name of a file in the file system, though in fact it can be any *connection* defining a connection between S and external sources of data—more about that later in this section. The second argument, `what`, defines what class of object the user wants `scan` to look for. If `what` is, say, any object of class `character`, then `scan` will interpret the data items as elements of a character vector. The possibilities are unlimited, because methods can be defined for `scan` to read data of any class at all.

The method for an ordinary S list, for example, expects to get items successively to add to each of the elements of the list supplied as `what`. Suppose we want to create a list having two elements named `x` and `y`, each being numeric vectors. We will read this in from a file. The file needs to have the first `x` value, then the first `y` value, then the second `x` value, and so on. We will be less likely to make errors if the file has two items per line, though `scan` doesn't care about that. Suppose the file is `"xyData"`, and the first few lines are

```
156 348
182 325
211 333
212 345
218 325
220 334
246 334
```

Then `scan` can read the data into the list we want.

```
> xy = scan("xyData", list(x = numeric(), y=numeric()))
> xy
$x:
 [1] 156 182 211 212 218 220 246 247 251 252 254 258
[13] 261 263

$y:
 [1] 348 325 333 345 325 334 334 332 347 335 340 337
[13] 323 327
```

There are other methods for scan, and some additional arguments. The most important of the other arguments is n, which controls the number of items to be read, if we don't want to read to the end of the file.

As we mentioned, providing the character string name of a file is only a special case. Any S connection object would do instead. Connection objects represent all the kinds of things in your computing environment from which S can read data, or to which S can write data. The connection objects can be manipulated in a variety of ways in S, when you want more control over how data is transferred.

The most important way to manipulate a connection is to open it. We didn't open "xyData", so scan opened it for us, and then closed it again before returning. Another call to scan with the same file would start reading again at the beginning of the file. Fine for this example, but in some cases we might want to read pieces of the data from the file, look at those, and then decide what to read next.

Suppose that the same data was written on another file, "xy2", in a different style. First the file contains the number of elements to expect in each of x and y (14 in this case), then all the x data, then all the y data.

The essential trick to scanning sequentially is to open the file first. Then, each successive call to scan will leave the connection open and pick up reading where the last call left off. The first time, we'll just read one integer number, say nxy. Next we'll read the x data, then the y data.

```
> xyfile = open("xy2")
> nxy = scan(xyfile, integer(), n=1)
> x = scan(xyfile, numeric(), n = nxy)
> y = scan(xyfile, numeric(), n = nxy)
> xy2 = list(x = x, y = y)
> xy2
$x:
```

```
 [1] 156 182 211 212 218 220 246 247 251 252 254 258
[13] 261 263

$y:
 [1] 348 325 333 345 325 334 334 332 347 335 340 337
[13] 323 327

> close(xyfile)
[1] T
```

As a final step, we explicitly closed the connection. This isn't required but it is good housekeeping, freeing up operating system resources and preventing any accidental use of the connection later on.

1.5 Writing S Functions

Programming in S begins for real when you start writing S functions. S tries to make it easy to get started with the process. A function can be constructed from the expressions you have already given as tasks to S, or it can be the result of wanting to change an existing function. Often, you find yourself writing the same, or similar, expressions over and over. It makes sense then to package these as a function.

1.5.1 Creating a Function Object

You can create a function by just typing in an expression that defines and assigns a `function` object. The idea is simple: take an expression you would use to compute something interesting, precede the expression by `function` followed by the parenthesized names of the objects involved, and that expression defines a function. Take the expression `log(x - min(x) + 1)` we used on page 14. To turn it into a function:

```
> logtrans = function(x) log(x - min(x) + 1)
```

Like any assignment expression, this tells S to evaluate the expression on the right of the "=" and associate it with the name on the left. The difference in this case is that the expression on the right, when it is parsed, defines a *function object*. The parser converts the reserved word `function`, followed by a parenthesized list of arguments, followed by an S expression into a function object. The expression is called the *body* of the function object.

A function definition can appear anywhere you want in an S expression. Most often, it appears on the right of an assignment, as in this example, which associates this function object with the name `logtrans`.

After the assignment, the S evaluator can take a call of the form:

```
logtrans(x1)
```

and evaluate it, using the same rules we discussed in section 1.3. Indeed, there is *no* difference between a function created by a user's program and one supplied with S.

Once the computations get a little more complicated, providing an in-line function definition as we did here becomes clumsy, or even impossible. Often, we need the function body to contain several expressions, including assignments. The technique is to enclose these expressions in braces and separate them by semicolons or new lines. As in many languages, the expressions in the braced list are evaluated one after another. In S, the key point is that the value of the whole braced list is the value of the last expression. Functions like this quickly get too complicated to type straight off. But perhaps we can manage something fairly simple. Suppose we want to turn the following sequence of tasks into a function:

```
> a = min(x1)
> b = max(x1) - a
> xx = (x1 - a)/b
```

The computation has the effect of shifting the numbers in `x1` to the interval `0` to `1`. Let's type the same sequence of expressions, but this time as the body of a function:

```
> shift = function(x) {
+ a = min(x)
+ b = max(x) - a
+ (x-a)/b
+}
```

The S parser keeps prompting for more input until it has a syntactically complete expression. The change made from the three tasks to the three expressions is typical: we replaced the third line by an expression without an assignment. This expression becomes the value of a call to the function.

The other key difference between tasks and function definitions is that ordinary assignments in the body of the function, such as those for `a` and `b`, now become *local* assignments in the evaluation of a call to `shift`. The assignments and any storage they require go away when the call is completed.

1.5.2 Turning Tasks into a Function

As an alternative to typing in the three expressions in the body of `shift`, we can ask S to create a function containing some recently typed tasks. To do this, you supply the name you want to give the new function as an argument to the function `history`, which will then turn the expressions it picks up into the body of a function. The arguments `pattern` and `max` in a call to `history` select only tasks matching some string pattern and limit the number of tasks selected. In the computations to shift `x1`, we typed three expressions, each of which contained the string `"x1"`:

```
> a = min(x1)
> b = max(x1) - a
> xx = (x1 - a)/b
```

To turn these into a function object named `shift`,

```
> history("x1", 3, function="shift")
Function object shift defined
and saved on file shift.S
```

The `history` call had two side effects: it created an object named `shift` on the database and it wrote a file `"shift.S"` containing an equivalent assignment expression. (Computations in a functional language such as S aren't supposed to have side effects, and we do try to avoid them. But S is willing to make exceptions to proper behavior when the result is a useful tool for the user, as it is in this case.)

The notion of capturing some tasks from the recent history is particularly attractive if we really just want to save typing. Consider the `plot` and `summary` example from page 8:

```
> plot(budget$Date, budget$Amount)
> summary(budget)
```

If I find I'm doing those two computations fairly often, it makes sense to turn them into a function, say `bshow`. This does it:

```
> history(function = "bshow", pattern = "budget", max = 2)
Wrote the S definition for function show to file bshow.S
> bshow
function
{
    plot(budget$Date, budget$Amount)
    summary(budget)
}
```

That's it: the function `bshow` is ready to use.

The tasks usually will not be exactly what you want in the function, just a rough approximation. In the `bshow` case, the function is fine if I always want to look at the same dataset. I might have wanted the function, though, to look at a number of similar datasets, with the dataset being the argument to the function. In the `shift` case that is most certainly the situation.

The technique is to take the file created by `history` and edit it. The file `bshow.S` starts off as this:

```
bshow = function()
{
    plot(budget$Date, budget$Amount)
    summary(budget)
}
```

All we need to do is edit the first line to be:

```
bshow = function(budget)
```

Once the file is edited, we can redefine the S object `bshow` by the S task:

```
source("bshow.S")
```

Your chosen user interface environment will determine how you do the editing, and may provide some alternative to `source` as a mechanism for loading the revised definition back into S. In our examples, we'll avoid assuming any particular interface and stick to standard S mechanisms. In the next subsection, we look at a standard way to do the editing.

1.5.3 Editing a Function

Nothing is perfect, and often you will feel that some existing function, written by you or by someone else, could be improved by a few changes. Since S functions are S objects, procedures for doing this are simple. We will show here how to dump a function to a text file and then source in the edited file to give you a new version.

The only hard part of this is understanding the function well enough to change it! If you wrote the function yourself, you hope that you still remember what it was supposed to do. If it's a standard S function or one written by someone else, your chances will depend on how complicated the job the function does and on how clearly the author codes, as well as your own experience with S, of course. If the function looks obscure, consider that often you can modify the behavior of the function without rewriting it by just defining a new function that calls it in a slightly special way.

Suppose, though, that you do want to edit the function. If it's one of your own functions, chances are that you will already have a copy of it on a text file. If not, or if it is a function someone else wrote, you can use `dump` to create a text file. Suppose I wrote the function `bshow` (see page 21), but you decide that instead of plotting the same two components each time, you want to plot all the pairs of components. You could just create your own version of `bshow`, but it is usually smarter to start off with a different name for the new function. That allows you to compare the old and new easily, just in case the new function doesn't work quite right. We copy the function object, say to an object named `myshow`, and use `dump` to dump the object to a file.

```
> myshow = bshow
> dump("myshow")
"myshow.S"
```

We didn't tell `dump` where to write the output, so it named the file using the object name, as `"myshow.S"`. The file contains:

```
myshow = function(budget)
{
    plot(budget$Date, budget$Amount)
    summary(budget)
}
```

All we need to do in editing is to change the plot call to be `plot(budget)`. Then

```
source("myshow.S")
```

creates the revised object `myshow`.

Different user interfaces may provide you with other ways to dump the definition of a function to a file, or to edit a function without being conscious of dumping it to a file at all. Choose whatever mechanism you find most convenient. As you do more programming in S, you will want to develop a systematic approach to organizing the functions and other software you own. Use the S chapter for this: in a chapter you can co-ordinate S functions, other S objects, documentation and, if you want to, related software in other languages. It's worth keeping in mind that S functions are objects: the process described here creates or modifies an S object in your chapter database. How you dump the function to a text file, what name you give the text file, how you source it back in, even whether you keep the text file around are all your decisions. S provides some tools but imposes no rules. You give S the tasks to dump and source: the S evaluator looks only for the S object corresponding to the function's name the next time you call it.

1.5.4 Debugging and Testing Functions

Once a function is defined or redefined, we need to see whether it does what we want. Once the new function is assigned, it is available for use, so we can proceed immediately to give it something to do. S is designed to encourage you to plunge in and try things: if they don't work, the reason will often be obvious. You can edit the function to fix the obvious problem and try again, all within a few minutes.

Sometimes, however, it pays to look a little harder at problems, particularly if the cause is not obvious. S provides some tools to help, and the design of the language lets you use and modify these tools in an unlimited way. Let's look at an example using the function `bshow` defined on page 21: after editing, we end up with the following function:

```
> bshow
function(budget)
{
    plot(budget$Date, budget$Amount)
    summary(budget)
}
```

Notice that naming the formal argument `budget` doesn't create any problem, even though there is also an object named `budget` on our database. Inside the function, `budget` is assigned and found locally.

The first test should be whether we can redo the tasks that inspired the function in the first place.

```
> bshow(budget)
  Transaction        Category          Amount
 Min.   :1439   Service   :3   Min.   :-368.00
 1st Qu.:1440   Withdrawal:1   1st Qu.: -50.56
 Median :1442   Telephone :1   Median : -20.24
 Mean   :1442   Tax       :1   Mean   :  12.37
 3rd Qu.:1444   Misc      :1   3rd Qu.:  33.50
 Max.   :1445   Interest  :1   Max.   : 652.50
 NA's   :   3   (Other)   :2
```

```
          Date
 Min.    : 1.0
 1st Qu.: 1.0
 Median : 2.0
 Mean   : 3.1
 3rd Qu.: 3.0
 Max.   :14.0
```

Good, now let's try it on another object we have around, `budget2`:

```
> bshow(budget2)
Problem: No method for plotting class "NULL" vs class "numeric"

Debug ? (y|n):
```

Not so good. In evaluating the expression, S has come to a point where some function decides there is a problem, and that it doesn't know what to do next.

The function then invokes S's error recovery, for example by calling the S function `stop`. You have control over what happens next, but by default S invites you to debug the problem interactively, through the function `recover`. This function sets up an interactive environment for you, in which you can type ordinary S expressions, plus a few commands special to `recover`.

The first important difference from ordinary evaluation is that you are working *locally* in a function call where the problem arose. Arguments and objects assigned locally in the function can be studied, even modified. To get started, just answer `y` to the invitation.

```
> bshow(budget2)
Error: No method for plotting class "NULL" vs class "numeric"

Debug ? (y|n): y
Browsing in frame of bshow(budget2)
Local Variables: budget
```

Whenever you don't know what's available, evaluating the expression `?` will provide some hints.

```
R> ?
Type any expression. Special commands:
'up', 'down' for navigation between frames.
```

```
'where' # where are we in the function calls?
'dump'  # dump frames, end this task
'q'     # end this task, no dump
'go'    # retry the expression, with corrections made
Browsing in frame of bshow(budget2)
Local Variables: budget
```

In this case, let's look at the data:

```
R> budget
             row.labels Transaction Amount Data
Magazine              1        1446    -20   15
    Misc              2        1447   -105   18
 Deposit              3          NA   1000   19
```

Staring at that for a bit shows that the last column is called "Data", not "Date", as we had intended. Seems likely to have caused confusion. We can, in this case, try modifying the data on the fly, to change the last name. Then, the command go will tell S to go on with the task.

```
R> names(budget)
[1] "row.labels"  "Transaction" "Amount"      "Data"
R> names(budget)[[4]] = "Date"
[1] "Date"
R> go
   row.labels     Transaction         Amount
 Min.    :1.00   Min.    :1446   Min.    :  -105.00
 1st Qu.:1.25   1st Qu.:1446   1st Qu.:    -83.75
 Median :2.00   Median :1446   Median :    -20.00
 Mean    :2.00   Mean    :1446   Mean    :   291.70
 3rd Qu.:2.75   3rd Qu.:1447   3rd Qu.:    745.00
 Max.    :3.00   Max.    :1447   Max.    : 10000.00
                NA's    :   1


       Date
 Min.    :15.00
 1st Qu.:15.75
 Median :18.00
 Mean    :17.33
 3rd Qu.:18.75
 Max.    :19.00
```

Very nice. To be honest, most debugging problems aren't quite so simple that we can just edit the data and go on. More typically, there is something wrong with one of the functions. Just use the command `q` to quit from `recover`.

The error recovery is only one way to use interactive browsing in S functions. Instead of waiting until an error occurs, you can call the function `browser`, which provides the same style of interaction.

You can call this function anywhere, but the simplest way to use it is to trace calls to a particular S function. No editing of the function is needed: a call to `trace` will provide a temporary version of the function you want to trace. A typical notion is to call the browser either on entering or on exiting a call to some function.

Consider the `shift` function defined on page 20. Suppose we're using it to rescale the output from some complicated computation. The results so far are in `bigX`, a large numeric object. We'll use `shift` on that and save the result.

```
> length(bigX)
[1] 400000
> xx = shift(bigX)
```

We should check that things worked as expected. But clearly we don't plan on examining 400,000 numbers. What to do?

Testing software is a hard and very important part of programming; we'll discuss it many times throughout the book. S has a variety of tools, and the *everything is an object* concept will turn out to be essential. Right now, we just want a way to look at the data; the function `plot` might be good, since it would show all the numbers on a single plot. For 400,000 numbers, though, a plot might take a little while. Let's settle for just looking at a few numbers.

```
> xx[1:10]
 [1] NA NA NA NA NA NA NA NA NA NA
```

The value `NA` in S stands for undefined numeric data (missing, or the result of some numeric computation whose result is undefined). Seems unlikely that all the initial values were undefined in `bigX`:

```
> bigX[1:10]
 [1]  0.6317972 -1.2800208 -0.6903045 -0.1199328  1.4295871
 [6] -0.3390379  0.1302444 -1.5837125  1.5309090 -0.7084244
```

But let's not just flail around: now we can use some of the debugging tools in S. Let's trace the computations in `shift`: if we say

```
trace(shift, browser, exit=browser)
```

then an interactive browser will be called from each call to `shift`, once at the beginning and again just before returning. In this case, there isn't much to see at the beginning so let's just use:

```
trace(shift, exit=browser)
```

Like `recover`, `browser` evaluates any expression you type, but in the context from which `browser` was called. To remind you where you are, it changes the prompt into a `b` followed by the name identifying the context. We look at the values of `a` and `b`:

```
> trace(shift, exit=browser)
> xx = shift(bigX)
On exit: Called from: shift(bigX)
b(shift)> a
[1] NA
b(shift)> b
[1] NA
```

They're both undefined: this clue should make us look up the documentation of `min`:

```
b(shift)> ?min
Title:
        Extremes
Usage:
        max(..., na.rm=F)
        min(..., na.rm=F)
Arguments:
  ...:  any number of numeric arguments.
  na.rm: a logical value, default 'FALSE'. If 'TRUE',
        missing values are ignored.

Value:
        the single maximum or minimum value found in any of
        the args; any 'NA's in the data produce 'NA' as a
        result unless 'na.rm' is 'TRUE'.
    etc.
```

Ah, so if `bigX` had any undefined values at all, the min and max will be undefined. The function `is.na` reports `NA` values in its argument; we don't want to see the 400,000 logical values, so we just ask if `any` of the values were missing.

```
b(shift)> any(is.na(x))
[1] T
b(shift)> q
```

That solves the mystery, and now we can decide what we want to do. We will edit the function `shift` to fix up the problem.

What *do* we want here? Either `NA`'s are unacceptable, in which case we should check and produce an error, or we only want to rescale using the defined values. It's easy to include a test, and the `any(is.na(x))` will do fine. The S function `stop` is the standard way to produce an error; it takes any number of arguments, pastes them together, and uses the result as an error message:

```
if(any(is.na(x)))
  stop("missing values not allowed")
```

The call to `stop` puts the user into a dialog with the `recover` function, just as happened somewhere deep in the plotting on page 25.

But is this really the right decision on our part? It puts the burden on the user to do something about missing values. If it were true that `NA`'s in the data made the `shift` computation fundamentally meaningless, such a decision would be reasonable. But the shift is just an application of arithmetic, and S arithmetic isn't bothered by missing values. Missing values in operands of arithmetic produce missing values in the result, without complaining, and that seems eminently sensible here.

S functions will be more useful if they don't introduce inessential extra requirements on the user. In the `shift` example, all we need to do to retain the basic generality of the arithmetic operations is to ignore `NA` values when computing the scalars `a` and `b`. The online documentation for `min` on page 28 already showed us how: include the `na.rm=T` argument.

Here is an implementation:

```
shift =
function(x)
{
  a = min(x, na.rm=T)
  b = max(x, na.rm=T)-a
  (x-a)/b
}
```

Notice that the arithmetic to do the shift itself doesn't need to take any account of `NA` values, since we're following the standard S method in dealing with them. Let's try out the new version:

```
> xx = shift(bigX)
> xx[1:10]
 [1] 0.5617603 0.2598438 0.3529725 0.4430463 0.6877482
 [6] 0.4084449 0.4825545 0.2118845 0.7037491 0.3501110
> min(xx, na.rm=T); max(xx, na.rm=T)
[1] 0
[1] 1
```

Now the answer seems reasonable.

The first instinct here might be to just throw away the `NA`'s, returning only the actual values. This is a bad idea, usually, because the user will likely be working with other data parallel to `x`. If we throw away the missing values, the user can't relate the scaled result to those other objects. Worse, we didn't provide any clue that the values were thrown away. As a result, some error or confusion is likely to result later on, making for a mystery.

So far we have concentrated on debugging: browsing interactively after we have discovered that something needs fixing. Other tools in S help to *find* things that need fixing. One useful tool is the function `all.equal`. You call this function with two arguments, `target` and `current`. The notion is that both are supposed to have come from equivalent computations: `all.equal` compares them and reports the nature of any differences. In general, this is a tricky business, requiring some notions of when things are "close enough" and of which pieces of a complicated object we want to regard as essential. The function can be extended by writing new methods, but it comes equipped to work on most common S objects. Testing with `all.equal` works best if there are some nice identities implied by our new function. In `shift`, for example, we should be able to get back the original data by rescaling and then adding on the minimum of the original.

```
> a = min(x1)
> b = max(x1) - a
> xx = shift(x1)
> xx = xx * b
> xx = xx + a
> all.equal(x1, xx)
[1] T
```

Not all the values of `xx` will necessarily be identical to those of `x1`, when we are doing numerical computations; for this reason, `all.equal` allows for some tolerance in deciding on equality. You may want to look at documentation for other S functions designed to assist in testing: try `?identical`, `?do.test`, or `?error.check`.

What general lesson do these examples show? A very important one: *Use S to Program in S.* The same S expressions we might use to study data interactively can be used, in exactly the same way, to test or debug our programs. When browsing either in error recovery or in a traced function, all the resources of S are available. These browsers let us work on objects in the function call frames, but these are ordinary S objects, just local rather than stored on a database.

1.5.5 Documenting S Functions

We have used the `?` online help operator to get information about S functions and other topics. Once we get into programming, we should think about documenting our own functions as well. Other people who use our functions will appreciate it, and even if we're only programming for ourselves, it helps to document functions in order to remember what they really were meant to do when we come back to them later on.

All S functions are self-documenting. The software that manages the S online help will look for explicit documentation, but if there is none, it will construct documentation from the S object itself. For function `shift`:

```
> ?shift
Title:
        Function shift
Usage:
        shift(x)
Arguments:
  x:    argument, no default.
```

Not very informative, but at least it tells us the arguments.

The first step in documenting a function is to include some descriptive comments in the function definition. Any comments at the top of the definition are interpreted by the self-documenting software in S as a description of the function. Let's add a couple of lines to the last version of `shift`:

```
shift =
# rescale 'x' to the range 0 to 1.
# No missing values allowed.
function(x)
{
  if(any(is.na(x)))
    stop("missing values not allowed")
```

```
    a = min(x)
    b = max(x)-a
    (x-a)/b
  }
```

With the comments added, users get some useful information:

```
> source("shift.S")
> ?shift
Title:
        Function shift
Usage:
        shift(x)
Arguments:
  x:    argument, no default.

Description:
        rescale 'x' to the range 0 to 1.  No missing values
        allowed.
```

Adding comments to functions is easy, and you should try to make it a habit whenever writing a new function to put a line or two describing it at the top of the definition. When you edit the function, glance at the comments to see if they should be changed to reflect the change in the function itself. These simple steps will go a long way to making your functions easy to use.

Eventually, you will likely want some more elegant or thorough documentation for functions that prove really useful. Section 9.3 discusses techniques for creating and editing S online documentation formally; for the present chapter, we can get along fine using comments in the functions.

1.6 Defining Classes and Methods

Methods in S define how a particular function should behave, based on the class of the arguments to the function. You can find yourself involved in programming new methods for several reasons, usually arising either from working with a new class of objects or with a new function:

1. You need to define a new class of objects to represent data that behaves a bit differently from existing classes of objects.

2. Your application requires a new function, to do something different from existing functions, and it makes sense for that function to behave differently for different classes of objects.

You may also just need to revise an existing method or differentiate classes that were treated together before. However it comes about, defining a new method is the workhorse of programming with objects in S.

Conversely, defining a new class is less common, but often the crucial step. Classes encapsulate how we think about the objects we deal with, what information the objects contain, what makes them valid. You will likely write many more methods than class definitions, particularly since each new class definition typically generates a number of new methods. But the usefulness of your project will depend on good design of the object classes, probably more than anything.

To begin, though, we will discuss the simpler project of designing methods. Let's take on a project to write a function that returns a "one-line" description of an object. We could just type the name of the object, of course, and S would show us that object. Also, the function `summary`, supplied with S, is designed to summarize the essential information in an object, usually in a page or so. Our project takes `summary` one step further, with the goal of a one-liner.

We will name the function `whatis` and give it just one argument, the object we're interested in. One definition might be:

```
> whatis = function(object) class(object)
> whatis(1:10); whatis(state.x91); whatis(whatis)
[1] "integer"
[1] "matrix"
[1] "function"
```

Okay, but not much of a programming contribution. What else might we want to know about the object? Something about how big it is, perhaps. We can use the S function `length`. So we might try, as a second attempt:

```
> whatis = function(object) paste("An object of class",
+    class(object), "and length", length(object))
```

The S function `paste` pastes strings from all its arguments into single strings. Let's try this definition on a few objects:

```
> whatis(x1)
[1] "An object of class numeric and length 14"
```

```
> whatis(xm)
[1] "An object of class matrix and length 42"
> whatis(whatis)
[1] "An object of class function and length 2"
```

Well, better but not great. The idea of length is fine for numeric objects, and generally for the `vector` objects we discussed on page 13. But for a matrix we would like to know the number of rows and columns, and for a function it may not be clear what we would like, but certainly the length has no obvious relevance at all. Let's go back to the simpler definition, `class(object)`, and think things over.

1.6.1 Defining Methods

Around now we realize that the generic purpose of the function (in this case, to produce an informative one-line summary) needs to be implemented by different methods for different kinds of objects. The class/method mechanism in S provides exactly this facility. We will define methods for those classes of objects where we can see a useful, simple summary. The existing function still plays a role, now as the *default* method, to be used when none of the explicit methods applies. For this purpose we will want to return to a simple definition.

For a definition of `whatis` for ordinary vectors of numbers, character strings, or other kinds of data, the `length` is quite reasonable. As mentioned on page 13, this is where virtual classes are so helpful. We don't need to implement a method for every actual vector class, just one method for all vectors.

The method is defined by a call to the function `setMethod`:

```
setMethod("whatis",
  "vector",
  function(object)
    paste(class(object), "vector of length", length(object))
)
```

We tell `setMethod` three things: what generic function is involved, what classes of arguments the method corresponds to, and what the definition of the method is. S uses the term *signature* for the second item: in general, it matches any of the arguments of the function to the name of a class. The definition of the method is a function; it must always have exactly the same arguments as the generic. This is the first method defined for `whatis`, so S just takes the ordinary function as defining the generic.

The `vector` method will do fine for those ordinary vectors, but for objects with more complicated classes, we can do more. Consider matrices, for example. We would like to know the number of rows and number of columns. What else should we include in a one-line summary? Well, matrices are examples of S *structures*: objects that take a vector and add some structure to it. So we might ask whether the relevant information about the underlying vector could be included. We decided before that the class and the length are useful descriptions of vectors, but in this case we don't need the length if we know the number of rows and columns. We can include the class of the vector, though, and this is useful since matrices can include any vector class as data. All that information can be wrapped up in the function:

```
whatIsMatrix = function(object)
   paste(class(as(object, "vector")), "matrix with",
         nrow(object), "rows and", ncol(object), "columns")
```

In order to make this the matrix method for `whatis`, we call the function `setMethod` again.

```
setMethod("whatis", "matrix", whatIsMatrix)
```

We can call `showMethods` to see all the methods currently defined for a generic.

```
> showMethods("whatis")
     Database    object
[1,] "."        "ANY"
[2,] "."        "matrix"
[3,] "."        "vector"
```

Three methods are defined, all on the working database (which happened to appear as "." in the `search` list of databases). The method corresponding to class `ANY` is the one used if none of the other methods matches the class of `object`; in other words, the default method.

At this point, there are some observations worth noting.

- We did not define the default method. Because there was an existing definition for the function, S assumed that definition should continue to be used when no explicit method applied.

- The call to `setMethod` clearly had some side effect, since the new method definition persisted. For nearly all applications, you don't want to worry about exactly *what* side effect; to be safe, just use tools such as `setMethod`, `dumpMethod` and `source` to set, dump, and redefine the methods.

- We remarked on the usefulness of virtual classes, not only `vector`, but also `structure`. There are a number of virtual classes in S, and you can define more; see section 7.1.5 for a discussion.

- In the `matrix` method, we created a special function, `whatIsMatrix`. That is usually a good idea, so we can test the function before explicitly making it a method. There are two ways to use the function: making its current definition the method, as we did here, or writing the method as a call to the special function. Either way is fine, but in the first case redefining `whatIsMatrix` will *not* change the method, because it was the value, the *object*, that was passed to `setMethod`.

If you want to edit the definition of a particular method, the function `dumpMethod` will write it out to a file. In fact, `dumpMethod` works even if no method has been explicitly defined for this class. For this reason, it's the best general way to start editing a new method:

```
> dumpMethod("whatis", "numeric")
Method specification written to file "whatis.numeric.S"
```

I asked for the method for class `numeric`; there is no explicit method, but since class `numeric` extends class `vector`, that method was written out. The file name chosen combines the name of the generic and the signature of the method we're looking at. The file `"whatis.numeric.S"` contains:

```
setMethod("whatis", "numeric",
function(object)
paste(class(object), "vector of length", length(object))
)
```

The `setMethod` call, when evaluated, will define the new method, but currently just contains the method for class `"vector"`. After we edit the file,

```
source("whatis.numeric.S")
```

will define the method. As an exercise, you might try writing a method for numeric objects: perhaps in addition to length, the `min` and `max` might be interesting, but take a look at page 29 first.

Let's look at a few calls to `whatis` with the methods defined so far.

```
> whatis(1:10)
[1] "integer vector of length 10"
> whatis(x1)
```

```
[1] "numeric vector of length 14"
> whatis(x1 > 0)
[1] "logical vector of length 14"
> whatis(letters)
[1] "character vector of length 26"
> whatis(xm)
[1] "numeric matrix with 14 rows and 3 columns"
> whatis(paste)
[1] "An object of class function"
```

The case of a function object still falls through to the default method, because a `function` object is not a vector. There is nothing particularly difficult in dealing with functions as objects, but you will need to find some tools to help. If you'd like to try writing a `whatis` method for function objects, see page 79 in section 2.6.1.

1.6.2 Defining a New Class

Designing a class is an extremely important step in programming with data, allowing you to mold S objects to conform to your application's needs. The key step is to decide what information your class should contain. What does the data mean and how are we likely to use it? There are often different ways to organize the same information; no single choice may be unequivocally right, but some time pondering the consequences of the choices will be well invested.

The mechanism for creating classes is fairly simple: you call one function, `setClass`, to define the new class, and then write some associated functions and methods to make the class useful. For many new classes, this includes some or all of the following:

1. software to create objects from the class, such as generating functions or methods to read objects from external files;
2. perhaps a method to validate an object from the class;
3. methods to show (print and/or plot), or to summarize the objects for users;
4. data manipulation methods, especially methods to extract and replace subsets;
5. methods for numeric calculations (arithmetic and other operators, math functions) that reflect the character of the objects.

We will sketch a few of those here, using a relatively simple, but practical, example. In section 1.7, a more extended example is given. See section 8.2 for more information about important functions, those for which you may want to write new methods.

Suppose we are tracking a variable of interest along some axis, perhaps by a measuring device that records a value y_1 at a position x_1, a value y_2 at x_2, and so on. The vector y is the fundamental data, but we need sometimes to remember the positions, x, as well. This example was developed by my colleague Scott Vander Wiel for an application with x being distance along a length of fiber optic cable, and y some measurement of the cable at that position. Clearly, though, the concept is a very general one.

How do we want to think of such data? Basically, we want to operate on the measurements, but always carry along the relevant positions and use them when this makes sense; for example, when plotting. What is the natural way to implement this concept? We could represent the data as a matrix, with two columns. This leaves the user, however, to remember when to work with one or the other column, or both. We could represent the data as a list with two components, but this has a similar problem.

S provides a class definition mechanism for such situations. We can decide what information the class needs, and then define methods for functions to make the class objects behave as users would expect. Users of the methods can for most purposes forget about how the class is implemented and just work with the concept of the data.

Classes can be defined in terms of named *slots* containing the relevant information. In this case, the choice of slots is pretty obvious: positions and measurements, or `x` and `y`. The S function `setClass` records the definition of a class. Its first two arguments are the name of the class and a representation for it—pairs of slot names and the class of the corresponding slot. The function `representation` constructs this argument. Let's call the new class `track`, reflecting the concept of tracking the measurement at various positions.

```
setClass("track", representation(x = "numeric", y = "numeric"))
```

S now knows the representation of the class and can do some elementary computations to create and print objects from the class. The operator `"@"` can be used to extract or set the slots of an object from the class, and the function `class` can be used to extract or set the class. To create a `track` object from positions `pos1` and responses `resp1`:

```
> tr1 = new("track", x = pos1, y = resp1)
```

The function `new` returns a new object from any non-virtual class. Its first argument is the name of the class, all other arguments are optional and if they are provided, S tries to interpret them as supplying the data for the new object. In particular, as we did here, the call to `new` can supply data for the slots, using the slot names as argument names.

Since S knows the representation of the class, an object can be shown using the known names and classes of the slots. The default method for `show` will do this:

```
> tr1
An object of class "track"

Slot "x":
 [1] 156 182 211 212 218 220 246 247 251 252 254 258 261 263

Slot "y":
 [1] 348 325 333 345 325 334 334 332 347 335 340 337 323 327
```

It's also possible to convert objects into the new class by a call to `as`. For example, a named list, `xy`, with elements `x` and `y` could be made into a `track` object by `as(xy, "track")`.

Most classes will come with *generating* functions to create objects more conveniently than by calling `new`. S imposes no restrictions on these functions, but for convenience they tend to have the same name as the class. Their arguments can be anything. For `track` objects, a simple generator would look much like what we did directly:

```
track =
# an object representing measurements 'y', tracked at positions 'x'.
function(x, y)
{
  x = as(x, "numeric")
  y = as(y, "numeric")
  if(length(x) != length(y))
    stop("x, y should have equal length")
  new("track", x = x, y = y)
}
```

From a user's perspective, a major advantage of a generating function is that the call is independent of the details of the representation.

```
> tr1 = track(pos1, resp1)
```

For some classes, we also want to be able to read the objects simply from text files, using the `scan` function (see section 1.7 and page 50 for an example). For now we'll assume that `track` objects are just generated from positions and measurements that are already available.

Once we can create objects from the class conveniently, we want to look at them. A method for the function `show` will be called when S automatically displays the result of a computation. This function takes only one argument, `object`, the object to be displayed. The display can be either printed (the usual choice), or plotted. Two other functions, `print` and `plot`, can also have methods defined for these two cases. These functions have a more general form, but the method for a new class of objects is usually called with just one argument. The function `summary` is expected to produce a short description, ideally to fit on a single page regardless of the size of the object.

Methods for `show` should display the objects so that all the essential information is visible in a simple, intuitive way. Easier said than done, in many cases! For `track` objects, we want to associate corresponding values from the `x` and `y` slots. It should be clear to users which is which, but we also want to remind them that this is a special class, and not a list or a matrix. You might want to stop reading at this point and think about or sketch out some possibilities.

At any rate, here is one line of thinking. A fairly nice way to pair the values is to make up a matrix with two rows or columns for the `x, y` values. But a two-column matrix will waste all sorts of space printing, so it better be a two-row matrix. We can get that by using the S function `rbind`, which pastes its arguments together as rows of a matrix. Let's try it:

```
> xy = rbind(tr1@x, tr1@y)
> xy
     [,1] [,2] [,3] [,4] [,5] [,6] [,7] [,8] [,9]
[1,]  156  182  211  212  218  220  246  247  251
[2,]  348  325  333  345  325  334  334  332  347

     [,10] [,11] [,12] [,13] [,14]
[1,]   252   254   258   261   263
[2,]   335   340   337   323   327
```

Not too bad, but we need to distinguish the rows by providing row names: `"x"` and `"y"` should be good enough. The column names are not too nice either, because they make readers think that a `track` really is a matrix, and when we get on to thinking about other methods, that is not the way we want to think of `track` objects. We could put in empty column labels, but it might be better to be reminded of how many points there are in the object, so let's use 1, 2, ... The row and column names of the matrix are set by replacing the `dimnames` (see the online documentation `?dimnames`).

```
> dimnames(xy) = list( c("x", "y"),
+   1:ncol(xy))
> xy
    1   2   3   4   5   6   7   8   9  10  11  12  13
x 156 182 211 212 218 220 246 247 251 252 254 258 261
y 348 325 333 345 325 334 334 332 347 335 340 337 323

   14
x 263
y 327
```

Good enough for now. We now package up the method as a function, and supply it to `setMethod`. The best way to do this, as usual, is to call `dumpMethod`:

```
> dumpMethod("show", "track")
Method specification written to file "show.track.S"
```

Whether or not a method has been defined explicitly, this will write out the current method, maybe the default method. Then we can edit the definition in the file to be what we want, in this case:

```
setMethod("show", "track",
  function(object) {
    xy = rbind(object@x, object@y)
    dimnames(xy) = list( c("x", "y"),
         1:ncol(xy))
    show(xy)
 })
```

Once we source in this file, the new method is defined.

```
> source("show.track.S")
> tr1
    1   2   3   4   5   6   7   8   9  10  11  12  13
x 156 182 211 212 218 220 246 247 251 252 254 258 261
y 348 325 333 345 325 334 334 332 347 335 340 337 323

   14
x 263
y 327
```

The style of this method is typical of many: we construct a new kind of object, and then apply the original generic function to this object. This allows us to reuse the existing method (in this case, for printing matrices), without having to worry about the details of that method. Use this technique liberally, it often provides the most elegant implementation of methods. Just be sure that when you recall the generic function you are getting a different method: otherwise an infinite loop will result. (S catches such loops fairly cleanly, but they still can be confusing to debug.)

For many classes, plotting the objects is as important as printing them. For `track` objects, this is certainly true. The first question is how to plot the objects by themselves: a method to interpret expressions such as `plot(tr1)`. A natural plot is just a scatter plot with the positions on the x axis and the measurements on the y axis. Obviously, this was partly what suggested the names of the slots in the first place. The second question is how to plot tracks against other objects: in this case we would usually just take the `y` measurements to define the object, and ignore the `x` values.

The function `plot` has arguments `x` and `y`, plus other arguments for supplying various parameters.

```
> args(plot)
plot(x, y, ...)
```

For `plot`, we want to provide methods that depend on both the `x` and `y` arguments. The method is set the same way as before, but now we give as the second argument to `setMethod` a signature with two arguments specified. For example, if we want to plot a `track` object on its own, the `y` argument to `plot` is missing. This is a perfectly legitimate class to specify.

```
setMethod("plot",
   signature(x = "track", y = "missing"),
   function(x, y, ...) whatever ...
)
```

There remains the definition of the method. Let's take a simple approach and just call `plot` again, with the two slots as arguments: `plot(x@x, x@y)`. To establish the corresponding method:

```
setMethod("plot",
   signature(x = "track", y = "missing"),
   function(x, y, ...) plot(x@x, x@y, ...)
)
```

As it happens, the same result is obtained from converting the object to its "unclassed" form: `unclass(tr1)` is a named list with elements `x` and `y`, and a plot method for such objects produces the same scatter plot. Either way, the only major drawback is that the labels for the plot are not very nice. From looking at the documentation of `plot`, we see that arguments `xlab` and `ylab` supply labels. We can edit the method by calling `dumpMethod`, for example, and editing the resulting file to supply labels:

```
setMethod("plot",
   signature(x = "track", y = "missing"),
   function(x, y, ...)
      plot(unclass(x), xlab = "Position",
           ylab = "Value", ...)
)
```

Evaluating this task redefines the method with fairly reasonable labels.

So how about plotting `track` objects against other objects? Here we want to specify the `x` argument, but not the `y` at all, if the `track` object is to be on the x axis. We can just omit the `y` argument from the signature. The method is simple again (aside from labels):

```
setMethod("plot",
   signature(x = "track"),
   function(x, y, ...) plot(x@y, y, ...)
)
```

Similarly, if the `track` is on the y axis:

```
setMethod("plot",
   signature(y = "track"),
   function(x, y, ...) plot(x, y@y, ...)
)
```

With these three methods, we are now prepared to plot using track objects in a pretty reasonable style.

Not perfectly, however. Notice that the labels in the plots will not show the expression defining the track object (e.g., `tr1`). Also, there are other plotting functions besides `plot`, to do things like drawing lines, adding scatters of points, or displaying text at co-ordinates defined by the data. It would be nice to include all these as well in the methods for `track` objects. All this can be accomplished and quite elegantly, but it is a little too far into the structure of S methods to include here. See section 8.2.3 if you want to go on to the next stage.

1.7 Classes and Methods: An Extended Example

When we take up a new programming language or environment, there often turns out to be a central, challenging aspect to any serious project. A great deal of satisfaction comes from mastering the central aspect, from feeling that our new software fits well with the language, maybe even that the software is starting to be elegant as well as useful.

For programming with data, certainly when using S, the single most central aspect concerns organizing data into objects, specifically by designing classes. The challenge includes making the class correspond cleanly to the key concepts about the data, providing enough methods to let users work easily with the data, while keeping users from misleading results or unnecessary problems. For serious applications, no perfect design is likely, but after some experimenting and thought, we can hope for a satisfying feeling that the design and the underlying application fit well together.

As an extended example, let's look at the device manufacturing application on page 2. Here the important challenge is think about the probe-test data that are central to understanding how well the process is working. What is the relevant information needed and how can we use it to provide insight to our users?

In the rest of this section, we will look at a few steps in providing software for this application, using the tools of S to design functions, classes, and methods. In looking at the example and especially in working with your own applications, I think you will find two general principles helpful:

1. We're writing this software for *users*, and simple steps along the way will often make for much more usable results. For example, a couple of lines of initial comments in functions and methods, aimed at explaining the function to *users*, provide for online documentation right away. Providing such simple assistance can soon become a natural part of your programming style.

2. But remember that S tries to move from ideas to software quickly. You rarely need to work for a very long time before coming up with an initial version. A few key functions, an initial design for a class, and some corresponding methods will let you and the first users do something, often with useful results. Gradual refinement will then add features, take care of problems you hadn't anticipated, and move the software toward a more mature form.

We will follow a few stages in how software for probe-test data *might* be designed using S. The first steps shown here wouldn't likely be your (or my) very first attempt to write the software, but after three or four quick rounds, something at this level would likely arise. Nor is the path we will follow in any way the only or necessarily the best approach, but it does illustrate how some ideas of what is needed get turned into S software. An actual, and very successful, library of S-based software for these applications is described in reference [4] in the bibliography.

The overall goal for this application is to design software to organize, visualize, and model the data describing the manufacture of electronic devices. Classes of objects will be needed to represent various sources of data during and after manufacturing such devices. Specifically, consider defining a class, `probeTest`, for the probe-test data taken at the end of manufacturing. The traditional results of these tests are single characters indicating the result of the test, for each device (each "site") on each wafer. The same sites appear on all the wafers, so the data can naturally be thought of as a matrix of, say, wafers by sites, the elements of the matrix being the characters defining the results. This is certainly not the *only* way we could represent the results; we might just think of a single string with all the results pasted together. In practice, we might want to play around a bit with both versions and eventually decide which is better. To keep this chapter from getting too long, let's fast-forward a bit over that experience. It turns out that a matrix structure has some key advantages. We can easily talk about different sites on the same wafer (elements in a row) or the same sites on different wafers (elements in a column). Also, we can go easily from the codes for the results to other information that keys off the codes (for example, colors in a plot); if we had used a string for each wafer we would spend a lot of time extracting characters to do computations on them. So a matrix of results it is.

The *levels*, that is, the possible results for the tests, are fixed: the testing machine is programmed to do certain tests, depending on the design or *code* being manufactured in this case. One or more characters indicate success; the remaining, failure of the tests at a particular stage, for a particular reason. The matrix needs also to be supplemented to capture the essential spatial nature of the tests. For understanding the data, we need to know *where* on the wafer each site is located. This information is also part of the code.

So the fundamental design challenge for us is to combine the probe test results with the information about the code that makes the results meaningful. Because the design and manufacturing process is so very challenging, there is potentially a huge amount of information about the code. Let's start with an extremely simple assumption; namely, that for each code we have some way, unspecified for the moment, to look up all the information we need. In fact, each code does have an identifier assigned during manufacture and much information is available once we know the identifier. The catch is that we will eventually have to organize that information for the code in some sensible way. It looks as if we've generated another class design problem, or at least a challenge to organize some more data, inside or outside S.

To get on with the design of the `probeTest` class, though, let's just assume the code identifier is stored in the `probeTest` object and that the methods for class `probeTest` will get any other information they need by calling a function that accesses some database of information about the various codes. Now we can define the class.

```
setClass("probeTest",
    representation(results = "matrix", code = "string"))
```

The call to `setClass` records the definition of the class. The `representation` function constructs an argument with pairs of slot names and the classes of the corresponding slot. In the example, there are two slots, with names `results`, and `code`. Every object from class `probeTest` will have these slots, with each slot having the class prescribed in the representation.

1.7.1 Creating Objects from the Class

A generating function with the same name as the class will as usual help users to create objects from the class without depending on the details of the representation. Generator functions tend to take a variety of arguments, some or all of them being optional, for situations where the user has varying amounts of information available at the time the object needs to be created.

In the `probeTest` class, we have two pieces of information: the code identifying the test and the test results themselves. The user of the generator might have the code only, the code and some results, or neither.

Given the code, let's assume that a function, `waferLayout`, can provide the other fixed information about this design, in particular, the number of devices per wafer. When we come to *redesign* the class, we'll want to think carefully about what information goes where, but for the moment let's make a pretty simple assumption: `waferLayout` knows about some S objects that define the various codes, and returns a named list with elements defining all the obvious things we need to know about the code. Included will be such things as where on the wafer each device appears, what characters are legal test results, etc.

We decided that the `results` slot would be a string matrix with columns corresponding to sites and rows corresponding to the wafers in the lot being tested. So if the user supplies the code, we can start off the matrix with the right number of columns, and 0 rows. If the user supplies `results` as well, we can check the supplied results against the definition of the design in a number of ways, but let's leave checking to the next round in the design of the class. One reason (aside from wanting to get the first version of the class going quickly) is that we're really talking here about *validating* an object from the class. That is usually best handled in one place, by a validation method, rather than locally in the generator; again, something that can reasonably wait until our second round of design.

Here's a definition for the generating function, taking two optional arguments for the code and the results.

```
"probeTest" =
function(code, results)
# generates an object from class 'probeTest'; if supplied,
# 'code' should be the character string identifying the design
# code.  If 'results' is supplied, it should be a string matrix
# with columns corresponding to the test sites for this code.
{
  if(!missing(code)) {
    design = waferLayout(code)
    nwafers = design$nwafers
    if(missing(results))
      results = matrix(string(), 0, nwafers)
    new("probeTest", results = results,
        code = code)
  }
  else new("probeTest")
}
```

Typical of such functions is the call to `new`, with arguments to set some or all of the slots in the new object. The three lines of comments provide some description for the online documentation of the generating function, what a user will see in response to

```
?probeTest
```

For getting probe-test data, a method for reading the data from a file is a more interesting and useful way to generate new objects than a generating function. Probe test equipment traditionally generates for each wafer some printed output, somewhat like Table 1.1. In the bad old days, humans were expected to scan many pages of such tables looking for patterns. A complete test output file would contain one table of this form for each of the wafers in the current lot. To scan in such data, we want to make each row of the results matrix from the characters in the table, after throwing away the dots used as filler. The `scan` method could figure out the number of rows and columns in the printed output, but since this also is defined by the design code, let's just assume the function `waferLayout` returns us the information we need for this particular code; among other things, the number of sites, the numbers of rows and columns in the printed test results and `levels`, the possible characters for the test results.

```
404899801
. . . . . . . . . . . . . . . . . .
. . . . . K C D C C E C E . . . . .
. . . B N K C J C C C C C D . . . .
. . . K C K K E J K C C C 1 1 D . .
. . C B C K K J C C K C L K C C . .
. C C C N K E K K K K K L 1 1 C 1 .
. 1 E L N E K K K K C N M 1 C 1 C .
. N 1 L C I C K K K K N L 1 1 1 1 1
D 1 E C C 1 C L N 1 L 1 1 1 1 1 C 1
D D 1 1 C N N M 1 1 D C 1 1 E C 1 1
B D M N 1 M C C N F 1 1 D C 1 D 1 C
. D E L N 1 E C 1 1 1 M D L N 1 1 B
. D 1 1 1 1 N N C C 1 1 1 1 1 1 1 .
. . E 1 1 C C C 1 C E L 1 1 C 1 . .
. . D C 1 C 1 1 C 1 1 E M C 1 1 . .
. . . E 1 L 1 1 1 1 1 1 1 1 M . . .
. . . . B 1 N L C 1 1 E 1 N . . . .
. . . . . . . . . . . . . . . . . .
```

Table 1.1: *A typical text file with the probe-test results for a single wafer. The first line identifies the wafer. The dot characters are just filler,* `1` *means pass, other characters are failure modes.*

Forgetting for the moment about methods and classes, let's write an S function that scans in this sort of data and makes a matrix of the test results, one row per wafer. The arguments to the function should include `file`, a file or other S connection to read from and perhaps `nwafers`, the number of wafers for which we expect data. If `nwafers` is unknown, we'll just read until there is no more data on the file. We'll call the function `scanProbeTest`.

One design challenge for `scanProbeTest` is knowing how much data to read for each wafer. Fortunately, this is something else we can figure out based on the code information. We know the number of columns and the number of lines in the input data (two more than the number of rows in the layout). The product of these is `nfields`, the number of fields including the dots used as filler. We will make this another argument to the function, which can then read one item to get the wafer identifier and `nfields` more to get the test results. It will then throw away all the filler items, leaving the test results for this wafer.

S has some standard functions we can use to build up the results for all the wafers: `c` will concatenate the wafer identifiers and `rbind` will combine rows of test results. A simple but adequate definition of the function is in Table 1.2 on page 50. Like all `scan` functions, it begins by opening the connection if it's not already open, and arranging to close it when the function exits. In the `while` loop, the function reads the label string and then `nfield` more results. The `scan` function and nearly all S input functions do not report an error if they reach the end of the file; they simply return fewer than the requested amount of data. That's an advantage of a functional, object-based language: there is no need to handle an error, when the calling function can simply examine the value of the call to `scan` instead. Our function detects the end of the file when it can't read the next wafer identifier and breaks from the loop.

Our function could, however, be more careful about looking for errors. For example, it doesn't check for an end of file when reading the test results, although it easily could. It also doesn't check the validity of the results, for example that they are single characters. These are details that we should look to once the basic method works. First, though, let's take the function and embed it in a method for `scan`. Even though no method has been explicitly specified yet, we can and should start off by dumping the method:

```
> dumpMethod("scan",signature(what="probeTest"))
Method specification written to file "scan.ANY.probeTest.S"
```

The method written will be whatever method S would currently select for this class of argument, probably the default method in this case. Starting

```
"scanProbeTest" <-
# read and interpret probe-test data from 'file'.  Each
# wafer comes as a single field giving the code,
# followed by 'nfields' each of 1 character, test sites
# plus some 'filler' to make a rectangular picture.
# The filler sites are thrown away.  Reads up to
# 'nwafers' such wafers, or to the end of the file.
# Returns a string matrix of the test results, one row
# per wafer.
function(file,  nwafers, nfields, filler = ".")
{
  if(!isOpen(file, "r")) {
    file = open(file, "r")
    on.exit(close(file, "r"))
  }
  nrow = 0; rowlabels = string()
  results = string()
  while(nrow < nwafers) {
    ## read the next wafer id
    rowlabel = scan(file, what = string(), n = 1)
    if(length(rowlabel) == 0)
      ## no more data
      break
    nrow = nrow + 1
    el(rowlabels, nrow) = rowlabel
    ## now read the test results
    chars = scan(file, what = string(), n = nfields)
    results = c(results, chars[chars != filler])
  }
  nsites = length(results)/nrow
  matrix(results, nrow = nrow, ncol = nsites, byrow=T,
       dimnames = list(rowlabels, 1:nsites))
}
```

Table 1.2: *A function to scan in probe-test data, such as that in Table 1.1.*

with this file rather than from scratch has the advantage that we have a correct method specification for scan, with the right arguments. Our job is then just to edit the body of the method, with fewer chances to get the boring details wrong.

The method in this case needs to get the information about the layout for the call to scanProbeTest. In fact, we could get this information heuristically from the file itself, just by knowing that one line with the wafer identifier is followed by a rectangular layout as in Table 1.1. It's simpler, though, and less error-prone in principle, to assume this information is maintained separately. We'll assume the function waferLayout looks in a table, based on the code, and returns the information we need: nrow and ncol, the number of rows and columns in the text file of probe-test results, as in Table 1.1. To install the method, then:

```
setMethod("scan", signature(what = "probeTest"),
  function(file, what, n, ...) {
     p = waferLayout(what@code)
     if(is.null(p))
       stop("code description for \"", code,
        "\" not found")
     if(missing(n)) n = 10000 # big enough
     data = scanProbeTest(file, n, p$nrow*p$ncol)
     what@results = data
     what
   }
)
```

A user would invoke the method by passing an empty probeTest object for the relevant code to scan.

```
> probe48998 = scan("wafers/48998.map", probeTest("48998"))
> dim(probe48998@results)
[1]  50 233
```

This lot contained 50 wafers, each with 233 sites, for a total of 11,650 devices.

There are many other ways to use the scan methods; for example, a separate function could take the file name only, infer the identifier and from that the code by reading the first line of the file, and from that get the probe layout, generate the empty probeTest object, and call scan.

1.7.2 Displaying Objects from the Class

The `show` method for a class of objects provides the automatic display of expression results. Usually, this will print the data in some form. With the probe-test data, however, it might be more convenient to display the results graphically. Plotting the probe-test data, particularly with interactive control of the plot, was a key contribution of the work on wafer data. A printing method, in contrast, generally ends up producing output that looks roughly like the files of input data in Table 1.1, which can be rather uninformative when there are many wafers in a single dataset. We can choose to make the `show` method plot or print, or even to choose between the two depending, for example, on the presence of a suitable graphics device. Let's go on to define methods for `plot` and `print` and assume `show` will choose between them.

The plots developed for the S-Wafers software used color coding for the different outcomes, plotting a filled rectangle at the co-ordinates of each site corresponding to the outcome. Each wafer then is plotted with all the devices colored in, as shown in Figure 1.2 on page 4, or in color on the cover of the book.

I have to emphasize what a major advance this simple technique was for the study of the data. The plots have two crucial advantages over the printouts such as Table 1.1 on page 48. First, *much* more data can be seen at once. The plot can be reduced to fit as many as fifty on a page and one can still see the patterns. Second, patterns from one wafer to another are much easier to see as colored (or even as grey-level shading) than they are when we have to match particular characters. The ability of the users to visualize the test results had just been substantially enhanced. Nor is that the end of the story: *interaction* can be added to the plot, giving users further insight into the data. We won't pursue interactive graphics here, but will mention it on page 90 in Chapter 2, as an example of an interface (in this case, from S to Java).

So, how would we implement such a plot? Each of the devices is a little rectangle on the wafer; we essentially want to draw and fill in these rectangles, with a color chosen to correspond to the result for the corresponding device. From some snooping through the available graphics functions (in Table 3.10 on page 103, for example), we find the function `polygon`. This draws one or more polygons and, conveniently, will optionally fill them if we supply corresponding colors.

So we need some information: the co-ordinates of the polygons for all the devices. But this information is fixed, based on the code. So, as with the information we used in the `scan` method, we can assume that the `waferLayout` function has this information predefined, as named components `xpoly` and `ypoly`. We also want to draw in the edge of the wafer, so these co-ordinates should also be available, as `xedge` and `yedge`. One more piece of information is required: we need to know the possible results of the tests. It's essential that the plots be consistent, both within a lot and from one lot to another for the same code. Comparing results under different circumstances will be a powerful source of insight, and the plots make such comparisons practical.

The possible results, then, will be one final piece of information supplied by `waferLayout`, the component `levels`. This is a vector of *all* possible test results; we will use colors corresponding consistently to those. You can specify colors in S by giving a *color map*, which provides some number of colors, defined any way that your graphics device understands (often, levels of red, green, and blue). Once the color map is given, colors in S graphics are specified by integers taken to be indexes into that color map. For the plotting example, we're going to assume that a color map has been defined for this code. We can then use color number 1 for the first level, 2 for the second, and so on (the colors don't need to be distinct, if we want to treat some test results as equivalent).

Now we can describe what the plot method should do: for each wafer, draw the polygons representing the devices, coloring each polygon in by the corresponding color in `levels`. Once again, let's start by defining a function to do the computation:

```
"waferPlot" =
# plot each wafer in 'object', showing the edge and
# filled polygons for each device, in color 'results-1'
function(object)
{
  L = waferLayout(object)
  oldPars = par(L$graphicPars)
  on.exit(par(oldPars))
  results = match(object@results, L$levels)
  for(i in 1:nrow(results)) {
    plot(L$xedge, L$yedge, type = "l")
    polygon(L$xpoly, L$ypoly, col = results[i,] - 1)
  }
}
```

As far as the graphics is concerned, nearly all the action is in the loop, over the wafers. Two S graphics functions are called for each wafer: `plot` and polygon. The plot call produces a new plot each time, so normally it would clear the screen or move to the next page of a postscript printout. It then plots the edge of the wafer, as a connected line—that's what the argument type="l" does. On top of this plot, polygon then fills in the polygons for all the devices, colored by matching the corresponding results to the levels. Subtracting 1 makes the first level into the background color, following the convention in Figure 1.2 of making the successful devices invisible.

There are a few details in the design typical of plotting methods in S. Graphical parameters in S are set by a call to the par function; they control general features of graphical output. The wafer layout information contains suitable parameter settings to plot all the wafers together, such as layout parameters (mfrow) to put multiple plots on one page. With multiple plots specified, each plot call moves to a new position on the same page, not to a new page. The use of on.exit makes sure that any changes made will be restored when waferPlot exits. Whenever a function changes options or parameters, good style requires restoring the user's view of the world on exit. A more sophisticated plotting method would have incorporated graphical parameters from the user's call (the "..." argument to plot) and allowed them to override the waferlayout values, for example by specifying the user's parameters after the waferlayout parameters.

The plot method itself is then very simple indeed:

```
setMethod("plot", signature("probeTest", "missing"),
function(x, y, ...)
  waferPlot(x))
)
```

This is the method for calls to plot with only one dataset provided, specifying "missing" for argument y. We discussed plot methods for class track on page 42, with either one or two arguments.

Let's turn now to a method for print, which should print essentially all the information in the object, in a form comprehensible to an end user. There may be a lot of information, and in fact there is in the probe48998 object. It's probably the best strategy, however, to print that information, not wasting space but not throwing data away either. The user's interface to S may provide ways to navigate over a substantial amount of printed output, searching or paging. The role of the method should be to provide that information, preferably without second-guessing how the interface will work.

We will reproduce roughly the printed layout of Table 1.1. This amounts to a character matrix for each wafer, with elements corresponding to sites filled in with the corresponding results and other elements left blank. To do this, we need to know for each site the corresponding row and column position it came from when the data was scanned. We could infer this empirically, but again it will be cleaner to assume `waferLayout` gives us back `rows` and `cols`, two vectors of length equal to the number of sites for this code. We want to fill the corresponding elements from the `results` data. The idea of replacing part of a matrix when you know the row and column position of each element comes up frequently, and S provides a shorthand way to do it. If `where` is a two-column matrix, with the appropriate row positions and column positions as *its* two columns, then

```
x[where] = value
```

inserts the elements of `value` in the right places. Just what we need for this example.

As we did when creating both `scan` and `plot` methods, it's best to start by writing a function to do what we want, and then worry about making a method out of it. Here's a function that starts with the matrix of results and the necessary layout information, and writes output similar to Table 1.1.

```
probeDisplayMatrix =
# print the matrix 'results' of test results to 'file'.
# Each wafer is printed as a character matrix 'nrow' by 'ncol';
# 'filler' is a character to put in where there are no sites.
function(file, results, nrow, ncol,
     rows, cols, filler = " ")
{
  x = matrix(filler, nrow, ncol)
  where = cbind(rows, cols)
  labels = dimnames(results)[[1]]
  for(i in seq(nrow(results))) {
    x[where] = results[i,]
    cat(labels[i], "\n", file = file)
    for(j in seq(nrow))
      cat(x[j,], "\n", file = file)
  }
}
```

Like `scanProbeTest`, the function `probeDisplayMatrix` does all the hard computations (well, not all *that* hard), without worrying much about how to extract the information needed to do them. It's likely to survive with little reprogramming needed when we revise the organization of the class, or even when applied to a different but related class. We did the printing here using a fairly low-level function, cat; not a bad idea when we know precisely what we want. It's a good idea, though, to look first for a related existing function; there are functions for printing matrices (the objects listed on page 16 include some), but they turn out to be somewhat inconvenient if we want to reproduce Table 1.1. Turning off features we don't want is in this case as hard as generating the desired output with a lower-level function.

The method for print gets the necessary information about the object and calls probeDisplayMatrix.

```
setMethod("print", "probeTest",
  ## the test results are printed in (approximate) positions
  ## in a two-way layout.  For more control of output, see
  ##  the function 'probeDisplayMatrix'.
  function(object, ...) {
    p = waferLayout(object)
    probeDisplayMatrix(stdout(), object@results,
      p$nrow, p$ncol, p$rows, p$cols, " ")
  }
)
```

The comment lines at the beginning of the definition for the method work like comment lines in function definitions: they provide some automatic documentation. In this case, the documentation is just for this method. Users can get detailed documentation for individual methods by preceding the "?" operator with "method", and following it by a call to the function we're interested in, but with the arguments just set to the class(es) for which we want documentation.

So, for example, to see the method documentation for function print when the first argument is of class "probeTest":

```
> method ? print("probeTest")
Title:
        Method print
Usage:
        print(object = probeTest)
Arguments:
  object: argument, no default.

Signature:
  object: "probeTest"

Description:
```

```
the test results are printed in (approximate)
positions in a two-way layout.  For more control of
output, see the function 'probeDisplayMatrix'.
```

The method itself produces printouts like Table 1.1 for each wafer, but filled around with blanks.

1.7.3 Manipulations and Modeling

The ability to visualize device test data, quickly and with interactive control, was a breakthrough for many of the people responsible for device manufacture. Tedious reports, bulky and inconvenient, were replaced by essentially instant views of the data. The software described in reference [4] became an indispensable part of the engineer's tools. Perhaps the most important benefit was simply the new attitude that it *was* feasible to view, study and model the data describing the manufacturing process.

Naturally, this led to an interest in more analytical techniques, and eventually to the application of models incorporating features such as spatial patterns in the test results. Some lots may exhibit, for example, more defects near the outside of the wafers, or in a region near one corner. To analyze such patterns, we need a convenient way to express them; specifically, some compact ways to extract subsets of the results defined by various spatial patterns. As our final example, let's describe one way to introduce methods to extract spatially defined subsets of the probe-test data.

The operator `x[i]` in S is the standard way to extract subsets of data. A convenient way to extract spatial subsets would be to use this operator on probe-test data with the expression inside the square brackets describing the region of interest. For example,

```
probe48998[x > y]
```

would extract the results for sites in the region defined by `x > y`, the sites below the line $x = y$.

For the programmer, there seems to be a big catch in such expressions. The positions `x` and `y` are specific to the data—they are computed from the `waferLayout` information for this particular code. So for them to be supplied as arguments to the subset expression, it appears that the user needs to precompute them for the particular dataset, making the whole computation extremely tedious.

The solution comes from an aspect of S that will be emphasized over and over throughout the book: *everything is an object.* In particular, not only is the computed value of `x > y` an object in S, so is the expression itself, before it is evaluated. An S function can capture such an expression and then evaluate it explicitly, in a context the function sets up.

Given this idea and the information returned by `waferLayout`, the solution to our programming problem is easy. Using the familiar technique, let's first write a function to do the extraction, and then put that into a method. We'll call the function `probeSelect` and give it two arguments: the `probeTest` object and the expression describing the region. Notice that we want the expression as an object itself here (we'll show how to get that object in the next stage). The definition of `probeSelect` is very simple, but definitely not obvious, until you have had some experience with using unevaluated S expressions. Here is the central idea. We won't get the results desired if the expression `x > y` is evaluated in the ordinary way, in the context of a standard S function call. Okay, so instead we will *create* a context in which `x` and `y` are defined the way we want, and evaluate the expression explicitly in that context.

Specifically, we create an S *evaluation frame* from the layout information. An evaluation frame in S is a temporary database in which objects, such as `x` and `y` in our case, are made available with whatever definition we want. One way to set up an evaluation frame is to call the function `new.frame`, giving it a named list as an argument. Each of the named elements of the list becomes an object in the frame, with the same name. Just what we need, since `waferLayout` returns a named list with, among other things, `x` and `y` as elements. Once we have created the frame, we call the S function `eval` to do an ordinary S evaluation, but explicitly.

```
"probeSelect" =
## the subset of the probeTest sites satisfying the spatial
## expression in 'expr', which should be an S expression object
function(obj, expr)
{
  frame = new.frame(waferLayout(obj))
  eval(expr, frame)
}
```

Although we expected the expression to be some sort of logical criterion, such as `x > y`, in fact it can be any computation using the information produced by `waferLayout`.

It remains to turn this into a method. The function `"["` corresponds to the subset operator. Aside from the special way in which we write expressions to call it, this is just an S function like any other. In particular, we can define methods for it.

```
setMethod("[", "probeTest",
function(x, ..., drop = T) {
  expr = substitute(...)
  sites = probeSelect(x, expr)
  x@results[, sites]
}
)
```

This method assumes that the user is supplying some expression to select sites. The expression `substitute(...)` returns the actual argument corresponding, *unevaluated*; for example, the object `x > y` rather than the result of evaluating that expression. Just what we wanted in `probeSelect`. The value of the call to `probeSelect`, in this case, is the logical vector telling us which sites satisfy the expression in `x` and `y`. Lastly, the method uses this to index the sites, that is, the columns of the results matrix, `x@results`.

Further refinements of the extraction method for probe-test data would certainly elaborate on our first version. Spatial patterns are one important kind of subset, but we probably would like to provide other kinds of subsets as well. We might want to select complete wafers, rather than arbitrary sites or results. Even when subsetting sites, it might be better to return whole wafers, to retain the `probeTest` class. (The unselected sites could be set to `NA`, perhaps.) The argument `drop` to `"["` is used to control similar decisions when subsetting matrices and multiway arrays, and could be used similarly for `probeTest`.

Distinguishing different kinds of subsetting can use a variety of techniques. One that requires just a little extra programming on the part of the user is to supply the spatial expression as a special class of object. Modeling software makes much use of the `formula` class to provide a symbolic definition of the model. If users were willing, we could apply the same idea here:

```
probe48998[~ x > y]
```

For the price of that extra character, `probeSelect` would no longer require the call to `substitute` in the method. In fact, `probeSelect` could itself be a generic function with the method for class `formula` corresponding to the original definition. Just one of many interesting directions to explore.

This brings us to the end of the example, and of the chapter. It's only the beginning of the application of programming with data to studying probe-test and related data, however. The S-Wafers software described in reference [4] shows what can be done. In our examples here, a modest amount of programming created initial versions of S software to visualize and compute with the test data. The visualization software itself is an important tool in practice. The other computations, especially the ability to represent spatial patterns, open the way to a deeper understanding of the data and of the processes that produce that data. Spatial patterns can now be incorporated easily in spatial *models* that can then be applied to the data. The models make it possible to test out ideas quantitatively about what effects may be at work. This is an example of how computing with data can help in science and its applications.

Chapter 2

Concepts

This chapter presents the key concepts of S, with the perspective of a user planning to do some programming in the language.

The concepts in this chapter underly all computing in S. The more you come to understand them, the more programming with S will become natural.

If the concept seems abstract or its exact meaning vague, don't worry too much about the precise meaning at this stage, particularly if the ideas are new. The concepts and the examples to be presented in this and other chapters re-enforce each other—each step in understanding the concepts makes the examples more natural, and each exercise with the examples helps clarify the concepts.

This chapter works well in conjunction with Chapter 1, which has more examples and details but less background.

2.1 S and Other Languages

The S language has a syntax that looks much like other "scientific" languages, especially the C family of languages (such as C, C++, Java, or awk). It has some uncommon features (e.g., very general kinds of assignments) and avoids some common ones (e.g., declarations). These differences of form are not very important and we can deal with them as they arise. Much more important, however, are differences in how you use the language.

You never write a program in S. Rather than users having the task of creating a program, users start an S session and give *it* tasks. So, S is a programming environment without programs. In place of programs, you use, create, and modify S *objects. Everything is an object.* S computes with objects and manages them for the user. This includes maintaining databases of objects. Objects hold data about the world outside, but they also are used to define everything in S itself.

S objects are much more general than those of the other languages mentioned, and farther from the hardware. Where single numbers and bytes provide the atoms for languages like C or Java, the atomic objects in S are dynamic, self-describing vectors of arbitrarily many numbers, character strings, or raw bytes. Similarly, where the most elementary computations in other languages are modeled on machine instructions (conceptually at least), it's best to think of the elementary S computation as a call to some subprogram (say, a C subroutine) that computes some well-defined result, without coming back to S for further information.

Numerical operations on atomic S objects are such computations, but so are sorting, summary computations, algebraic transformations, and many others. Even matrix computations can be regarded this way, although they deal with slightly more structured objects. Programmers can add their own elementary computations, just by interfacing S to a routine in C or Fortran. The essential "atomic-ness" of the computations comes because S turns control over to the routine; until the routine exits, the S evaluator and object management exert no influence.

In terms of objects and computations, then, S is a higher level language than, say, C. This makes many experienced programmers nervous—are they less in control? Yes, in a sense, and deliberately so. Remember our main goal: to turn ideas into software, quickly and faithfully. S takes over much of the low-level organization, so that each conceptual step the programmer takes can do more, and get us closer to the results we need. Occasionally, the computational overhead of this approach slows down the mechanics of the computing importantly. This is a serious point. There are good ways to deal with the problems, and we will discuss them. But the best advice is this: don't worry about such issues until you *really* need to. S works hard (especially the current version) to organize large computations sensibly. You will nearly always do better to concentrate on the big picture, what you want to learn and how that relates to the data. Particularly as you get used to this big thinking, you will likely find that S objects and functions fit your concepts, allowing you to postpone worrying about the details, either forever or at least until you know the particular computation is really important, and worth sweating over those details.

Everything that goes on in S can be thought of in terms of tasks and objects. Tasks are the computations and objects are the things that the computations manipulate. Tasks and objects each rely especially on one organizing principle. For tasks, this is *functions*: each step S takes in carrying out a task is likely to involve evaluating a call to a function. The early stages of programming in S should lead you to write some new functions, or to modify some existing functions. These at first will likely do nothing more than make it easy to present the same task to S repeatedly, perhaps changing something about the details of the task each time.

As tasks are organized by functions, objects are organized by *classes*. Each S object has a class that describes what the object is, what information it contains, and how it relates to other classes. In most cases (and always in your initial use of S) you don't need to know the class of objects: just assume the class is chosen to be natural for the object: `"numeric"` for numerical results, `"logical"` for the results of tests, and so on. The class helps the object to be interpreted correctly when used in later tasks. Just as programming will lead you to define or modify functions, it will often also lead you to define new classes of objects, though this is a more advanced kind of programming, which you will likely come to later.

Functions and classes come together in *methods*. These describe how function calls should be evaluated, based on the classes of objects appearing as arguments. A function with different methods is a *generic* function: the actual definition to be used for the function is selected by S for each call. For example, `summary` is a generic function that produces an object that summarizes the object supplied as an argument. So if `x` has class `"numeric"` or class `"matrix"`, `summary(x)` will be computed using the method for that class. When we write the call to `summary`, we don't need to specify this information. S objects are always self-describing, so S finds the class information and anything else it needs to know by examining the object itself. Methods can depend on more than one argument; for example, the expression `match(x, table)` matches the data `x` against the values in `table`; the evaluator may use the class of either argument to choose a method. S will also apply its knowledge of how classes are related to make the widest sensible use of a method; for example, a method defined for all multiway arrays would apply automatically to class `matrix`, which specializes class `array`.

For the programmer, methods provide an important way to organize computations. Instead of having to maintain and understand the whole range of computations for a function, we can work on particular methods, for particular classes of objects. I can write a new method for `summary` for a class of objects that interests me, without touching the other methods or even needing to know what methods exist.

The natural progression in using S as a programming environment is a gradual one. The first step is to write some simple functions to carry out tasks of interest to you. As you become more acquainted with the language, these functions are likely to become more ambitious. You will want to organize them and make them more convenient for you and others to use.

Before long, you are likely to decide that your applications involve special kinds of data. To deal with your data effectively, you will want to design classes of objects that reflect this specialness. Often, only a little new software is needed to do this, if the new classes are an extension of existing classes. The essential programming steps at this stage are to design the class and to provide some fundamental functions and methods to work with it.

These are the key concepts in any programming with S. In the rest of this chapter, we will examine them in more detail.

2.2 Communicating with S

The first question about any tool is "How do I use it?" To get things going, you start a *session* with S. You then communicate with S by giving it a *task*; S tries to perform the task and then waits for the next task. The session continues through successive tasks until you decide to quit. Performing a task means evaluating some S expression, in several steps; typically, each step is a call to some S function.

So the communication with S has three layers: the session, the tasks within the session, and the evaluations within each task. The mechanics of starting and quitting the session, and of sending individual tasks to S, will vary depending on the particular user interface you are using. The interface is *not* fixed by S itself; in fact, S provides tools to design or modify the user interface. A common interface is one in which the user is prompted to type an S expression. When the expression is complete and the user tells S to go ahead (by the `Enter` key, say), S takes this as a task. Specifically, S parses the typed text, evaluates the resulting expression, and perhaps shows the result to the user. Once this task is done, S goes back to waiting for the next task.

This is the user interface we will assume for most of our examples.

```
> plot(wafers)
```

The interface prompted the user with the character string `"> "`, the user responded by typing `"plot(wafers)"` followed by `Enter`. This is a complete expression, so S took it as a task (in this case, the evaluation of the task produced some graphical output).

It's worth re-emphasizing something that is missing from these concepts: a *program*. S is a programming language and programming environment, but unlike more traditional environments, the user is not asked to write a "complete program" that has to be "run" to produce results. The communication with S reverses much of the traditional emphasis: instead of users having the task of creating and running programs, users give S tasks that S carries out. This style of communication is more typical of word processors and other interactive software than of programming languages. All the same, S is a real programming language, but without the concept of a user-written "program".

The distinction is not simply abstract. S aims to shorten the time between an idea and the software that implements the idea. Before you start to implement a new idea, you will likely be relying on some simple tasks that you can give S to get the information you need. You will often type a short expression that says something you want to know; for example, "What does a plot of the dataset `wafers` look like?" Eventually, we have to go beyond such very simple tasks, but all along the way the emphasis is on building up software by giving S a sequence of tasks.

2.3 Data: Objects

The concept that permeates S more than any is that of an *object*. The name `wafers` in the expression `plot(wafers)` tells S to find an object associated with that name: S maintains databases of objects that can be accessed by name. But in fact every evaluation in S, and every step of every evaluation, produces an object as its result.

Functions, such as `plot`, also correspond to objects. And the expression `plot(wafers)`, before it is evaluated, also corresponds to an object. Indeed, this brings us to a key concept in understanding S.

Everything in S is an Object

This maxim will keep recurring throughout this and other chapters. Look under *object* in the index for some examples. Essentially, S is prepared to use any object anywhere in a task. The language can operate on objects that represent functions or expressions just as well as on objects that represent numeric data. The maxim has practical importance, not just abstract interest. Using S to manipulate S expressions, for example, is a powerful technique that underlies nearly all the programming and debugging tools, among many other applications.

Of course, not every function can make sense of every object: doing arithmetic on a function object won't (usually) work. None of these constraints, however, are intrinsic to the programming environment itself. Many computations in S really are universal: they can be trusted to give *some* result when applied to any object. This may be true of computations we design as well as of those supplied with S. An example is a little function called `whatis`, which we will design and elaborate on throughout the book. The value of a call to `whatis` is a character string, a brief description of the object supplied as the argument:

```
> whatis(letters)
[1] "character vector of length 26"
> whatis(plot)
[1] "Object of class function"
> whatis(1:10)
[1] "integer vector of length 10"
> whatis(Quote(1:10))
[1] "Object of class call"
> whatis(xx)
[1] "integer matrix with 3 rows and 4 columns"
```

Everything is an object, and all objects are, deep down, similar. Functions will make distinctions among objects and try to do the right computations in the particular case. The fundamental tools for making these distinctions are the *class* of objects, and the *methods* associated with functions.

2.4 The Language

The next concept we need to grasp is the language itself, the material with which we express computations. Looking at S expressions, you will see a language that will be familiar in most respects to anyone used to programming in C, C++, Java, or other similar languages. To illustrate, let's take a function defined in section 1.7.2:

```
"waferPlot" =
# plot each wafer in 'object', showing the edge and
# filled polygons for each device, in color 'results-1'
function(object)
{
  L = waferLayout(object)
  oldPars = par(L$graphicPars)
  on.exit(par(oldPars))
  results = match(object@results, L$levels)
  for(i in 1:nrow(results)) {
    plot(L$xedge, L$yedge, type = "l")
    polygon(L$xpoly, L$ypoly, col = results[i,] - 1)
  }
}
```

We've chosen this as a typical example of medium size. You can look up the discussion on page 53 to learn what it does, but don't worry about understanding the computations, just notice the style: lots of function calls, some assignments, an iterative expression, and braces to group expressions together. Comments start with the character # and continue to the end of the line. If you're unfamiliar with S syntax and want an overview, see section 3.2.

Most of the language consists of function calls; it is only a slight exaggeration to think of the evaluation of a task as equivalent to a sequence of evaluations of function calls. In this sense, S is a *functional* language: that is, evaluation of a function call is the essential step when S carries out a task.

2.4.1 S as a Functional Language

Programming in S usually follows this rather simple model: functions are built up by combining calls to other functions, using the value of one call as an argument to another, and so on. Of course, nearly every programming language has this concept: the distinctive feature of S is that its functional style and the generality of objects in S allow the concept to be used very extensively. Languages that, for example, can only treat certain data types as the value of functions will have to use some other mechanism to build up and pass around more general objects. But in S everything is an object (remember?) so whatever can be computed can be returned as the value of a function.

Nearly every example in this book illustrates this concept, but we include a few simple examples here. Suppose we want to subtract a from all the values in some dataset x, and then divide the result by b. The expression in S is exactly the obvious one:

```
(x - a)/b
```

(That the function calls here are written as arithmetic operators is just a convenience provided by the parser. Both "-" and "/" are S functions in the same sense as log, say.) The application of the general concepts becomes clear if we ask what happens when this expression is evaluated. The function "/" is called with two expressions as arguments: x-a and b. To evaluate the first argument, the function "-" is called with x and a as arguments. Evaluating this call produces an object, and this object is the evaluated first argument to "/". So, in a particular case suppose x is the sequence of numbers from 1 to 10, a is the smallest value of x and b is the largest value minus the smallest.

```
> x = 1:10
> x
 [1]  1  2  3  4  5  6  7  8  9 10
> a = min(x)
> b = max(x) - min(x)
> (x-a)/b
 [1] 0.0000000 0.1111111 0.2222222 0.3333333 0.4444444
 [6] 0.5555556 0.6666667 0.7777778 0.8888889 1.0000000
```

In this example, where everything is defined in terms of x, we could dispense with a and b:

```
> (x - min(x))/(max(x) - min(x))
 [1] 0.0000000 0.1111111 0.2222222 0.3333333 0.4444444
 [6] 0.5555556 0.6666667 0.7777778 0.8888889 1.0000000
```

The function calls are more deeply nested in evaluating this expression, but they are still evaluated the same way.

One more example, this time to illustrate how powerful a functional language can be in expressing computations. (This example involves slightly more abstract notions; you can skip it without missing any key practical tools.) There are many algorithms for the task of *sorting*, that is, ordering some data values so that each value is less than the following value, in some sense of comparison. The most famous algorithm, and perhaps the most studied algorithm in all of computing, is called *quicksort*. It can be described very simply in words. Given some data x, say, we choose one element, called the *fence*. We split x into three parts, according to whether the values are less than, equal to, or greater than the fence. We apply the same algorithm, recursively, to the first and third of these parts and concatenate the results. With a rule for handling very small objects (so the algorithm doesn't loop forever), and a rule for making the comparisons with the fence, the algorithm is defined.

In S, we can express this algorithm in exactly the way it is described in words.

```
quicksort =
function(x) {
  if(length(x) < 3) q2(x)
  else {
    fence = sample(x,1)
    c(quicksort(x[x < fence]),
      x[x == fence],
      quicksort(x[x > fence]))
  }
}

q2 =
function(x)
if(length(x) == 2 && x[1] > x[2]) x[2:1] else x
```

The function quicksort does the choice of the fence and the recursive splitting. The auxiliary function q2 handles small objects (objects of length less than 3). The first line in the body of quicksort checks whether x is small enough to use q2. If not, we select the fence as one randomly sampled element of x and, in one expression, compute the value of the function. The call to the function c concatenates the three parts we described, the first and third parts being returned by recursive calls to quicksort.

The quicksort algorithm has been implemented in just about every major computing language (C, Pascal, Fortran, you name it). Donald Knuth, in *The Art of Computer Programming* (Chapter 5), even shows an implementation in his abstract assembly language. At 62 steps, this is rather long but even if you look up an implementation in one of the other languages, I assert you will find it longer, less clear and/or more limited than the S function.

Why? S brings three of its key features to bear on this computation:

1. Everything is an object, so to put together the key recursive computations, you just put together the three objects.

2. Objects are self-describing, so `quicksort` is not cluttered up or restricted by having to worry about what kind of data is in `x`.

3. Just as the quicksort algorithm depends only on the notion of comparing two values, so the S function depends only on the comparison function (the `"<"` operator) and not on the actual method.

The third point is the central notion of functions and methods, which we will examine in section 2.7. Whatever the class of `x`—whether supplied with S or designed by you—so long as a method for comparison exists, `quicksort` will apply *without any change* to that object. Most languages either require the user to supply information to allow the algorithm to work on particular classes of data, or else restrict the algorithm to one or a few predetermined classes.

2.4.2 S Expressions as Objects

A pure functional language would, in principal, have nothing but functions. S does not go quite this far. The language has more to it and, more importantly, S supports facilities such as persistent databases that require some evaluations to have side effects. Section 2.5 takes up the database concept in S. In the present section, we look at the language as a whole. This can be done best by considering expressions as S objects, in two steps: first, as objects *before* being evaluated; and second, the process of evaluation itself. Keep in mind that all this is legitimate S programming discussion, not something remote or "privileged". All the language objects are accessible to any S programmer and while parsing and evaluation usually take place automatically, they are also available explicitly as programming tools whenever you need them.

When you communicate a task to S, you usually type some text. The text is parsed; once that happens (and assuming S can make syntactic sense of what you typed), an S object is created corresponding to the text. Normally, S then goes on to *evaluate* that object, but right now we want to talk about the unevaluated expression. To look at examples, it is convenient to suppress the evaluation; in particular, the function `Quote` takes an expression and returns it unevaluated.

Typing the 12 characters `plot(wafers)`, followed by the `Enter` key, causes S to evaluate the expression and produce a plot as side effect. Typing

```
> plotCall = Quote(plot(wafers))
```

creates an object of class `call`. All the function calls and operators are also objects of class `call`:

```
> class(Quote(x + y))
[1] "call"
> class(Quote(1:10))
[1] "call"
> class(Quote(-x))
[1] "call"
```

Assignment operators are an exception among infix operators; they are special enough to have their own class.

```
> class(Quote(a = 1:10)); class(Quote(a <- 1:10))
[1] "="
[1] "<-"
```

The two assignment operators have the same semantics; `"<-"` is more typing but useful because it can't be confused with named arguments, which also use `"="`. Another specialized function-like part of the language is the expression `return(expr)`, which causes an immediate return from a function call; it also has its own class.

The remaining parts of the language provide control. Objects of class `if` have a test condition and one or two possible expressions to evaluate based on the logical value of the test.

```
> class(Quote(if(length(x) < 3) a(x) else b(x)))
[1] "if"
```

Objects of class `for`, `while`, and `repeat` express loops; each object contains a *body*, an S expression that will be evaluated some number of times when we ask to evaluate the whole loop. Within any of these loop expressions, an expression can consist of the reserved word `next` or the reserved word `break`, to instruct the evaluator to start the next iteration of the loop or to break out of the loop. Finally, several expressions enclosed in braces and separated in input text by new lines or semicolons, are grouped together into an object of class `"{"`:

```
> class(Quote({ x = y - x; y = sqrt(y)
+ x + y}))
[1] "{"
```

All these objects are `language` objects, meaning that when the object is evaluated, something nontrivial happens. Evaluation often takes place automatically, but it can also occur explicitly by calling the function `eval`. For a language object `expr`, `eval(expr)` will generally be different from `expr`. The virtual class `language` groups all the language objects together, to allow methods to be written for them all. Each of the language classes extends class `language`.

Evaluation in S can be understood as a process driven by the class of the object being evaluated. Again, function calls are the central step. S evaluates a function call by finding the function definition corresponding to the function called, matching the actual arguments in the call to the formal arguments in the definition, and then evaluating the body of the function, in the context defined by the matched arguments. If the function is a generic function, evaluating the body will involve finding a *method*, that is, a particular definition of the function associated with the classes of the arguments in the call. The concepts of functions and methods are examined further in sections 2.6 and 2.7.

The evaluation of conditional expressions, loops, and braced expressions are defined in terms of evaluating their elements. The elements of a `"{"` expression, for example, are just the individual expressions grouped together. The `"{"` expression is evaluated by evaluating the subexpressions in order. The last one evaluated becomes the value of the whole expression.

The expression

```
if(length(x) < 3) a(x) else b(x)
```

has three elements: the test expression in parentheses, the `if` expression immediately following, and the (optional) `else` expression at the end of the expression. S evaluates the first element and requires it to produce an unambiguous `TRUE` or `FALSE` value. Based on this value S then evaluates the second or the third element, whose value becomes the value of the whole expression.

Loops are evaluated by evaluating the body so long as the loop remains valid. Most often, the loop's body is a braced expression. The three kinds of loops differ in how they decide to iterate. A `for` loop, such as

```
for(i in names(data)) {
```

```
y = data[, i]
print(summary(y))
plot(x, y, ylab = i)
}
```

assigns the name given in the parentheses to each element of the `in` expression in turn. Each time it then evaluates the body of the loop. A `while` expression evaluates the test expression, in parentheses when the expression is parsed, and interprets the result as a single logical value. So long as the value is `TRUE`, the evaluator then evaluates the body of the loop, followed by evaluating the test again. A `repeat` loop just keeps on evaluating the body. Any loop can be exited immediately by evaluating a `break` expression or a `return` expression; a `repeat` loop can only be exited this way.

A distinction in S that may produce some surprises at first comes between general logical expressions and single tests for branching. The expression

```
if(length(x) > 0)
    xMax = max(x)
```

uses the value produced by the > operator as a single logical value to control whether the following expression is computed or not. This is conceptually quite different from an expression such as

```
posTest = (x > 0)
```

The second example is standard S computation; depending on what `x` contains, the value can be an object of any length, or a structure, or anything else, depending on the class of `x`. But the first example depends on getting a single `TRUE` or `FALSE` value from the comparison; anything else will produce an error because the S evaluator will not know what to do next. The same requirement applies to the condition in a `while` loop.

Once you are accustomed to working in S, this distinction will cease to be a puzzle, but it does have some practical implications. Don't count on logical comparisons to be valid in conditional tests, unless you build them from functions guaranteed to produce single values of the right class. For example, a very common mis-statement is to use the `"=="` operator in a condition:

```
if(x == NULL) x = rnorm(n) # Wrong!
```

This is sure to be a mistake: the `"=="` operator doesn't have a method defined for class `NULL`, and even if it did, it would be likely to produce something surprising when, say, `x` was a vector of length `n`.

S provides some tools to do what you probably meant in these cases. In the example above, what we *meant* was "if the whole object x is identical to the NULL object". The S function identical is designed for just this case:

```
if(identical(x, NULL)) x = rnorm(n)
```

The call to identical produces TRUE if the two objects are entirely identical, and FALSE otherwise. (The function is also a rather quick way to test total equality, but it's the clean definition that is really important here.)

Some other useful tools in these situations are: any and all, which reduce a vector of logical values to one value; is, which tests an object for being in or extending a class; is.na function to test for missing values; the "&&" and "||" control operators; and functions such as length and class that are pretty much guaranteed to produce single values. Even some of these functions have to be used with caution: any and all require their argument to be a logical vector and if there are NA's possible in the values, these have to be explicitly ignored by supplying na.rm=T in the call.

2.5 Databases and Chapters

The S concept of a database is essentially that of a set of *names* (distinct, non-empty character strings), each name associated with an *object*. Databases and the functions defined for them permeate S, for both temporary and permanent storage. They give S a unique approach to assignment expressions and related computations. A duality between objects with named elements and databases with named objects provides some important programming techniques as well.

2.5.1 The S Model for Databases

Like most programming languages, but unlike purely functional languages, S includes assignment expressions. An assignment such as

```
logRev = log(revenue - min(revenue) + 1)
```

always has the same meaning in S: "Evaluate the expression on the right side of the assignment and associate it with the name on the left side in the appropriate database." After this expression is successfully evaluated, the local database now has associated the object evaluated on the right of the assignment with the name "logRev". Think of an S database as a collection of names, each associated with an S object. Databases are used for many purposes, two being particularly important:

1. All the *persistent* objects in S (the objects that persist from one S task to another, or from one S session to another) reside in some database. S sets up these databases (typically, when a user invokes the `S CHAPTER` command) and provides a variety of functions to work with them, using or modifying the objects they contain.

2. Each call to an S function creates a temporary database while the call is being evaluated. All the arguments to the function, and any ordinary assignments that are included in the definition of the function, will be kept in that database.

When we are talking about the model for evaluation in S, the temporary databases generated during the evaluation of a function call are called *frames.* An assignment expression occurring in a function definition, such as

```
results = match(object@results, L$levels)
```

in the example on page 67, saves the value of the expression on the right of the assignment in the frame, associated with the name (`"results"`). Assignment expressions typed by the user at the top level are also assigned in a frame, frame 1, associated with the task itself. Assignments in this frame, however, are special in that they are committed to the persistent database, known as the working data, appearing in the first position in the S search list. If there is an error or interruption of the task, the commitment is not done, leaving the working data unchanged.

When we say that *all* S objects reside in databases, this includes the objects we create and manipulate in ordinary S expressions, but also objects that describe S itself (what classes and methods exist, for example) and objects that provide the online documentation that you see in response to `?topic` expressions. The `S CHAPTER` command sets up three databases under the current directory for these three purposes, but you should always think of a chapter as a single entity, the directory in which the command was executed.

When you want a model for how S implements databases, imagine a table of objects stored by name, with some S primitives to look up a name, to associate a name with an object, and to remove a name and the associated object. Database operations are quite finely tuned, since they occur so often. It is good programming style to make use of them whenever possible, by organizing your own needs for associating names with objects as S database operations. In particular, databases with quite a large number of objects can be maintained without serious slowdown when you are looking for a particular object. S provides facilities to organize large libraries of objects for more efficient storage and retrieval.

This idea of named objects defining a database leads to a dual concept: the whole database or frame as a single object, with named elements corresponding to each object in the original database. This duality can be exploited when programming in S, by extracting the objects in a frame as a named list, for example. The functions in Table 3.13 on page 106 provide other similar information.

The duality occurs most often in the opposite direction: treating an object as a database, either by attaching it, or by creating a frame from the object.

```
attach(fuel.frame)
```

The search list now contains a database, in which each named element of the S object becomes an independent object. After the attach, S expressions can be evaluated that refer to these elements by name, just as they would objects in an ordinary chapter database. S expressions can get, assign, or remove any object by name, and the changes will remain available throughout the session; furthermore, when the database is detached, the user can arrange to save the modified database as an ordinary object again.

The same idea is useful for temporary frames as well: a named object provides the initial values in a frame, and further S expressions evaluated there can modify the frame. The S models library uses this idea very extensively:

```
fit1 = lm(Fuel ~ Weight, data = fuel.frame)
```

The `data` argument is turned into an evaluation frame, with the named elements of `fuel.frame` (`Fuel` and `Weight`, in particular) as objects. Computations done in this frame by `lm` lead to fitting the linear model.

2.5.2 Dumping Objects; Sharing Chapters

Data organized in persistent databases can be shared among users with access to the local file system; users only need to call the S `attach` function with the path of the chapter directory. If the assignment of `logRev` at the beginning of the section was in chapter `"/home/jmc/financial"`,

```
attach("/home/jmc/financial")
```

will allow you to refer by name to `logRev` and all the other objects in that chapter in any S expression.

What about sharing S objects and databases between *different* computer systems? If the two systems have the same hardware and operating system, we could just package up the whole chapter directory and ship it, using our preferred tools for archiving, compressing, and shipping. More often, though, we can't count on compatibility between the systems. For this purpose, S supports a text-only dumping format for objects, usually referred to as "data dump" format. Objects, and even whole chapters, can be dumped in this format, and then restored on the same machine or any other.

Because the dumped file only contains ordinary text, the two machines need not be compatible, other than both supporting S. Even this restriction need not limit us totally: the structure of the dumped data is relatively simple, so programs in other languages can read or even write files in this format. The data dump format is used, then, to communicate not only among S programs but also with other software. (For example, an experimental interface between S and Java, described in reference [2], used Java methods to read and write data dump format.) The format is presented in detail in Chapter 5.

Data dump format can be used for either whole chapters or individual objects. The chapter level is particularly useful, not only for sharing chapters, but for making backups and keeping successive versions of a chapter. The function `dumpChapter` carries out a sequence of operations that produce dumps of all the objects in a particular chapter. These dumps are on files, such as `all.Sdata`, which contains dumps of all the ordinary objects in the chapter. A file `DUMP_FILES` has a list of all the dumped files. By using `DUMP_FILES`, the programmer can make an archive of the chapter or use a version control system to keep track of changes to the chapter.

At the level of individual objects, the S function `data.dump` takes the names of any number of S objects on a database and produces a file containing all the data in all those objects. The function `data.restore` inverts the process, reading in such a file and assigning the corresponding objects in the receiving database.

Data dump format guarantees to put all the information in the object onto the dump file, so it is the right way to exchange data between machines and programs. The format is not ideal for humans to edit, however, being hard to read and unforgiving of mistakes. To edit S data by hand, the function `dump` uses a different technique. This function, like `data.dump`, produces a text file, but this time the text file contains one or more S expressions, each typically an assignment. When the S expressions are parsed and evaluated, for example by using the `source` function, the dumped objects are recreated on the working data. The typical case, editing a function definition, has been shown several times in earlier sections of the book. The technique is not limited to functions, however. Most importantly, analogues to `dump` exist for dumping the definitions of classes, methods, and documentation objects (`dumpClass`, `dumpMethod`, and `dumpDoc`). Because these are special objects, the reverse process of getting them back again after editing is a bit different, but the essential approach works the same way: dump the current definition, edit, and execute the edited version.

The approach also works for ordinary objects *other* than functions, but the dumped file may not always be easy to edit. A method can be defined (specifically, for the function `dput`) to say how to dump a particular class of objects. The goal is a file that can be sourced back in, while having a format clear enough for a human to edit. The designer of a particular class of objects should try to provide such a method. (Try, for example, `dump` on a matrix object to see a partially successful example.) Designing the `dput` method can be a real challenge, and for some classes, there may be no format for dumping arbitrary objects that is both fully general and readable by humans.

Another approach is to program the `scan` function with methods to read files for particular classes of objects, or to use some more specialized functions instead. In this case the idea is not precisely the `dump` and `source` pair but class-specific S expressions to create the file and to scan it in again after editing. In the case of a two-way table, for example, the functions `write.table` and `read.table` provide special-purpose functions.

2.6 Functions

Since S expressions mainly consist of function calls, programming in S largely reduces to designing what these functions will do. A function is an object (yes, again, everything is an object). A typical call to a function, say:

```
myscale(newData)
```

first asks the S evaluator to find the function object corresponding to `myscale`. If we print this object, instead of calling it, it might look like:

```
> myscale
# rescale 'x', by default to the range '0,1'.  If supplied,
# 'location' is a central value for 'x'.
# After subtracting this off, the result is divided by 'scale'
function(x, location = min(x), scale = max(y))
{
  y = x - location
  y/scale
}
```

Because `myscale` has been assigned on a database, we can call it by just supplying the name:

```
> x2 = myscale(log(1 + x1))
```

The essential concept, though, is more general: S takes two objects, the call to the function and the definition of the function, and uses both to evaluate the call.

2.6.1 Function Objects

A function object contains two kinds of information. The *formal arguments* tell S how to interpret calls to the function, and the *body* is the expression whose value is returned from the call. In the definition of `myscale` above, the body is the braced list of two expressions. When the S evaluator has matched the actual arguments to the formal arguments, it evaluates the body and returns the resulting object as the value of the call. In `myscale` the value of the last expression in the braced list, `y/scale`, is the value of the body, and so of the call.

Because users can call S functions with named and unnamed arguments, both the position and the name of an argument in the function definition is relevant. We say "formal" arguments when we want to distinguish the arguments in the function definition from the "actual" arguments in a call to the function.

Besides the names of the formal arguments, the definition can include assignment expressions to create *default* values for arguments that are needed in evaluating the body, but that are not supplied in the call. In `myscale`, arguments `location` and `scale` have defaults, but `x` does not.

Functions, like everything, are objects. S provides a number of tools to compute with function objects. The function `functionArgNames` returns the names of the arguments to any function object, while `functionBody` returns the S object that is the body of the function.

```
> functionArgNames(myscale)
[1] "x"        "location" "scale"
> functionBody(myscale)
{
  y = x - location
  y/scale
}
```

When the evaluator encounters the name `myscale` in these expressions, it's irrelevant that a *function* object was assigned to that name. S just lookes for the object by name, and supplies it to be used in evaluating the call to `functionBody` or `functionArgNames`. Nor does it matter that the argument was a name; any other expression that evaluated to a function definition would do as well.

Evaluating a function call is *the* core of S, so a concept of how it works will help greatly in understanding the language. Starting with the object representing the call, for example,

```
myscale(log(1+x1))
```

S finds the definition of the function and matches the actual arguments to the formal arguments. In the example, there is only one actual argument, unnamed, corresponding to the expression `log(1+x1))`. S matches this expression to the formal argument `x`. What does "match" mean here? The expressions in the actual call are matched to the formal arguments: by name if the actual arguments were named, and then by position (for example, the first unnamed argument matches the first unmatched formal argument.) See section 4.4.4 if you want all the details.

When S is evaluating the body of the function and, for the first time, needs the value of the argument `x`, it will look for a matched expression; if there is one, that expression is evaluated and assigned locally as `x`. If the argument was missing, but there was a default assignment in the function definition, this expression will be evaluated instead. In practice, there are many more programming possibilities than this suggests: because calls are objects, S programmers can test whether an argument is missing and do other manipulations on the call itself.

Two details of the evaluation are important. First, notice that an argument is evaluated when S *needs* the value; this technique is known as *lazy* evaluation and is essential to the generality in treating arguments. Some special S functions take the name of a formal argument and do computations that do not need the value. For example, `missing(x)` evaluates to `TRUE` if the formal argument `x` was not matched in this call.

Second, it matters in S *where* an evaluation takes place. The body of a function call is evaluated in an evaluation frame that S creates. Local assignments are in this frame; so, for example, the assignment of a new value to `x` in the body of `myscale` does not overwrite any object called `x` in another frame or on a database. The evaluation of an actual argument takes place in the frame of the caller. In this example, when argument `x` is first needed, S will look in the local frame of the calling function for `x1` when it evaluates the corresponding actual argument. The default expressions for missing arguments, however, need to be evaluated in the current frame so they can refer to other arguments and local assignments. For example, the default for argument `y` in the following function is defined in terms of argument `x`:

```
lcor = function(x, y = lag(x)) cor(x, y)
```

If `y` were missing in a call to `lcor`, S would evaluate `lag(x)` and look for `x` in the frame of the call, not in the frame where that call originated.

2.6.2 Defining New Functions

A function object can be created by just assigning an expression of this same form, or by copying an existing function, or by special tools provided by S. Once an initial version of the function exists, it can be edited and redefined. See Chapter 1, page 22 for an introduction. S tries to make it simple to create and edit functions for simple applications. The style we encourage is to turn an idea, suggested by something just tried directly perhaps, into a function in a few simple steps. Try out that function; it won't likely be quite what you want, but a few more steps will revise it, ready to try again.

There are many ways to get started, the simplest being just to type in a function definition. For an example, let's go back to the expression in section 2.4 on page 68:

```
> (x - min(x))/(max(x) - min(x))
 [1] 0.0000000 0.1111111 0.2222222 0.3333333 0.4444444
 [6] 0.5555556 0.6666667 0.7777778 0.8888889 1.0000000
```

This can be turned into a function object, with this expression as its body:

```
> myscale = function(x) (x - min(x))/(max(x) - min(x))
```

Another way to start off a function definition is to copy the definition of an existing function and edit in some changes. The function `dump` can be used to create a text file with the definition of the function on it.

```
> myscale = scale
> dump("myscale")
[1] "myscale.S"
```

The value returned by `dump` is the name of the file written. At this point we can edit the contents of this file to make `myscale` differ from `scale` in whatever way we want, and then evaluate the task

```
> source("myscale.S")
```

The call to `source` will parse the file and evaluate the parsed expression as a task. In particular, the call to `dump` wrote an assignment expression for `myscale` onto the file; the call to `source` will carry out this assignment, and redefine object `myscale`.

A third way to create a function is to capture some of the expressions from recent tasks, using the S function `history`. To capture the last expression involving `min` and make it into a function called `myscale`:

```
> history("min", 1, function="myscale")
Function object myscale defined
and saved on file myscale.S
> myscale
function()
{
    (x - min(x))/(max(x) - min(x))
}
```

Now `myscale` is an S object that contains the function definition. The function created by `history` has no arguments; we would probably like to edit it to make `x` an argument, which only requires putting the name `x` between the parentheses. Editing the file `myscale.S` to make this one change gives:

```
myscale = function(x)
{
    (x - min(x))/(max(x) - min(x))
}
```

and then:

```
> source("myscale.S")
```

will parse and evaluate the file.

The details of how you dump objects to a text file, edit the file, and redefine the objects from the edited version will differ depending on your chosen user interface for S. The concept to keep in mind, though, is that new functions are objects that you create to do tasks. Most of us do this by following a typical S *programming cycle*. After initially creating a function object, we repeatedly try it out, edit it, and source the new version in. Each iteration of the cycle can be short in your programming time, and it should be, at least initially while you're experimenting with a new idea. The goal, as always, is to turn your ideas into software, quickly and faithfully.

2.7 Methods

The definition of a function says nothing about what classes of objects might be supplied as arguments: there are no declarations in S. When you start out with a new function, you're usually just thinking of a few applications. If the function turns out to be useful, you or somebody else will start to apply it more generally. Chances are that some of those applications will not fit very well to the original definition.

To make the function more general, you can just edit the definition, making the computations conditional on the properties of the arguments to the function. A better style, in most cases, is to define *methods* separately for particular classes of arguments. Then instead of one function definition, you can develop as many different methods as make sense. Each method can still be simple (relatively simple, anyway), since it only has to worry about the particular classes of objects for which it was designed. The effect for users may be essentially the same, but the software organization will benefit, with less clutter and a clearer relation between the objects and how they are treated.

Let's take as an example the little function `whatis`, which is supposed to return a one-line description of an object. We used this function on page 66. Notice that when we used it on a numeric vector, it gave back both the class and the length. Applied to a matrix, it told us the number of rows and columns; applied to a function, it gave the number of arguments. A single definition that handled all these cases sensibly would start to become unwieldy. Worse, it would gradually come to know something about every class of object to which it would be applied.

Suppose we started out with the simplest definition of `whatis`:

```
> whatis = function(object) paste("Object of class", class(object))
```

Now, we decide to make the function smarter, for vector objects, by returning the length of the vector as well as the class:

```
setMethod("whatis", "vector",
  function(object)
    paste(class(object), "vector of length", length(object))
)
```

The call to `setMethod` specifies three things: the name of the function, the class that the argument should match, and a new function definition to be used as the *method* if the match succeeds. Otherwise, the behavior of the function is unchanged: the original function definition is retained as the method to use when the match fails—the default method, as we now call it. Functions for which methods are defined are called *generic* in S, the idea being that the function now defines generally what you want it to do, but individual methods can make explicit how that generic purpose is achieved.

There are three key concepts here. First, we organize software by a conceptual two-way table: by generic function, roughly corresponding to *what* we want to do; and by methods indexed by classes, corresponding to *how* we will do it. This is a much clearer organization than the use of conditional expressions within one large function. The second concept, which follows from the first, is that of isolating information about classes. Only the methods related to the class need to know about that class, and then only the properties of the class relevant to the purpose of the generic.

The third concept is that of gradual refinement. Because S keeps definitions of functions, methods, and classes in permanent chapters, you can query such definitions and use them as the starting point for further refinements. This approach has practical advantages, since S can produce an expression that defines the current version of a method, say, leaving the programmer only the task of editing in the refinements. A collection of `dump` functions produces the current versions of objects, methods, or classes. The function `dumpMethod` does this for a method, given the name of the function and the class or classes to match.

You can *always* get the current method definition, so long as the function itself is defined at all, even if no method has been explicitly defined for these classes. S will always select a method to evaluate the function; initially, it just selects the single definition of the function. Once methods have been defined, either one of these will be chosen or if none applies, the default, original definition. No matter which, dumpMethod will select that and dump it as the method, ready to edit.

Suppose we want to go on refining the whatis function to do something special for matrix objects.

```
> dumpMethod("whatis", "matrix")
Method specification written to file "whatis.matrix.S"
```

No method exists for class matrix, but a method for class vector was defined, and matrix extends this class:

```
> extends("matrix", "vector")
[1] T
```

If we look at the file written out, we can see that dumpMethod wrote out a call to setMethod that used the vector method definition:

```
setMethod("whatis", "matrix",
  function(object)
  paste(class(object), "vector of length", length(object))
)
```

The file is ready to edit, but we only need to refine the one line to produce something more suitable for a matrix object.

If I seem to belabor the gradual refinement concept, that is both because it has significant advantages in programming with data and because it differs from the approach of programming languages that expect programmers to produce a full and formal definition of the computations to be done. S is both a programming language and an interactive environment. The philosophy of the language is to exploit the resources of the environment to make programming a more gradual, evolutionary activity. So far as possible, each step leads gently to the next, from non-programming interaction through increasing involvement with the language.

2.8 Classes of Objects

So far, we have used the notion of an object's class informally, essentially as the name for "what" the object is. That is a good place to start, but if you want to design classes of objects (a very powerful tool in S), you will need this concept fleshed out somewhat.

Every S object has a class, a character string attached to the object that labels it. To each unique class name there may belong some description, stored in one or more S databases. For example, the standard libraries that are attached whenever you run S contain information on a large number of different classes, including the standard ones such as `"numeric"`, `"logical"`, and all the other kinds of S vectors. User-defined chapters can also contain information on classes; in fact, there is essentially no special behavior for "system" libraries in S.

What information does S actually "know" about a class? Sometimes, nothing except that this is the name of a class. Methods can be written for objects with any class. Effective methods, though, usually want to assume something about the objects from a particular class that distinguishes them from other objects. Two kinds of information about a class are commonly stored by S: the representation of the class, and the relation of the class to other classes. Representation information says something about what data an object from the class must contain, to be valid. This can be done in two different ways: classes may be defined as a list of named *slots*, each slot defined to contain an object of some other class. In this way, relatively complicated objects can be defined in terms of simpler ones. The specification of a class can also include a *prototype*, a single object of some other class. When an object from the class is generated by a call to `new`, the value returned will be a copy of this prototype, with its class set to this class.

Either or both of the representation and prototype can be supplied. Here are three examples:

```
setClass("track",
  representation(x = "numeric", y = "numeric"))

setClass("matrix", representation("array"),
   prototype = list(NULL, c(0, 0), NULL))

setClass("sequence", prototype = numeric(3))
```

The first example defines class `track` with two slots, both numeric. In the second class `matrix` is defined to include the same slots as class `array`—this is the effect of including an existing class with no slot specified. A prototype is also provided, so that `new("matrix")` will return an object like the prototype. If no prototype had been provided, `new` would be applied to the class of each of the slots. For example, `new("track")` returns an object with each slot containing `new("numeric")`.

If a prototype but no representation is given, the new class specializes the prototype to a different purpose but has no explicit slots. The `sequence` class, for example, uses three numbers to represent a general sequence. We don't want functions to treat those numbers as an ordinary numeric vector. Classes defined from prototypes are important in preventing confusion between what an object physically contains and how we want to think about the object.

Interclass relations in S are described by what we call `extends` or `is` relations. Such relations allow methods to be defined for one class but then applied to all classes that extend that class: a method defined for class `vector` can be applied to `numeric`, `logical`, or even `matrix` objects. A class $\mathcal{B}$ extends a class $\mathcal{A}$ if the representation of $\mathcal{B}$ explicitly includes class $\mathcal{A}$. More generally, an object, b, of class $\mathcal{B}$ is a class $\mathcal{A}$ object if b can be used in any method where an object of class $\mathcal{A}$ was called for. The `is` relations can usually be inferred from the representation of the class, but they can be explicitly set as well. The relations can be conditional, in that not all objects of class $\mathcal{B}$ need have the property. Also, they may or may not involve some computations in order to treat b as if from class $\mathcal{A}$.

The concept here is one of substitution or coercion. Given the object b and a method for class $\mathcal{A}$, evaluation can proceed because the S evaluator knows how to coerce b to become an object from class $\mathcal{A}$; after the coercion is done, the computation of the method can proceed, using the resulting class $\mathcal{A}$ object. The formal definition of the classes and of their relations makes this process explicit; its application is a key part of evaluating methods.

Class representations and `is` relations are at the heart of what makes S classes useful: they allow much computation to be organized automatically by S, based on general information about classes and methods. Programmers usually don't have to worry about such things, except when they want to design new classes. It does help to keep in mind, though, that representation and `is` information does exist and can be studied.

A class may be specified to be a `virtual` class. Such classes express what various other classes have in common. For example, the `vector` class expresses what numeric, logical, and other S vectors have in common: roughly, the concept of a number of elements that can be indexed from 1 up to the `length` of the vector. Virtual classes are a powerful concept. Methods can be written for them that apply to all the actual classes.

Newly defined classes can include a virtual class; they then automatically have access to these methods. For example, new classes can be defined to be vectors, if that makes sense, and be eligible to be used where existing "built in" vectors could be used. Virtual classes may not imply anything about the representation (as `vector` does not), but they may have a representation themselves. In this case, they not only link the actual classes via methods, but they enforce some common aspects of the representation. The class `structure` is an example of such a class: matrices, time-series, and other structure classes share both methods and part of their representation through this virtual class.

One subtle point often arises in software systems that deal with objects and classes: is a class an object? (Remember we promised that everything is an object. But if so, what is the class of a class?) Some systems introduce the notion of *meta-objects* for similar situations. S uses its notion of a database instead: information such as class descriptions are kept in ordinary S objects, but the objects are stored in *meta databases.* These are just ordinary S databases (directories usually), but they have been designated to be used for special purposes. The S `CHAPTER` command creates databases for method and class descriptions, along with the database for ordinary S objects. S functions that set method and class information perform assignments in the meta database. All the standard S software is available to manipulate objects in the meta database, but there are no conflicts with names for ordinary S objects and ordinary computations will not overwrite anything in the meta database. In order to have a compact term, we will refer to *metadata* throughout the book, but this always means ordinary data stored on meta databases.

S uses the same mechanism to store documentation objects—objects that contain the online documentation for S functions, classes, etc.

2.9 Interfaces

The final major concept of S to be discussed is the *interface.* S explicitly assumes that not all computations and not all data will exist only in the S language itself. Data exists on files and other similar entities in the file system. Computations may be done in the shell (the operating system's command line evaluator), or in compiled subroutines. Providing useful interfaces to such computations in a language like S requires more than just a way to "run" computations. Because of the functional and object-oriented nature of S, the interface needs to be defined through function calls and through the structure of the objects communicated through the interface, in both directions.

The interface to files and the like is defined through a virtual class, `connection`. The interfaces to the shell and to subroutines are defined via functions, each providing an interface and a way to communicate data.

2.9.1 Connections

Much computation in S requires data to be read from or written to some external source. Common examples are files in the file system and windows on the display. The data being transmitted either way is usually thought of as characters, divided into lines by newlines. S objects from the `connection` classes connect the S session to such sources, providing a unified view for the programmer.

Specific connection classes correspond to things in the operating system that provide useful, but different facilities: `file`, `pipe`, and `fifo` objects. In addition, `textConnection` objects treat S character vectors as connections. The `display` class refers to windows or the user's display as a connection. All connection classes extend the virtual class `connection`.

Connection classes provide several facilities for the programmer, by comparison to the underlying C-style concepts.

1. They provide a uniform mechanism for many S functions that need to read or write data. Where it makes sense, any S connection can be supplied to these functions.

2. During an S session, underlying software keeps track of currently active connections, allowing mixed reading and writing and other operations that are difficult or error-prone at a lower level.

3. Much of the programming detail needed in doing input and output can be hidden from the programmer by the support software.

If computations can be expressed in terms of connections generally, the result is usually more convenient, reliable, and efficient input and output.

2.9.2 Interfaces to the Shell and Other Programs

A command shell is an interpreter that takes lines of text and interprets them as commands in a (fairly unstructured) language. Derived historically from the `sh` command in Unix, command shells exist in various forms and on Unix, Linux, and Windows systems. A command shell provides fundamental facilities to deal with files, directories, and user communication; in addition, many other tools and languages expect to be invoked as commands from the shell.

Basic to the shell's model of computation is the notion of *standard input* and *standard output*: for many uses of the shell, lines of text are read from the file or stream of standard input and the command then writes lines of text on standard output. The S function `shell` uses this model to provide an interface to the shell. The function takes as arguments a command to execute and optionally an S object, `input`, to act as the standard input to the command.

In the `shell` interface, lines of text are mapped from and to elements of a character vector. The S object `letters`, for example, has 26 elements containing `"a"`, `"b"`, and so on. If supplied as the `input` argument, the command would get the corresponding 26 lines of standard input. Similarly, the standard output of the command can be returned as the value of the call to `shell`: each line of output becomes a character-string element of the value. (Some commands may not produce useful output; instead, we may want to know the termination status of the command; in this case `shell` is called with `output=F` and returns the status as its value.)

Uses of the interface may be as simple as turning a simple shell into an S function, such as the function `date`:

```
date = function() shell("date")
```

In other examples, the interface may be used to prepare for S output to files:

```
shell(paste("mkdir", path), mustWork = T, output = F)
```

In this case, S makes a new directory, whose path in the file system is specified in `path`. The optional argument `mustWork=T` says that an error should be generated if the command fails.

Another interface to shell commands comes through the `pipe` connection class. This interface anticipates starting up a shell command that will get its input not from a single S object, but from successive output from S, via functions such as `writeLines` and `cat`. Similarly, data flows back to S in this case via input functions such as `readLines` and `scan`.

A third, slightly less obvious interface to other systems uses `fifo` connections. These look like files in the file system but have the property that data is always written to the end of the file and read from the beginning of the file. Once the data is read, it disappears. Also called named pipes, `fifo`'s are an adequate mechanism for programs to communicate with each other: one program writes to the file, another waits for input to be available and then reads it. S can function on either side of this communication, as a writer or (via `setReader`) as a reader.

For a less mundane example of interfacing to other programs, we can look at the example of plotting probe test data, discussed in section 1.7.2. In that section we designed an S plotting function to display test results, coded by color, with a separate plot for each of the wafers in a lot. A valuable plot, but one that could be made even better by allowing the user to interact: to look at one or all failure modes; to see the composite behavior over all wafers in the lot, etc. You could program all that interaction in S; user interaction is something S does fairly well. But other systems are designed expressly for a combination of user interaction, graphics, and window management. One notable example is the Java language.

You don't need to know anything about Java to follow this discussion. Just take my word that a Java program can be written to display probe test data such as shown in section 1.1 and on the cover of the book. In fact, Mark Hansen and David James followed up their S-Wafers software, [4], with Java code that did just this. It also provided interactive control as suggested above. You can see the Java plot in action by going to the web site,

`http://cm.bell-labs.com/stat/project/icmanuf`

When first designed, the Java plot was considered stand-alone, but in fact it's an excellent candidate for the S concept of interfaces.

Two approaches make sense: a rather simple one based on the shell interface and a more ambitious one using `fifo`'s. Let's begin by assuming the Java program has been written, in the first case, as a command, `JavaWafers`, defined on our own computer. This command reads the probe test results from its standard input and then runs the Java plot, with the results displayed on the user's screen and the user able to select various plotting modes by clicking on that plot.

The shell interface is then going to be a plot method for the `probeTest` class of data. The plot method in S starts up the `JavaWafers` command, with the data from the S object as its input. We need a function in S that strings out all the information needed by the command: the number of sites, number of wafers, the row and column positions for the sites, and then the results themselves. Let's just assume that the function `probeTestForJava` does this, without worrying about the details (they are in fact very much like the `print` method on page 55 in section 1.7.2).

The method itself is then very simple indeed:

```
setMethod("plot", signature("probeTest", "missing"),
function(x, y, ...)
  shell("JavaWafers", probeTestForJava(x))
)
```

To be more precise about the shell command, we want it to spawn the Java program in the background and return. Java will then provide further interaction directly with the user's display.

Much the same graphics and interaction could be used in quite a different way, communicating via `fifo`'s. In this scenario, there is some additional software involved that manages communications between an S process and the Java software running the example. This software is a *message server*; it receives and delivers messages for its clients. In this example, the S process as a client writes the information for the wafer plot to a `fifo`. The message server notices the data on the `fifo`, reads it, interprets it as a message for the Java program, and passes it along. The Java program reacts to the message as before, by generating the wafer display for the user.

Communicating messages via a message server requires more substantial software, but has several advantages. An experimental implementation of this approach, with the message server implemented in Java, was written to explore communication between S and Java (see reference [2]). Where the shell command essentially restricts the Java program to run on the same machine as S, the message communication approach can be (and in fact, was) implemented with a Java *applet*; that is, with the Java code invoked across the web. Also, with the communication going on in messages, it makes sense to see the interaction as a continuing dialog, with both systems listening for messages as well as originating them. In the future such interaction is likely to be done at a higher level, with communication handled by a general approach rather than *ad hoc* code. However, the concept of the *interface* as central to open-ended programming with data will only become more important.

2.9.3 Interfaces to Subroutines

The functions `.C` and `.Fortran` provide interfaces to subroutines in C and Fortran. These routines need to have been compiled and linked dynamically to S. If so, the S evaluator will generate a call to the routine. The arguments to the S interface function, in addition to the name of the routine, must match the arguments to the subroutine in number and type. To each of the permitted argument types there corresponds an S class; for example, class `numeric` corresponds to the data type `double *` in C and to `double precision` arrays in Fortran.

It's assumed that the subroutine will produce some useful results by overwriting some of its arguments: this is the standard C or Fortran approach to many computations. S is a functional language, so nothing in the S objects should ever be accidentally overwritten; instead, the value returned by the interface function is a list corresponding to the arguments, but with any overwritten data reflected in the corresponding element of the list.

For example, the Fortran routine `chol` takes a square numeric matrix, the number of rows, a scratch vector of that length, and two other arguments to control singularities. The routine writes its result, a matrix representing the decomposition of the input matrix, back on top of the first argument. Then a simple interface to this routine would be as follows:

```
"chol" =
function(x)
{
  p = nrow(x)
  cc = c("numeric", "integer", "numeric",
         "integer", "integer")
  z = .Fortran("chol", x, p, numeric(p),
         0, 0, CLASSES= cc)
  z[[1]]
}
```

The special argument `CLASSES` to `.Fortran` tells the interface what classes the arguments should be to conform to the declarations in the subroutine. The interface makes sure the arguments get coerced to the right class.

The returned value is a list; we only want the overwritten matrix as the result of the S function, so it's the first element that gets returned as the value. The actual interface to this routine should be more careful; for example, Fortran has no way to check that the matrix is really square, so a check for `nrow(x)==ncol(x)` should be done in S.

For detailed information on the `.C` and `.Fortran` interfaces, see section 11.2. There is another interface to C, designed for the bold soul who wants to manipulate the structure of S objects in C (section A.1). It's recommended *only* for those who really need it, and certainly not at any early stage in your S work. Lastly, there is an interface through a call to `.Internal`, mostly derived from earlier versions of S. Don't use this interface for new programming.

The concept underlying all the interfaces to C and Fortran is that the subroutine's computations are *atomic* to the S evaluation model. In a formal sense, the evaluator gives up control to the routine and has no knowledge of what happens until the routine returns. In understanding the computational model for S, such routines play the role of individual "machine" operations in other languages. They are as close to the machine as the S computational model gets.

This concept has some practical consequences. First, the atomic computations are nearly always defined in terms of arbitrarily large objects; arithmetic, for example, is defined for vectors of arbitrary length. Second, the set of atomic computations is neither small nor fixed. Many computations are included, such as mathematical computations, matrix decompositions, and table lookup, that are in no sense "close" to any typical hardware instruction set. But because these computations typically do nothing internally with the kind of generality that evaluations in S imply, they are conceptually atomic and often much faster than an iterative S calculation to achieve the same goal.

The efficiency issue illustrates the value of an extensible concept of atomic computations. Where new, useful computations can be added to the set of such routines, the S evaluation model can expand its notion of what computations are atomic. Substantial effort can go into making these computations efficient if the new atomic computation can be used in many ways; if, for example, it applies to all numeric data, or even to all vectors. Such redefinition of the language's atomic computations is a highly recommendable way to extend the system.

Chapter 3

Quick Reference

This chapter gives a quick reference to a number of topics. S has a very large number of functions, classes, and methods. Finding the right one for a particular need can be difficult. Browsing here may enable you to find a function or other technique that does what you want. The tables list a large number of functions and other tools; for further information, you can then look up the topic in the index of the book or with the online help in S.

This chapter plays the role of a rather long reference card, situated in the middle of the book, rather than being provided separately. You will find it useful to flip back here when searching for the right function to use. This chapter works well in conjunction with the index, particularly the entries under "hint".

3.1 The S Session

The way you start and end an S session may depend on the user interface you prefer; here we describe the classic interface from a shell program. In this interface, you start the session by typing the line `S` in response to the shell prompt. Quit as in the table: in cases of desperation you can kill the S process, for example by sending it a *quit* signal.

Tool	**Purpose**
`S -i` *file*	Use *file* as a source file to initialize the S evaluator. *Warning:* this overrides the standard S initialization, including the rest of this table.
`.S.init`	File in your login directory, of S tasks to be evaluated on startup.
`.S.chapters`	File in your login directory, with chapters to be attached on startup.
`.First`	Object in working data to be evaluated on startup.
`.Last`	Evaluated before quit.
`record(file)`	Make a detailed record of the session.
`interactive()`, `setInteractive(on)`	Test, set whether the session is interactive.
`q()`, `exit(error)` `terminate(...)`	Three ways to quit: normally, right away (with status `error`), and with a terminal error.

Table 3.1: ***The Session.*** *Tools to customize the S session. The* `-i` *file is an S source file;* `.First` *and* `.Last` *are functions or S language objects.*

In the standard S session, you will normally be prompted to provide input. The evaluator then waits for an event, such as a line of typed input. If the line is a complete expression, the evaluator invokes the standard task: roughly, the text is parsed, the resulting expression is evaluated and the result, if it is not an assignment or other special object, is shown to the user. If the current expression is incomplete, the evaluator will save the partial expression, issue a continuation prompt, and wait for the next event.

The behavior of the S evaluator can be controlled on starting the session, and on quitting from the session, by the tools shown in Table 3.1. The session can be customized at initialization, or later, by specifying event handlers (see Table 3.23 on page 117), and by attaching or detaching databases (see Table 3.14). For a fuller discussion of the S session, see section 4.1.

3.2 The S Language

The language for the standard S parser has a fairly typical C-style grammar, with a few extra features. Table 3.2 on page 98 summarizes the grammar. The syntactic type in the first column is defined as a pattern in the second column. When the parser successfully matches input in this form, it creates a corresponding object, whose class is given in the third column. For example, the `call` object is created when the parser matches an arbitrary expression (an Expr) followed by a left parenthesis, followd by an argument list followed by a right parenthesis. Quoted material in the definition appears literally as shown (the parentheses, for example). Other syntactic types either themselves appear on the left of this table or in Table 3.3 on page 99, if they are defined in the language. The syntactic forms in the second table do not correspond to distinct classes of objects; their role in the object depends on the context.

Types not appearing in either table are lexical types, defined as sequences of string patterns, similar to regular expressions. These are summarized informally in Table 3.4. When a lexical type appears on its own, an object is created of the corresponding S data class. In all three tables, text in *italics* is used to give informal, verbal equivalents to standard syntactic forms.

Type	Definition	Class
Call	Expr "(" ArgList ")"	"call"
Infix	Expr Op Expr Unary Expr Expr "$" ExName Expr "@" ExName	"call"
Subset	Expr "[" ArgList "]" Expr "[[" ArgList "]]"	"call"
Assign	Expr AssignOp Expr	AssignOp
Conditional	"if(" Expr ")" *and optional* "else" Expr	"if"
Iteration	"for(" Name "in" Expr ")" Expr "while(" Expr ")" Expr "repeat" Expr	"for" "while" "repeat"
Flow	"break" "next" "return(" Expr ")"	"break" "next" "return"
	"(" Expr ")"	"("
	"{" ExprList "}"	"{"
Literal	Integer *or* Numeric *or* String *or* Name	
Function	"function(" arglist ")" expr	"function"

Table 3.2: ***The Language(I).*** *Syntactic types in S. The type* Expr *can be any of the types defined by a row of the table. See section 4.2 for a more leisurely discussion.*

Type	**Definition**
ArgList	*0 or more of* Arg *separated by* ","
Arg	*Optional (* ExName "=" *) and optional* Expr
ExprList	*0 or more of* Expr *separated by (* ";" or new line)
ExName	Name *or* String

Table 3.3: ***The Language(II).*** *Syntax in S other than* Expr.

Type	**Definition**	**Class**
Integer	*Optional (* "+" *or* "-" *) and one or more digits*	`"integer"`
Numeric	Integer Exponent *or* *(Optional* Integer *followed by* "." *and optional digits and optional* Exponent *)*	`"numeric"`
Complex	*Optional (* Numeric *(* "+" *or* "-" *))* Numeric "i"	`"complex"`
Op	"+", "-", "*", "/", "^", ">", "<", ">=", "<=", "==", "!=", "&", "&&", "|", "||", "!", ":", "?", "$", "@", "~", "*%anything%*"	
AssignOp	"=" *or* "<-"	

Table 3.4: ***The Language(III).*** *Lexical rules. A function named* "*%anything%*" *can be used as an infix operator; e.g.,* "%*%".

3.3 Computing with S

S has a vast number of functions. Tables 3.5 to 3.9 list some of the functions you are likely to find helpful for general-purpose computations in S. For details on any of them, use the online help; for example, `?max`, or `?":"`, with the quotes needed for operators that have special characters in the name—these are quoted in the table as well.

Functions	**Purpose**
`"+"`, `"-"`, `"*"`, `"/"`, `"^"`, `"%%"`, `"%/%"`	Arithmetic: Returns an object with numeric data, but with structure inferred from the arguments. `NA`'s and `Inf`'s produced or passed on where it makes sense.
`exp`, `log`, `log10`, `logb`	Numeric and complex exponential transforms; `logb` takes a base as argument.
`sin`, `cos`, `tan`, `asin`, ··· `sinh`, ··· `asinh`, ...	Numeric and complex trigonometric and hyperbolic functions.
`abs`, `round`, `floor`, `trunc`, `signif`	Numeric transforms: See also the modulo operator, `"%%"`, and integer division, `"%/%"`.
`">"`, `"<"`, `">="`, `"<="`, `"=="`, `"!="`	Comparisons: Returns an object with logical data values, but with structure inferred from the arguments.

Table 3.5: ***Numerical Computations.*** *The functions preserve structure (e.g., of arrays) in the arguments; operators generally return an object like the more structured of the arguments.*

Functions	**Purpose**
`max`, `min`, `range`, `mean`, `median`, `var`	Summarize numerical data in one or two numbers.
`rnorm`, `runif`, `rexp`, `rgamma`, `rlnorm`, `rcauchy`, `rbeta`, `rf`, `rlogis`	Generate random numbers. The letters after the `"r"` define the distribution. To find quantile, probability, and density functions, replace the `"r"` with `"q"`, `"p"`, or `"d"`.
`sample(x, size)`	Sample from object `x`.
`apply`, `lapply`, `sapply`, `rapply`, `tapply`	Apply functions: a compact way to express loops over arrays, lists, and levels.

Table 3.6: ***Utility Functions.*** *Some S utilities for quantitative computing.*

Functions	**Purpose**
`"&"`, `"|"`	Elementwise logical *and*, *or*. In `if` and `while` expressions, use the operators and functions below.
`"&&"`,`"||"`	Combine single `T`, `F` values. Only the left operand is evaluated if it determines the result.
`any`, `all`	Summarize by single logical values.
`identical`, `all.equal`	Testing: Use `identical`, not `"=="` to test two objects for equality.
`is.na`	Test and set `NA` values.

Table 3.7: ***Control Functions.*** *Some S functions for testing and controlling computations.*

Functions	Purpose
`"["`, `"[["`, `el`, `elNamed`	Extract and replace subsets, elements.
split, c, list, named	Split apart or glue together objects.
`sort`, `rank`, `sort.list`, `order`	Sorting: To put other data into the order defined by `x`, use `sort.list(x)`.
`match`, `pmatch`, `regMatch`	Matching: For large objects, use `match` with `string` objects (see section 5.1.3).
`diff`, `cumsum`, `rev`, `rep`	Ways to convert numbers, find break points, repeat patterns.

Table 3.8: ***Data Manipulation.*** *Some S functions for manipulating, sorting, matching, and converting data in objects.*

Function	Purpose
`format`	Numeric data to fixed-width strings.
`rawFromHex`, `rawFromAscii`, `rawToHex`, `rawToAscii`	Raw bytes from and to their coding as character strings.
`deparse`, `deparseText`	S language objects to character strings or to a single string.
`unlist`	List-like objects to the simplest vector holding its elements.
`unclass`	Any object to the list of its slots or to its prototype's class.
`cut`	Numeric data to ordered levels.

Table 3.9: ***Data Conversion.*** *Functions for converting the structure or contents of data, other than by the general* `as(object, Class)` *methods.*

Table 3.10 gives some standard S graphics functions and Table 3.11 shows some graphical parameters that can be supplied, either as arguments to the functions or separately through the function `par`.

Functions	**Purpose**
`plot(x)`	Plot the object `x`.
`plot(x, y)`	Plot of `y` vs `x`.
`boxplot`, `hist`, `qqplot`, `qqnorm`	Show the pattern of numeric data.
`lines`, `points`, `text`, `segments`, `polygon`, `abline`	Add to an existing plot or build up a special plot. To set the co-ordinates use `par(usr=)` or `plot(type="n")`.
`identify`, `locator`	Interactively select data or positions on a plot.
`title`, `axis`, `legend`, `mtext`	Add to the margins of the plot.

Table 3.10: ***Graphics Functions.*** *The basic S graphics computations..*

Parameter	**Purpose**
`pch, col, lwd, lty`	Specify the appearance of graphics: plot symbol, color, line width and type.
`adj, srt, cex, csi`	Text: control the position and size.
`usr`	The co-ordinate range for `x, y`.
`mar, xaxs, xaxt, yaxs, yaxt, lab, las`	The size and style of the margins around the plot and the axes and labels in them.

Table 3.11: ***Graphical Parameters.*** *Name and meaning of useful parameters supplied either as named arguments to graphics functions or to the function* `par`*: see* `?par` *for details and more parameters.*

Table 3.12 lists commands that are recognized by the interactive browsers, both the `recover` function invoked from the standard error recovery and `browser` itself, recommended to be inserted in a computation by calling `trace` for a function or `traceMethod` for a method.

Browsers allow standard interactive S expressions, with the distinction that local objects and assignments apply to the current function call frame. The prompt will typically include the name of the function corresponding to this frame. A few special functions are defined while the browsers are running. These are shown in Table 3.12; there is the additional shortcut that the parentheses and argument list can be omitted from these, so long as there is no local object with the same name.

Command	**Function**	**Purpose**
`where`	`where()`	Print the call traceback.
`up`	`up(n=1)`	Move up `n` levels of calls.
`down`	`down(n=1)`	Move down `n` levels of calls.
`q` or `c`	`q()`	Quit from the browser.
`stop`	`stop(...)`	Stop the task, no error handling.
`dump`	`dump()`	Dump the frames and quit.

Table 3.12: ***Browser Commands.*** *Names that can be typed as shortcut commands during the interaction in* `recover` *and* `browser`*. If the command name is also a local variable, you must use the functional form.*

The `browser` function and the function `debugger` can also be used to examine a named S list, treating the elements of the named list as objects. This technique is used to dump the S evaluation frames and examine them later on, but with limited interaction, since the lists are no longer S evaluation frames.

Some additional functions are useful tools in debugging: `debugPrint` prints an object with information about its class, etc., but clipping long vectors to save space; `traceback` tells you where you are, like `where` in the browser; and the functions in Table 3.13 provide pieces of the evaluation state.

Table 3.13 deals with computations using the S evaluator explicitly. The `sys.` functions and `eval` operate with the model that the evaluator has a list of frames, one for each currently open function call, and that each of these frames is a named list of objects.

Function	**Purpose**
`sys.frame(n)`	The named list of the objects in frame `n`.
`sys.frames()`	The list of all the evaluation frames.
`sys.call(n)`	The S `call` object that generated frame `n`.
`sys.calls()`	The `expression` object containing all the calls currently in progress.
`sys.function(n)`	The `function` object defining frame `n`.
`sys.parent(n=1)`	The frame number of `n` generations of calls back.
`sys.parents()`	The frame numbers of the parents of all the current frames.
`sys.status(n)`	A named list of frames, calls, functions, and parents for all the frames up to `n`.
`Quote`, `substitute`	Generate S language objects without evaluating them.
`eval`	Explicitly evaluate an S language object.

Table 3.13: ***The S Evaluator.*** *Programming with the evaluator, to extract information from the evaluation model (the* `sys.` *functions), or to control expressions and evaluation.*

3.4 Databases

An S database is a way of associating names with S objects; in the abstract, a database is a lookup table whose entries are S objects. The S search list is the set of databases in which S looks for functions and other objects. The first database in the search list is the *working data*, the database where S puts top-level assignments. The next two tables provide functions to deal with the databases on the search list (Table 3.14), and to extract or set information in individual databases (Table 3.15).

Function	Purpose
`search, searchPaths`	The names or paths of attached chapters.
`attach, library`	Attach a database, return a unique `attached` object. See section 5.3.
`detach`	Detach a database.
`synchronize`	Synchronize a database for external changes.

Table 3.14: ***The Search List.*** *Examine and modify the list of searched databases.*

A set of database tools allow you to examine and set the essential information about individual databases. By convention, the names of these functions begin with `"database."`. Table 3.15 lists them. The `where` argument to these functions can be a position, the string corresponding to the name of the database in `search()`, or an `attached` object, as returned by `attach` or `database.attached`. The third choice is the only one guaranteed to have a constant meaning, regardless of databases being attached or detached.

Function	**Purpose**
`database.attached`	The object of class `attached` corresponding to the arguments.
`database.position`	The database's position in the search list or `NA`.
`database.status`	The current status of the database, e.g., `"read"`.
`database.type`	The character string describing the kind of database, e.g, `"directory"`.
`database.name`	The character string name by which the database appears in the search list.
`database.object`	The object attached as this database, or the path string for a directory.
`databse.attr`	An old function used to include attributes with a database object.

Table 3.15: ***Database Functions.*** *S functions returning information about S databases. All take arguments* `where`, *identifying the database of interest, by position, name, or as an* `attached` *object and, optionally,* `meta`, *to use the corresponding meta database.*

Taking S objects from databases, possibly doing something with them, and then restoring them, requires functions to read the objects and then to write them again. The operation can take place on one object or on several. S provides three formats in which the operations can take place: the binary format used in S databases, a symbolic dump format, and a format suitable for the S parser. Table 3.16 shows, in the style of an S 3-way array, the relevant functions. For example, to **Read One** object in **Binary Format** call `get`. See the functions' online documentation for details and arguments.

Binary Format		
	One	**Several**
Read	`get`	
Write	`assign`	
Symbolic Dump Format		
	One	**Several**
Read	`dataGet`	`data.dump`
Write	`dataPut`	`data.restore`
Parse Format		
	One	**Several**
Read	`dget`	`dump`
Write	`dput`	`source`

Table 3.16: ***Moving Objects Around.*** *Functions to help move S objects, one or several, between databases.*

Here are some general points about the operations in Table 3.16.

1. Binary format is the natural way to move objects around during a session; symbolic dump format is for moving between systems and for archiving or version control (section 5.5); parse format is for humans to look at and possibly edit.

2. Files, or more generally S connections, that we read or write with either symbolic dump format or parse format will never contain anything other than ordinary text characters.

3. The reasons to use symbolic dump format, rather than parse format, to ship and archive objects are that the restore process is much faster and that there is a possibility of losing information for some objects with parse format.

4. Shipping a whole S chapter requires creating a set of files that define *all* the objects in the chapter, plus any other information such as source for C or Fortran programs used by the chapter. See section 5.4.2.

3.5 Programming

Programming starts as the writing of S functions and may go on to programming with classes and other parts of S. There is a fundamental programming cycle in S: dump the current version (even when there isn't one—see below); edit; source the edited version back into S; test it out interactively. This cycle iterates, of course, until the programmer is sufficiently satisfied or time runs out. The cycle applies not only to functions but to the whole of S programming: classes, methods, and documentation as well. Since everything is an object, the cycle amounts to dumping, editing, and recreating the relevant objects in each case.

Table 3.17 lists some of the functions supplied as programming tools by S for this cycle. Here are some notes on the rows of the table:

Dump: It's generally better to dump before even trying to define the object seriously, especially for methods and documentation, since this gets the form of the source file right. Then you can work on the contents.

Inspect: These produce more readable descriptions than just looking at the source code, usually via methods for `show`.

Source: Except for documentation, the files dumped are just S expressions; use `source`, or sink input, or treat the file as standard input to an S process. The argument to the source functions is always the name of the dump file. See the online documentation for the functions on the second line for details of what happens.

`meta=`: The actual S objects are read and written to the corresponding meta database.

	Functions	Classes	Methods	Help
Dump	dump	dumpClass	dumpMethod dumpMethods	dumpDoc
Inspect	getFunction	getClass	selectMethod, showMethods	?
Source ··· **calling**	source	source setClass	source setMethod	sourceDoc
Debug	trace	traceMethod		
Remove	remove	removeClass	removeMethod, removeMethods	removeDoc
Exists	existsFunction	existsClass	existsMethod	existsDoc
Find	findFunction	findClass	findMethod, findMethods	findDoc
meta=		"methods"	"methods"	"help"

Table 3.17: ***The Programming Cycle.*** *Some essential tasks in the cycle of creating S functions, classes, methods, and documentation, with tools for each. See page 110 for some notes.*

All the tools working with objects (everything except the **Source** row) take an argument with the character string name of the function, class, or documentation topic. They also generally take an optional `where=` argument to specialize the action or search to a particular database.

The functions such as `dumpMethod` in the **Methods** column take the name of the function and the signature of the method as their first two arguments. Where a second function is given in this column, it tries to work on all the methods for a particular generic function at once. See Table 3.20 on page 114 for more tools for methods.

The **Help** column deals with documentation objects, but if no formal documentation exists, tools such as `dumpDoc` will generate a documentation object from the self-documentation comments in functions (`selfDoc` gets self-documentation even if a documentation object exists). There can be several topics per documentation object. Any one of them will do to find or dump the documentation, and `sourceDoc` will reset them all.

The tools in Table 3.18 are used either in generating messages to the user (the first row in the table) or are ways to control what happens when warnings or errors occur. The four functions in the first row all take arbitrarily many arguments and paste them together, with no added white space, to form a message. The `message` function just prints the message; the other functions produce some measure of error handling (see section 6.4.2 for details).

Tool	Purpose
`message`, `warning`, `stop`, `terminate`	Make a message of all the arguments, then increasingly severe error action.
`options(error=`*expr*`)`	Set the error handler, usually for the rest of the session.
`.onError`, `.onError`N	Additional error actions, usually temporary.
`options(warn=`n`)`	Set the warn level; 2 makes warnings into errors.
`recover`, `dump.frames`, `dump.calls`	Suitable functions to be the `error=` option.

Table 3.18: ***Error Handling.*** *Writing error messages; arranging actions in case of error.*

3.6 Classes and Methods

Tables 3.19 and 3.20 deal with creating classes and methods.

Functions	Purpose
`getClass`, `setClass`, `removeClass`	The definition of a class.
`extends`, `is`, `setIs`	Examines or sets `is` relations between classes.
`validObject`, `setValidity`	Uses or sets a method to test that an object conforms to its class.
`getClasses`	The classes currently on a database, or anywhere.
`"@"`, `slot`,	Slots in objects.
`getSlots`, `slotNames`, `representation`, `prototype`	Parts of the class definition.
`setAs`, `showAs`	Explicit conversion between classes.
`getClassVersions`, `setClassVersion`	Version control for classes.
`getOldClass`, `setOldClass`, `removeOldClass`, `oldClass`	Deal with classes from previous versions of S.

Table 3.19: ***Tools for Classes.*** *Some S functions for dealing with class definitions. For the basic programming cycle, see Table 3.17. For a general discussion, see Chapter 7.*

Functions	**Purpose**
`setMethod`	Set a method for a particular function and signature.
`existsMethod` *and* `hasMethod`, `getMethod` *and* `selectMethod`	Test and get methods. The second in each pair includes inherited methods.
`getMethods`	The metadata object defining the methods for a specified function on a specified database.
`getGenerics`	The functions defined as generic functions, on the specified database, or anywhere.
`callGeneric`, `callMethod`, `evalMethod`	Explicit use of a generic function or a method for it.
`getGroup`, `getGroupMembers`, `setGroup`	Generic functions belong to a group generic function.
`oldMethod`, `UseMethod`, `NextMethod`	Tools for dealing with old-style S methods.

Table 3.20: ***Tools for Methods.*** *Some S functions for dealing with method definitions. For the basic programming cycle, see Table 3.17. For a general discussion, see Chapter 8.*

3.7 Documentation

The documentation file written by dumpDoc and read by sourceDoc is in the SGML language, a general markup language used, for example, in the HTML documents on the web. The documentation *objects* in S are of class sgml. The body of the documentation is built, recursively, from objects of class sgmlTag, each corresponding to a *tag*, or command, in SGML. When dumped, these objects turn into standard pieces of SGML, starting with the the tag name in angle brackets ("<s-value>") and ending with the tag name preceded by the character "/", again in angle brackets ("</s-value>").

There are many ways to use and edit SGML files; this book only deals with a very few. See the web site (page vi in the Preface) for more information. If you plan to edit the documentation objects directly, some sections (e.g., the usage section) should be altered in the S objects, because the SGML is generated in very stylized form. Others contain plain text, which can be edited easily in the SGML file. Table 3.21 summarizes the special top-level tags appearing in the dumped documentation file, with some hints on editing them. Chapter 9 gives further details.

Name	Prints as	Purpose; Editing
s-name		A topic name.
s-title	Title	The title of the help.
s-usage	Usage	The calling sequence of the function(s). Don't edit.
s-args	Arguments	Subsections for each of the arguments.
s-value	Value	The value returned by the function.
s-examples	Examples	S code for examples.
s-keywords	Key Words	Key words, if you want them.
s-see	See Also	Related topics, with discussion.
s-signature	Signature	The method signature.
s-slots	Slots	Descriptions of the class's slots.
s-contains	Contains	The classes this class extends.

Table 3.21: ***Documentation Sections.*** *Top-level* SGML *tags generated for S documentation. The last three rows apply to method or class documentation only. See Chapter 9 for a general discussion.*

3.8 Connections; Reading and Writing; Events

Connections are the objects that connect the S session to files and other things in the operating system from which you might read information or to which you might write information. The information is usually read and written as text, and reading is usually done in terms of lines of text (separated by newline characters), but neither of these restrictions is strict and neither should normally get in your way. Table 3.22 shows some of the tools available for dealing with connections. See the online help for individual functions and Chapter 10 for the overall information.

Function	Purpose
`file`, `pipe`, `fifo`, `textConnection`, `terminal`	The connection classes and corresponding functions (except `terminal`).
`open`, `close`, `isOpen`	Opening and closing connections; see section 10.5.2 for options.
`scan`, `readLines`, `parse`, `parseSome`, `dget`, `dataGet`, `readRaw`	Reading connections, expecting fields, or other special format.
`cat`, `writeLines`, `dput`, `dataPut`, `writeRaw`	Writing to connections.
`pushBack`	Push text back to be reread.
`seek`	Position a `file`.
`sink`	Divert standard output, input, or messages.

Table 3.22: ***Tools for Connections.*** *Connection classes and functions for them.*

The S session is driven by an *event loop*, in which the S evaluator waits for potential events. When an event is discovered, S does the corresponding action as a task, then resumes waiting for more events. If more than one event is waiting, S picks one (don't assume anything about which one). Events can be scheduled and controlled using the tools in Table 3.23.

Readers	
Function	**Purpose**
`setReader`	When input is waiting on the connection, call an action function to read it.
`dropReader`	Stop looking for events for this tag or connection.
`showReaders`	Print a table of current readers, with their tags.
`isReader`	Is there a reader on this connection?
`standardReader`	The action to mimic S's standard behavior.
`parseEvalReader`	The action to just parse and evaluate tasks.
Monitors	
Function	**Purpose**
`setMonitor`	Call the action function after specified waiting time.
`resetMonitor`	Called in the monitor's action to make it wait again.
`setSubEvents`	Start or stop looking for monitor events during the task.

Table 3.23: ***Event Handling.*** *Functions controlling readers or monitors.*

3.9 Interfaces

Table 3.24 summarizes interfaces from S to other software. The first two interfaces treat a character string as a command for the command shell to run; through these, any executable command in the operating system can be initiated. The remaining interfaces all relate to compiled code (subroutines) in C, C++, or Fortran (or in principle other languages that produce compatible object code). The three `"dyn."` functions supplement the automatic dynamic linking of S chapters; for most purposes, you can get along without these and should. See Chapter 11 for more details on interfaces.

Function	**Purpose**
`shell(command, input)`	Run `command` in the shell, with `input` as standard input.
`pipe(command)`	A shell connection. Write to it to give the shell command input. Read from it to get the shell's output.
`.C(NAME, ...)`, `.Fortran(NAME ,...)`	Call the `C` or `Fortran` routine `"NAME"`. Special arguments `CLASS=` and `COPY=` can control coercion, copying.
`.Call(NAME, ...)`	Interface to C for S-dependant C code; see Appendix A.
`dyn.exists`, `Fortran.symbol`, `C.symbol`	Check for an entry point; use `Fortran.symbol` to get the right argument.
`dyn.open`, `dyn.close`	Link in compiled files *other* than S chapters, which are done automatically
`.Internal`	An old-style interface to C; don't use it.

Table 3.24: ***Interfaces.*** *Functions in S that run commands or subroutines in other languages.*

Chapter 4

Computations in S

This chapter is about how to use S (as opposed to how to program in S), discussing the S session and the fundamental computations on data that S provides: data manipulation, numerical computations, comparisons and searching, computations to organize objects, graphics, models, and other utilities. Computational efficiency and the S evaluation model are also discussed. Chapter 5 is a companion chapter discussing objects and databases. Together they form a reference on the built-in computations in S, the existing building blocks with which you can program. If you have used S in the past, much of this may be familiar, and you may choose to come back here only when you need something specific.

4.1 The S Session

The user begins an S session by starting up an S process, from a shell or some other user interface. Just how that happens depends on the user interface chosen; classically, the user invokes S from a shell, by typing a command, typically `Splus` or `S`. This invokes a shell script that in turn starts up an executable version of S. Actually, the shell script will run any command given as an argument to it. One of the useful such commands is:

```
Splus SHOME
```

This prints on the standard output the directory in which S resides; you can look there for some files that tell you about your local version of S.

The traditional shell-level interface to S is far from the only one or the most convenient. Special graphical interfaces are likely to be more convenient, when they are tailored to the window manager or other general user interface. For users of `emacs`, the `s-mode` or `ESS` interface is a general and convenient interface. Each of these special interfaces likely provides control features in addition to those described in this section. See the online documentation for some help in finding out. The control features described here depend only on S itself, and should be available regardless of what else is provided.

The S session continues until the user quits, until there is no further source for possible tasks, or until something very nasty happens that causes the S process to exit abnormally. Let's hope that you never encounter the third possibility. The usual way to quit is to call `q()`, or just to come to an end-of-file on the standard input. There are lots of other variants and you can control the details of what happens when you quit the session. See section 4.1.4 for a discussion.

4.1.1 Customizing the Session

The S user can customize the interaction with S in a variety of ways. The most important is to control the set of S databases available when looking for functions or other S objects. By including or excluding specialized libraries, the S session can be specialized to handle the problem areas of concern to you. For example, the library `models` includes extensive software for fitting and analyzing statistical models.

Another form of specialization is to set various S `options`, used by S functions to decide such questions as how to print output, what to do in case of error or warning situations, and many other details of behavior (see `?options`). Beyond these standard ways, S can be set up to have all sorts of tools to hand, such as particular connections, tables, or individual objects.

You can introduce these customizations at any time but often you would like them to take place automatically when the S session starts. You may want the specializations to be very general, applying to all S users on your system, or to all your sessions, or only to some of your sessions. There are two fundamental levels at which you can customize startup: the `.First` object and the S initialization file.

The initialization of the S process proceeds in stages:

1. Some very basic, fixed initialization brings the the evaluator manager to the point of being able to evaluate expressions.

2. S then looks for an initialization file. This is a text file containing S tasks that constitute the script for the session. By default, S uses the file `"$SHOME/S.init"`, but the initialization file can be specified explicitly (see page 122). The file is an S source file; when evaluated, it takes the remaining steps in this list.

3. S looks for a file defining the chapters to be included initially. The file `"$SHOME/.S.chapters"` lists the libraries that S itself includes by default. If the file `".S.chapters"` exists in your login directory, this is interpreted as the names of the S chapters you want attached in addition. The file should consist of names of libraries or chapters, one per line.

4. A working data (usually a directory containing an S chapter) is chosen. See section 4.1.2 for the rule used. With the working data chosen, an audit file is opened, if possible.

5. S looks for an object `.First` in any of the databases attached at the start of the session. If it finds one it (or its body, if it is a function) is evaluated in the top-level frame, so that assignments in `.First` are permanent.

6. S now looks for a file `.S.init` in your login directory. This can contain any additional initialization steps you want to apply.

If you provide your own initialization script, you can choose whether to *add* to the standard initialization, so that you retain the use of the `.S.chapters` file and `.First` object, or to *replace* the S initialization entirely. Unless you have some very definite ideas in mind, the first approach is recommended. Three tools can be used: the `.S.chapters` file to specify libraries or user chapters to be included; the `.S.init` file to supply general additional initializatio; and the `.First` S object in a particular S database. Another related technique is to provide initialization actions (`.on.attach`) in a particular library or chapter. Section 5.3.3 discusses attach actions.

Let's turn to some examples and describe the effect of the initialization steps in more detail. Suppose I want to include the S library `trellis` and a chapter in directory "S" under my login directory, no matter where I was actually running S. My ".S.chapters" file would be:

```
$HOME/S
trellis
```

Notice that the first line is expressed in terms of a shell variable, `$HOME`. As with calls to `library`, S will expand such variables to find the actual path of the chapter. The second line has only a name, with no "/"s. S interprets this if possible as a library under S's own directory (see `?library`—this is also consistent with how that function works).

Starting S up with this file, and no other special control, generates a search list as follows:

```
> search()
[1] "."  "$HOME/S" "trellis" "documentation" "main"
```

We got some databases we didn't ask for: the first one, the working data, is the current directory, ".". The selection of the working data is discussed in section 4.1.2. In addition, other S libaries were included. The last database is S library `main`. This is not an option, S won't run without it. The other libraries come from the "$SHOME/.S.chapters" file. They represent what you (or whoever owns S on this machine) conidered to be the essential S libraries. If you have a reason for *not* including one of these libraries, you can detach it in your `.First` or `.S.init`.

Next, let's consider the `.S.init` file and the `.First` function. A difference between these is that the initialization file applies no matter where you run S, while the `.First` only gets invoked if it is on the search path. Suppose we want the S prompt to be the current date, so we can identify session records, and want this to be true regardless of where we run the session. We could create a `.S.init` file containing the following single line:

```
options(prompt=shell("date '+%b %e: '"))
```

The option will be set *after* the standard initialization is performed.

Finally, suppose we want to take over the initial actions entirely. The best way to do this is to invoke the underlying S executable program directly, with an intialization script as a command-line option. To do this you need to have set the shell environment variable `SHOME` to the directory where the S-Plus or S system resides on your machine. (The `S SHOME` shell command tells you this.) In the standard version the program `$SHOME/cmd/Sqpe` is the executable version of the language. To use, say, file `specialS` in my login directory as the initialization script, invoke the shell command:

```
$SHOME/cmd/Sqpe -i $HOME/specialS
```

Remember that this overrides all the standard initialization. You will start with a minimal search list (a temporary database as working data and the required S libraries only).

4.1.2 Selecting the Working Data

When the S session begins, the evaluator needs to select a working database in which to assign persistent results of S expressions. This can be any valid chapter or S database on which you have write permission, and you can change the working database during the session by calling `attach` with `pos=1`. For convenience, it's best to start off with the arrangement you would prefer. There are several ways to guide S in picking the initial working database. Chances are you will have a preferred style and will nearly always work the same way: the following are the possible choices.

When S starts up a session it makes a sequence of attempts to pick the working database. Here they are in the order tried, with one exception that you can use if you don't like *any* of the following:

1. If the current directory is an S chapter, use that. The working directory will have name `"."`.

2. If there is a subdirectory `S` of the login directory that is an S chapter, use that.

3. If the login directory itself is an S chapter, use that.

4. If all these fail, S will create a chapter in `/tmp` for you to use, with a constructed name to make it almost certainly not currently used.

Actually, all of these are quite reasonable. My least recommended choice would be to use the login directory, because cluttering that up is not generally a good idea. However, that used to be the S default choice, after the local directory!

If none of these choices appeals, or if you feel the need to unconditionally control the location of the working database, that is also possible. Set the shell environment variable `SWORK` to the directory you want to use, before starting the S session. S will use this as the working database, regardless of other considerations. This is a forcing choice: if for some reason the directory pointed to by `SWORK` is not an S chapter, S jumps directly to creating the temporary directory. (This means that you could force the use of a temporary database by setting `SWORK` to, say, `""`.)

4.1.3 Keeping Track of the Session

S provides several tools to keep track of what has happened in a session. The *audit file* is an ordinary text file, kept in the S chapter directory used for the session. S records here the tasks given to the evaluator. You make use of the audit file by calling S functions that examine it; in particular, the function `history` will go back to find previous tasks. Calls to `history` can be used to rerun earlier tasks or to collect some tasks together to construct a function or a source file. The function `again`, which uses `history`, is a quick way to rerun a particular task:

```
again("lm")
```

says to rerun the latest expression in the audit file that contains the character string `"lm"`. The more general `history` function will search for all the occurrences of a string, up to a maximum number that is also an argument. If there are several matches, `history` will prompt you to select one. Alternatively, you can tell history to write all the matched expressions to a file and/or to turn them into a function. The latter is an easy way to start off a function that you want to encapsulate some tasks you have been giving S explicitly.

You can make a more complete record of an S session, for later reference or to share with others, by

```
record(file)
```

All the input you type and all the output S types back at you will be recorded, essentially just as you see them, on `file`, which may be any writable connection. An optional second argument to `record` is a character string to be written at the start of each line of user input. Post-processing using this string allows you to separate input from output, useful to make the record to look nicer. Recording is turned off by `record()`.

The audit file itself is the file `".Data/.Audit"` in the chapter used as the working database for your S session. Since the working data must always be writable, S is guaranteed to be able to create this file. If you look at the file, you will see S expressions and patterned S comments; these are used by functions such as `history` to interpret what has happened. You don't want to write things directly to the audit file yourself under any reasonable circumstances; the function `stamp` will put a message on the audit file for you. The `stamp` function is also a good way, as its name suggests, of putting a time stamp on printed or plotted output. Additional information about S permanent assignments will be written to the audit file if you set the `audit` option to any positive value:

```
> options(audit=1)
```

After this, stylized comment lines will be written to the file to mark any permanent assignments (such lines allow tracking the evolution of S objects in the database). If you want to get the name of the audit file (to communicate to a function or command that will read it, for example), call the function `audit.file`. You can also get the audit file as a connection; this might be useful to get a position on the file, again in order to communicate this information to a tool designed to read the file.

You can suppress audit output entirely by removing write permission from `".Data/.Audit"` before starting the session. Users occasionally run multiple S processes in the same working data, in which case you might want to suppress audit output from the later processes. Notice that making the audit file unwritable *after* starting a session does not stop the current S process from writing to the file, at least in the operating system versions I'm aware of. Once the file is open for writing, permissions are not checked again. Neither making the audit file unwritable nor running multiple S processes in the same chapter is recommended in general, because both tend to confuse the picture of what is happening in S and multiple processes may need to use `synchronize` frequently to ensure they all see the same view of the objects in the working database.

4.1.4 Quitting from the Session

Most S sessions end by the user deciding to quit: from the command-line user interface, by evaluating the S task `q()`, or by having no source for any further tasks (e.g., if S encounters an end-of-file on the standard input). There are a number of variations: you can end the session in other ways and you can control what happens when the session ends.

When the S evaluation manager quits the session, it looks in the search list of databases for an S object named `.Last`; if it finds one, it interprets that as an action to take on quitting. Like the `.First` object, to which it corresponds, it can be a function that will be called with no arguments, or an S expression to be evaluated.

S will end the session in response to four occurrences:

1. a call to the function `q`;
2. a call to the function `exit`;
3. a call to the function `terminate`;
4. no more possible sources for tasks.

The three functions `q`, `exit`, and `terminate` differ mostly in how normal the exit is intended to look. The functions `q` and `exit` both exit in a happy way as far as S is concerned, but `exit` takes an argument that, if non-zero, is used as the exit status of the S process. If S is being run from some other interface, for example a shell script, this information can be used to control what happens after the S session. Also, `q` quits at the end of the top-level task, while `exit` quits immediately.

The function `terminate` is for use when things are going very badly, and the S programmer wants to be sure the session ends. The function will try to exit gracefully (for example, it will try to evaluate the `.Last` action), but it *will* exit. If you think you might use `terminate`, or that you might encounter some terminal conditions such as memory faults, here are some hints that may be useful.

- If you want to use a `.Last` action, you may want to check whether S is exiting happily or unhappily. This can be done by checking the error level, which `terminate` will increment. If the call `error.level()` returns `0`, this is a normal exit, otherwise not.
- If `options(core.dump)` is set to `TRUE`, then a call to `terminate` will attempt to generate a core dump from the S session. In this case, the normal exit actions will *not* occur, so that the S process will be dumped more or less as it was when the extremely bad thing happened. Technically, in this case the S manager sends itself a quit signal.

Speaking of signals, those signals that S treats as terminal (such as quit) effectively generate a call to `terminate`, so the above remarks all apply as well to such signals, whether they come from outside the S process or from a call to `send.self`.

4.1.5 Interactive or Not?

The typical S session is *interactive* in the sense that the user is typing and/or using some other combination of input devices to send tasks to S, then waiting to see what happens.

An S session can be non-interactive, usually because S is invoked from the shell with a standard input that is not interactive (e.g., a file or the output piped from some other process). Most of what happens in S is the same for both interactive and non-interactive use, but S makes some decisions based on whether the session (more precisely, the current evaluator) is interactive. For example, the parser prompts by default for interactive use only. The choice of what to do with graphic output by default is another example: interactive graphics goes to a window interactively but to an offline postscript process non-interactively.

Your own S functions can test whether evaluation is interactive by calling `interactive()`. As an example, here is the default error option, the expression S evaluates when a problem or signal interrupts evaluation:

```
if(interactive()) recover()
else dump.calls()
```

If the current evaluation is done interactively, S calls `recover`, a function that offers the user the opportunity to examine and compute interactively with the S evaluation frames current at the time of the interruption. This function is not useful in a non-interactive situation, so in this case S will call a function that simply saves the function calls current at the time.

For programming interactions with the user, we may need some more precise information than `interactive` supplies, such as whether current standard input is coming from the user's terminal. See page 390 in section 10.4 for the corresponding tests.

Programmers can tell S to behave in an interactive or non-interactive way; this is sometimes useful when running S with a nonstandard script. The call `setInteractive(on)` sets the interactive state to `TRUE` or `FALSE` according to `on`, and returns the previous state.

4.2 The Language

The S programming language will be familiar in appearance to anyone used to programming in C or the other languages that share a similar style. It shares the control structures, assignment expressions, function call and operator syntax of this family of languages (roughly, "scientific" style programming). To that are added some extensions to function calling, some additional operators, and a much more general form of assignment expressions. The most important difference, however, is not syntax. The fundamental S concepts of task and evaluation cause the *effect* of the expressions to be very different from that in other languages. Also, the ability to deal with the language itself as an S object provides new ways to program.

4.2.1 Syntax

A semi-formal table for the S language syntax is given in Table 3.2 on page 98 in section 3.2. Here, we will discuss the language less formally, but in somewhat more detail.

An S expression is *complete* when it satisfies the syntax of an expression as described in Table 3.2 and more fully in the online documentation `?Syntax`. Complete expressions can be combined by separating them with newlines or the `";"` character.

S expressions consist largely of function calls. Functions are called by following the definition of the function with a parenthesized list of arguments. The definition is usually entered by typing the name under which the function is stored in a databse, as in the following examples.

```
mean(wafers1); c(0, x1, rep(y, rep(3, length(y))))
plot(wafers1, title = "Friday's Results")
```

Function calls look much as they do in other languages, but there are several syntactic generalizations in S. Any other S expression that evaluates to a function definition could replace the name preceding the argument list:

```
get("sqrt", where = "myLib")(1:10)
(function(x)x+1)(c(0, x1, y))
```

The call to `get` is evaluated, then the result used as a function definition. The second example uses an inline, constant definition of the function. The parentheses around the definition are needed because of precedence deficiencies in the parser.

Arguments in function calls differ from some languages in that they may optionally be named, as in the call to `plot` above. To see what arguments a function expects, either look at the full online documentation by typing `?remove`, for example, for function `remove`, or just `args(remove)` to see only the argument names:

```
> args(remove)
remove(list, where=, frame=, meta=F)
```

Named arguments only have to include enough characters to be unique. In a call to `remove`, the argument `frame` could be supplied as just `f=`

```
remove("tempX", f = n)
```

The S evaluator will complete the argument name.

Syntactic *name* expressions are usually references to an object associated with the corresponding character string in some database. The syntax for names allows names to be made up of letters, numbers and the character ".". Names cannot start with a number and generally *should not* start with "." (because several such names have special meaning to the S evaluator).

There are also infix operators for the usual arithmetic, comparison, and logical computations:

```
leg1Times/(leg1Times +leg2Times)
yield(wafers1) > .9
```

A few nonstandard infix operators are supplied as well. The most important, maybe the most useful of all operators, is the `?` operator, which invokes online help. It is most commonly used as a unary operator, followed by a name or a character string. S responds by printing the online documentation for the topic named:

```
?mean; ?"+"; ?Syntax
```

You need the quotes if the topic you are asking about is not syntactically a name, since the S parser will get an error without the quotes. The `?` operator can also be used as a binary operator; in this case, the first argument is interpreted as a special class of documentation, such as `method` or `class`. For example, `?array` asks for documentation on the `array` function, while

```
class ? array
```

asks for documentation on the array class. As a task all on its own with no arguments, `?` provides documentation on itself.

There are other operators to extract pieces of objects: the subset operator extracts part of an object:

```
wafers[1:5]
```

A similar operator with double square brackets, e.g., `x[[i]]`, extracts single elements of objects. Both operators take, syntactically, an arbitrary argument list with commas separating arguments. The "@" operator extracts a named slot from an object:

```
outCon @ description
```

This requires `outCon` to have a class with a `description` slot. A superficially similar operator, the "$" operator extracts a single element from an object with names: `lot1$GAS` evaluates as the element in the object `lot1` associated with the name "GAS". The two operators are really very different: the names for slots are defined from the class definition, not from the particular object; and "$" is only syntactic sugar to simplify, for example, `lot1[["GAS"]]`.

All the extraction operators can also appear on the left side of an assignment, in which case they are interpreted to mean replacing the corresponding subset, element, or slot.

Another commonly used special operator is ":", which evaluates to a sequence from its first to its second argument:

```
> 1:5
[1] 1 2 3 4 5
```

The sequence goes in steps of ± 1. We will deal with these and other operators as they become relevant. A list of them is included in the detailed documentation for topic `Syntax`.

S has assignment operators also. Assignments in S often look like assignments in other languages:

```
sum1 = summary(wafers1)
```

They mean something different, though. The essential concept is that the object, `summary(wafers1)`, is stored in a temporary or permanent database and associated with the name "sum1". Syntactically, they also differ in that the left side of the assignment can be an arbitrarily complex expression, built up of function calls in a recursive way (see section 4.5.2).

Several complete S expressions can be combined into one, by enclosing them in braces, and separating them by newlines or semicolons.

```
{ plot(wafers1)
  sum1 = summary(wafers1)}
```

The call to `plot` and the assignment would each have been a complete task if typed by itself and followed by `Enter`. With the braces, the whole thing is treated as a single expression.

The S language also includes expressions for iterating: `for` iterates over all the elements of an object; `while` iterates so long as an expression is true; `repeat` iterates forever. Special expressions `break` and `next` are available to break out of the iteration, or to start the next iteration.

```
for(i in length(wafers1)) plot(wafers1[[i]])

while( median(abs(fit$resid)) > eps) {
    fit = reweight(fit, weights)
    weights = reScale(fit$resid, resRange(fit))
}

repeat {
    fit = reweight(fit, weights)
    if(median(abs(fit$resid)) <= eps) break
    weights = reScale(fit$resid, resRange(fit))
}
```

Braced expressions are essential in most iterations to ensure that all the relevant computations are included.

S has the usual `if` ... `else` style of conditional expression, as we saw in the `repeat` example, or in:

```
if(median(abs(fit$resid)) <= eps) plot(fit)
else "Not done yet"
```

Comments are introduced in S input by the character `#`; any text starting with this character (if it isn't inside a quoted string) and ending with the end of the input line is taken to be a comment line. Comment lines are not (usually) thrown away; instead, they are saved up and incorporated with the next allowed subexpression. The subexpression is then parsed as a `"comment expression"` object. The evaluator reduces that object to its non-comment part for computation, but the comment part can be extracted, and is, for example, in making functions self-documenting.

4.2.2 The Language as Objects

Everything is an object, remember? A range of programming tools and techniques come from using the expressions being evaluated as S objects themselves, manipulating them and altering them. Functions such as `recover`, `browser`, and `trace` are not written in C and generally use few special constructions. Instead, they manipulate S function definitions and other S expressions, in S, to allow the user to examine details of evaluation.

If you want to do some manipulations of S language objects, you should be aware of several sets of tools to do so. The function `Quote` does nothing but return the expression you give it, unevaluated. In case this seems of little value, suppose you wanted to trace "^", but not to stop on each call, just to print out the class of the argument `e1`. We could construct a function specifically to do this, but you don't need to. Instead you can provide the `Quote`'d expression directly; naturally, you need the expression to be unevaluated until the traced function is called:

```
trace("^", Quote(message("Class of e1: ", class(e1))))
```

The complete call to `message` is inserted, *unevaluated* into a locally assigned version of the function object "^".

The function `substitute` also returns a version of the expression you give it, but in this case names in the expression will have an object substituted from the list in the second argument.

```
messageCall = substitute(message(MSG),
named(MSG=readLines(mFile)))
```

Like `Quote`, `substitute` does not evaluate its first argument; give it the argument `evaluate=T` if you want the first argument evaluated. Its second argument, however, is treated in a standard way: in the example, a call to `readLines(mFile)` will provide the object to substitute for `MSG`.

If the second argument to `substitute` is omitted, names are interpreted from the local function call.

```
trace("^",
    substitute({message(msg); browser()}))
```

If `msg` is an argument or locally assigned object in the function calling `trace`, its value will be substituted as the argument to `message`.

A set of tools lets you extract or replace pieces of function definitions: `functionArgNames` returns the vector of the functions formal argument names; `functionArgs` returns a named object whose names are the argument names and whose elements are the default values for the arguments; `functionBody` returns the S object that is the expression to be evaluated when the function is called; `functionComments` returns the character vector made up from those comments that provide the self-documentation for the function.

All these functions take an S function object as their only argument. All of them can also be used as replacement functions on the left of an assignment. For example, the following expression would add a line with the current date to the end of the description of a function called `checkProbe` stored on the working directory:

```
> functionComments(checkProbe) =
+   c(functionComments(checkProbe), "",
+   paste("Modified", date()))
```

The extra `""` puts a blank line on the documentation before the date.

4.2.3 The S Evaluator

S provides functions to give you a window into the evaluation as that evaluation goes on. Section 4.9 discusses the semantic model for S evaluation, but many of the tools can be used without worrying too much about the model.

As you can see from typing `where` in `recover` or `browser`, the S evaluator will typically have a number of evaluation frames open, as the evaluator calls one function from inside a call to another, etc. Functions are available to get essentially all the information in any of these calls, from inside an S function. By an arbitrary convention, the names of these functions all begin with `"sys."`. For example, `sys.call` returns the function call currently being evaluated, `sys.frame` returns a named list with all the objects in that current evaluation, etc. Table 3.13 on page 106 lists the available functions. These functions are useful in building interactive tools, such as `browser`. Knowing about them may help you understand existing tools and also let you modify the tools if you want to.

The function `eval` evaluates an S expression in a specified evaluation frame, not necessarily the current one. For example, here is a loop that evaluates expressions in the frame of the caller to the current function, until one of the expressions evaluates to `NULL`.

```
frame = sys.parent()
for(i in seq(along=exprs)) {
  value = eval(exprs[[i]], frame)
  if(identical(value, NULL)) break
  results[[i]] = value
}
```

Unlike `Quote`, `eval` evaluates its first argument, here `exprs[[i]]`, in the standard way to get the language object that will *then* be evaluated in the specified frame.

The distinction between evaluated and unevaluated arguments is unfortunate, but does make some common expressions in S easier to type. Linguistic purity would have had just one function, `Quote`, that did not evaluate its argument, requiring programmers to use that function to pass unevaluated arguments to other functions.

Play around with the various uses of `Quote`, `substitute`, and `eval`. The distinctions should gradually become more natural. In the case of `eval` in the example above, the `i`-th element of the `exprs` vector is selected, and presumably is itself an S language object, which then gets evaluated in `frame`. The `Quote` style might also make sense, but it has a different meaning.

```
eval(Quote(exprs[[i]]), frame
```

This says to evaluate the expression `exprs[[i]]`, for its own sake, in the context of the evaluation frame `frame`.

4.2.4 Control of Computations

S has fairly conventional expressions for conditional and iterative computations. The `if` and `if ... else` expressions evaluate a condition, expecting a single `TRUE` or `FALSE` value as the result. Depending on this value the first or second optional expression will be evaluated.

```
if(length(y[[i]]) y[[i]] else 0
if(min(x) > 0) x = log(1+x)
```

A related parallel construction is `ifelse`. This function takes three arguments

```
ifelse(cond, true, false)
```

For all the `TRUE` elements of `cond` the result has elements from `true`, for all `FALSE` elements arguments from `false`:

```
ifelse(x > 0, x, NA)
```

Notice that `ifelse` evaluates all three argument expressions, while the conventional `if ... else` will only evaluate one branch.

The most commonly used iteration in S is the `for` loop; this assigns each element of an object to the name supplied in the expression, and then evaluates the body of the loop.

```
for(var in names(data)) {
    x = data[[var]]
    cat("\n", var,"\n")
    print(summary(x))
    plot(x, abs(resid), xlab = var)
}
```

Loops, especially `for` loops, have a reputation for being bad in S because they operate slowly. The number of separate S function calls obviously grows proportionally to the number of times through the loop; if each iteration of the loop does a tiny computation, things can indeed be slowed down. In this case, it may be possible to rethink the computation in whole-object terms (see section 4.8). But loops are perfectly sensible if they express the computation naturally, especially if the individual steps of the loop do some serious computation, as in our example.

More general, but less frequently used iteration expressions are `while` and `repeat`. The `while` loop continues to evaluate the body of the loop so long as the single condition value is `TRUE`:

```
while(!fit$converged)
  fit = update(fit, x, y)
```

The `repeat` loop is just the special case of a `while` loop with a condition that is always `TRUE`, and not included:

```
repeat {
    fit = update(fit, x,  y)
    if(fit$converged) break
}
```

The `repeat` loop is more convenient if there are multiple ways to exit at different points in the loop, but it is generally the least clear when reading the code.

The body of any of the loop expressions is formally an arbitrary S expression. In practice it is usually a braced sequence of expressions, separated by semicolons or newlines. Within the body of the loop, S has two special control expressions: `next` to go to the next iteration of the loop and `break` to break out of the loop.

The `return` control expression causes the current function call to return, with the argument to the expression evaluated as the value of the function call:

```
if(all(x <= 0) return(NA)
```

A `return` is not needed if the function body is a braced sequence of expressions and the last expression is the value you want to return from the function. In functional languages, `return` expressions are rather deprecated, in that they can make it harder to see the behavior of the function. In fact, it's technically true that you are never *required* to use a `return` expression.

You can just put all the code that would have followed the `return` into a conditional expression.

```
if(all(x <= 0) NA
else {
  # whatever would have followed the return
}
```

However, when the condition prompting the `return` is buried deep in other computations, avoiding it may cause more complications than it saves.

The operators `&&` and `||`, discussed on page 148, may help express the conditions in `if` and `while` expressions:

```
if(min(x) > 0 && sum(log(x)) > zz)
```

The point about these operators is that the left-hand expression acts as a "guard" to the right-hand one, which will not be evaluated if it is not required. S programming rarely makes a special distinction between single values and general objects, but the conditional expressions in `if` or `while` are one case where a single value, and only `TRUE` or `FALSE`, must be produced. In addition to the operators above, functions `any`, `all`, and especially `identical` are useful, and operators `"=="` and `"!="` are to be avoided. See page 73 for some examples.

One final control structure is the `switch` function. In S, this is not a special construction but just an ordinary function. It evaluates its first argument and then, usually, evaluates one of the remaining arguments as the value of the call to `switch`. In most cases, the first argument is expected to evaluate to a single character string. The remaining arguments will have names that are possible values for this string. If any of them matches (exactly), that is the argument that `switch` will evaluate. If none matches and there is one unnamed argument, that is evaluated as the default value for the call to `switch`.

```
fit = switch(method,
    qr = lm.fit.qr(x, y, ...),
    chol = lm.fit.chol(x, y, ...),
    svd = lm.fit.svd(x, y, ...),
    stop(paste("Method", method, "not defined"))
)
```

If several names correspond to the same action, all but the last can be given as empty arguments: after matching a name, `switch` will continue until it gets a non-empty argument expression.

```
ok = switch(answer,
    yes = , YES = TRUE,
    no = , NO = FALSE,
    NA)
```

The strings "yes" and "YES" cause switch to return TRUE, the other two strings, FALSE; any string that doesn't match any of the four names results in NA being returned.

The first argument to switch can also evaluate to a single integer; in this case switch will evaluate the first, second, etc. of the following arguments. The integer form is generally not as useful; it's harder to see what is happening and there is no default action.

4.2.5 Assignment Expressions

S assignment expressions consist of an assignment operator, with an expression on the left and another on the right. The expression on the right is the *value* of the assignment; the expression on the left is either a name or a function call.

```
cp = crossprod(x)
diag(cp) = dr
```

The "=" in these expressions is the assignment operator, as in other C-like languages.

S also recognizes the operator "<-" as an assignment operator. When an assignment occurs *inside* an argument to a function, the operator "<-" must be used to distinguish the intended assignment from a named argument (see page 139). Otherwise, the operators behave identically. The arrow form requires a little more typing and is less familiar to programmers from other languages, but it expresses the asymmetric relation of the left and right side in an assignment, where "=" does not. In using "<-", though, you must not put a blank between the two characters. Both the following expressions are legal:

```
x <- 3
x < -3
```

but the first is an assignment and the second tells us which elements of x are less than -3.

The simplest assignment expression has a name on the left side.

```
cp = crossprod(x)
```

This is always evaluated by evaluating the right side expression and assigning the resulting object in the local frame, associated with the name on the left. Top-level assignments only differ in that the assignment is copied from frame 1 to the working data, when the corresponding top-level task completes.

```
> {start = fit$coef; new = update(fit, x, y)}
```

If the call to `update` completes, both `start` and `new` will be copied to the working data. If there is an error in the call, no copy of `start` will take place. But it will be in frame 1 for inspection from the browser during error recovery.

In the next simplest assignment expression the left side is a function call, and the first argument of the function call is a name. This is always evaluated as a simple assignment of a call to the *replacement function* corresponding to the function on the left. For example,

```
diag(cp) = dr
```

is evaluated as the simple assignment expression

```
cp = "diag<-"(cp, value=dr)
```

The function named `"diag<-"` is the replacement function corresponding to `diag`. It is expected to have `value` as its last argument and otherwise to have the same argument list as `diag`.

The full set of valid replacement expressions then comes from making the same substitution recursively, replacing a name in the simple case by a function with its first argument being a name.

```
diag(fit[["Variance"]]) = dr
fit[[i]]$residuals[ wgts < .001 ] = 0.
```

When some of the functions appear as operators, it may be a little hard to see the recursive nesting, but intuitively the generalization is simple: you can replace a piece of an object, or a piece of a piece of an object, or $\cdots$, where in S "piece" has a completely open-ended definition, via the replacement functions. See section 4.5.2 for the full definition and the implementation.

Assignment expressions also appear whenever you name an argument in a function call, though they are just called *named arguments* in this case and are interpreted somewhat differently. The call to any function associates each of the actual arguments with the name of one of the formal arguments to the function. S lets you omit names or supply just enough of the name to match, but the end result is the same as if all arguments had been named. The S function `match.call` lets you see this explicitly. It returns the fully specified call corresponding to a function and a call object.

```
> x = match.call(get,
+   Quote(get(nn[i], w=1)))
> x
get(name = nn[i], where = 1)
```

The `call` object stored in `x` has matched the two arguments to the formal arguments of `get`, added a name to the unnamed first argument, and completed the name `"where"`.

When the evaluator needs the value of one of the objects in a call, it will evaluate the right side of the corresponding assignment, in the frame of the calling function. In evaluating the call to `get` shown above, the first time that `name` is needed in evaluating the body of function `get`, the expression `nn[i]` will be evaluated and the result assigned as `"name"` in the frame of the call to `get`. I should emphasize that you rarely need to worry about the computations in such detail, but the model is there and can occasionally be useful when planning or understanding subtle computations.

Assignments can appear in the argument list in the function *definition* as well, to specify default expressions for arguments. In the definition of function `get`, for instance:

```
> get
function(name, where, frame, inherit = F, immediate = T,
    meta = 0, mode = "any")
{
    ....
```

The assignment of `inherit`, for example, will be evaluated when the value of `inherit` is needed and there was no corresponding actual argument.

All these assignments to the name of a formal argument take place in the local frame of the function call being evaluated. But what if you wanted the argument *itself* to be an assignment expression? In S terminology, you want the evaluation of the actual argument to have the side effect of doing an assignment in the calling frame. For example,

```
xTrim = trim( xUntrim = rnorm(1000), trVar) # Won't work!
```

The intention was that evaluating the argument to `trim` would have the side effect of assigning `xUntrim`. The evaluator, however, will look for an argument named xUntrim in the definition of trim, and probably won't find one.

If you do want to write such an expression, you need to either use the assignment arrow or enclose the expression in parentheses to remove ambiguity:

```
xTrim = trim( xUntrim <- rnorm(1000), trVar) # ok, if you must
xTrim = trim( (xUntrim = rnorm(1000)), trVar)
```

However, the general advice about arguments with assignment side effects is: *Don't use them.* They are bad programming style in any language, because they tend to hide the assignment from anyone reading the code. Assignments are just the expressions you don't want to hide, since they have side effects. In S such expressions are especially dangerous. For one thing, lazy evaluation of arguments means that you don't know when, or even if, the assignment will take place.

Assignments of this form are not dangerous, however, in a few functions that don't evaluate their arguments, such as Quote. These functions will try to allow assignment expressions in the relevant argument.

4.3 Numeric Computations

S provides numeric computations with the usual arithmetic operators, mathematical functions, and transformations:

```
+ - * / ^
```

are the standard arithmetic operators (^ for raising to a power). Additional operators include x %% y for the remainder of x modulo y, and x %/% y for the truncated division of x by y.

Mathematical functions include the usual exponential, trigonometric and hyperbolic functions, plus a slew of other functions of interest to the user community of S:

```
exp log log10 logb
sin cos tan asin acos atan
sinh cosh tanh asinh acosh atanh
gamma lgamma
```

Numeric transformations include absolute values, rounding and truncating, and similar functions:

```
abs round floor ceiling trunc signif
```

Numeric summaries include max, min, and range, plus typical statistical summaries of central value, such as mean and median; var computes variances. Notice that max and min produce a single number; S has also useful functions pmax and pmin to take any number of arguments and produce a "parallel" object with the maximum or minimum values of each argument. So,

```
pmax(0, temp)
```

has all the positive values in temp, and 0 wherever the corresponding value in temp was negative.

All these operators and functions look and behave much as you would expect from other languages. But the arguments are S objects, which are much more general than similar-looking arguments in most other languages. An important step to programming well in S is to think in terms of whole objects. If x is a numeric vector of length, say, 8567, it's key to think of a numeric computation such as log(x) as a single operation, not 8567 separate operations. Section 4.8 discusses this concept and other issues of strategy in using basic computations.

Numeric computations in S are designed to work with any numeric data (see section 4.3.2). If given non-numeric data, they usually complain:

```
> objects(1) +1
Problem in objects(1) + 1: Non-numeric first operand

Debug ? (y|n): y
Browsing in frame of objects(1) + 1
Local Variables: e1, e2

R> class(e1)
[1] "character"
R> is(e1, "numeric")
[1] F
```

If given numeric data that includes values outside the numerically valid range of the particular function, the corresponding values in the result will be represented as NA, and the function may warn you.

```
> temp
 [1]  11.9   3.4   2.9  -0.1  11.9   5.7   7.8  11.0   8.4   0.1
> log(temp)
 [1]  2.476538  1.223775  1.064711        NA  2.476538
```

```
 [6]  1.740466  2.054124  2.397895  2.128232 -2.302585
Warning messages:
  NAs generated in: log(x)
> 0/0
[1] NA
```

The underlying model for numbers in S includes `Inf` and `-Inf`. These stand intuitively for "infinitely large" and "infinitely negative".

Values generating these are considered part of the normal range for numeric functions and don't generate warnings:

```
> logTemp = log(pmax(0,temp))
> logTemp
 [1]  2.476538  1.223775  1.064711      -Inf  2.476538
 [6]  1.740466  2.054124  2.397895  2.128232 -2.302585
```

Although special, `Inf` and `-Inf` can be used in most ways just like any other numeric values:

```
> logTemp == -Inf
 [1] F F F T F F F F F F
```

This is definitely *not* the case with `NA`; you should think of this as a description that says we have no idea what the numeric value is. Beware of using `NA` in comparison operations.

```
x == NA  #Wrong!
```

Although the expression might seem sensible, it is not. Any value compared to `NA` is `NA`—how can we know whether a value that is missing is equal to another value or not? What this expression undoubtedly meant was: "an object like `x` with `TRUE` where the elements of `x` were missing and `FALSE` everywhere else." This result is the value of the expression `is.na(x)`.

```
> y = log(x)
Warning messages:
  NAs generated in: log(x)
> is.na(y)
[1] F F F T F
> x[is.na(y)]
[1] -0.8
```

To repeat the point, `y==NA` would evaluate to a vector of 5 `NA`'s.

You can set values to `Inf`, `-Inf`, or `NA` by using the three symbols in the replacement data:

```
> x[ x < -3] = -Inf; x[x > 3] = Inf
> x[ x == 0] = NA
```

You can also set `NA`'s by using `is.na` on the left of an assignment:

```
> is.na(x) = c(1,3)
```

The right side is interpreted as a subset of the object, and this subset is set to `NA`. For special classes of objects, this may be a safer way to set `NA` values, since the class can define a method for the replacement function of `is.na`.

4.3.1 Operations on Vectors and Structures

Mathematical functions and transformations operate on a numeric object to produce another numeric object of the same size and "structure". Structures in S are classes of objects that add to ordinary vectors some notion of the values being organized in space or time. Matrices, general multiway arrays, and time series are the most commonly used structures.

When we perform a numeric operation on one or more vectors or structures (or in fact any computation that operates element-wise on the values), the value retains the structure of the argument or arguments. A vector argument produces a vector result. A transformation of a structure returns a new structure like the argument, with the data part of the object replaced by the transformation. So if `x` is a matrix, `log(x)` is a matrix of the individual logarithms, and `abs(x)` a matrix with the individual absolute values. In fact, this concept extends beyond structures to any object that `is` a numeric vector in the S sense of substitution, if the function has a suitable method for class `vector`; we will deal with these ideas in Chapter 8.

Operators taking two arguments (operands) follow the same philosophy as much as possible. If one operand is simpler or has less structure than the other, S tries to coerce the simpler to look like the other and the value of the expression then looks like this more structured object. This philosophy is not pushed beyond the point where it seems unclear what the user intended.

If one operand has length 1, as in

```
x+1
```

S always acts as if the single number was expanded to the size and structure of `x`. If the two operands have the same length, then the value has that length also. In other cases, S has more trouble understanding what the user wants. There are two additional heuristic rules applied. First, if the length of the longer is an exact multiple of the length of the shorter, the operator implicitly expands the shorter by replicating it. Second, if one of the operands has length zero, the result has length 0. If neither of these conditions applies, the evaluator generates an error.

```
> length(x)
[1] 10
> x
 [1] -0.6 -0.7  1.3  0.2 -0.4  0.2  0.8  1.0  0.3 -0.7
> x + .1
 [1] -0.5 -0.6  1.4  0.3 -0.3  0.3  0.9  1.1  0.4 -0.6
> x + c(.1, .2)
 [1] -0.5 -0.5  1.4  0.4 -0.3  0.4  0.9  1.2  0.4 -0.5
> x + c(.1, .2 ,.3)
Problem in x + c(0.1, 0.2, 0.3): length of longer
operand (10) should be a multiple of
length of shorter (3)

Debug ? (y|n): n
```

Thus, `c(.1, .2)` was replicated to look like

```
> rep(c(.1, .2), 5)
 [1] 0.1 0.2 0.1 0.2 0.1 0.2 0.1 0.2 0.1 0.2
```

Replicating the operand of length 3 was considered to be less sensible, and the operator function generated an error. (I won't defend this exact rule too strongly, but it is based on computations for matrices or time series that sometimes have this form.)

The heuristic behind the treatment of zero-length arguments is that these usually result from computations involving some selection process, with a zero-length object resulting as a special case. For example, consider:

```
outerSquare = sum(x[abs(x) > eps]^2)
```

This looks innocuous, but what happens when none of the values in `x` is larger than `eps`? The argument to `sum` is then equivalent to

```
numeric(0) ^ 2
```

Pretty clearly, the user expects the result of this to be a vector of length 0, and that as a result `sum` will return 0. This is in fact what S does, although logical consistency (and earlier versions of the language) would extend the operand of length zero to match the longer operand.

If one of the operands has structure and the other doesn't, but S can interpret the vector as having the same length as the structure, the expression works and returns an object with the structure. This is the most general rule. Designers creating new classes of objects can and should devise more intelligent methods to interpret arithmetic operations when one operand comes from this class and the other is a vector.

If both operands are structures, the standard S operations apply a version of the same rule to the data parts. Again, this is not a very defensible general rule. Well-designed classes of objects should devise some methods that compute a result where the intention is clear and generate an error otherwise.

We are describing in this chapter how operators and functions behave on basic classes of S objects, in particular, on vectors and structures. They are not restricted to such classes, however. The designer of any class of objects is free to decide how arithmetic operators and mathematical functions should behave, and to implement methods accordingly. Chapter 8 shows examples of such implementations. The descriptions in the present chapter can be taken as paradigms, models suggesting how the functions would be expected to behave on other classes of objects.

4.3.2 Different Classes of Numeric Data

The class `"numeric"` refers to double-precision floating-point data; specifically, to elements of data represented by the data type `double` in the C language. Other kinds of numeric data can be created as well: class `"integer"` stores signed integer data (specifically, of type `long` in C) and class `"single"` stores data of type `"float"`. S users rarely need to worry about this: computations in S are designed to deal with whatever numeric data makes sense in the context.

You can coerce data explicitly to a particular class, for example, `as(x, "single")`. However, note that when S converts non-integer numbers to integer, it applies a rule that the result is `NA` unless the value is "near" an integer value. The definition of near is taken to be "within `.01`" of the nearest integer.

```
> as(x, "integer")
 [1] NA NA NA NA NA NA NA  1 NA NA
Warning messages:
  9 missing values generated coercing from
    numeric to integer in: as(x, "integer")
```

The lesson is that *automatic* coercion should be used only when you really believe the computations produced integer values, regardless of how they happen to be stored. If you want to force integer versions of numbers, specify the conversion by first calling one of the functions `round`, `trunc`, `floor`, or `ceiling`. These always produce integral values that convert to integer without generating `NA`'s.

S also includes numbers from the complex field, a slightly more exotic class but useful in special applications. Complex constants are written as a numeric, `+` or `-` a numeric followed by the letter `i`:

```
1+.5i; 1+0i; 0 -2.0i; 1i
```

If the first number is missing, it is assumed to be `0`. The two numbers are the real and imaginary part of the complex numbers; you can also generate complex numbers in various ways by using the arguments to the function `complex`; see `?complex`. Numeric data are coerced to complex numbers with imaginary part `0`.

```
> as(x, "complex")
 [1] -0.6+0i -0.7+0i  1.3+0i  0.2+0i -0.4+0i
 [6]  0.2+0i  0.8+0i  1.0+0i  0.3+0i -0.7+0i
```

Complex numbers have their own arithmetic and mathematical functions.

```
> y = as(x, "complex"); y + 1i
 [1] -0.6+1i -0.7+1i  1.3+1i  0.2+1i -0.4+1i
 [6]  0.2+1i  0.8+1i  1.0+1i  0.3+1i -0.7+1i
> (y+1i)^2
 [1] -0.64-1.2i -0.51-1.4i  0.69+2.6i -0.96+0.4i
 [5] -0.84-0.8i -0.96+0.4i -0.36+1.6i  0.00+2.0i
 [9] -0.91+0.6i -0.51-1.4i
```

The exponential and trigonometric functions have special definitions for complex arguments. Other functions are specialized entirely to complex numbers: `Arg`, `Mod`, `Conj`, `Re`, and `Im`. The other classic function for complex numbers is the (fast) Fourier transformation, `fft`. See the online documentation for all of these.

4.4 Testing and Matching

S provides all the usual operations in a programming language for comparing values and expressing logical tests, plus some less usual functions for related computations. Because S computes whole objects from whole objects, these operations generally operate "in parallel" on all the data in one or more objects. In addition, there are some interesting comparison questions that *only* arise because of the whole-object view; S has some special functions to address these. Because S objects are self-describing, the operations are computed appropriately for the kind of data being compared; that is, there are *methods* for these operations supplied with S, and additional methods can be defined to extend them.

4.4.1 Comparisons; Tests of Equality

S has the usual comparison operators, plus some special equality-testing functions. The S comparison operators,

```
< <= == != > >=
```

operate to compare the operands element by element:

```
temp < 0
x <= log(1+y)
fft(z) != 1i
```

The result of the computation contains logical values reflecting the truth or not of the comparison. These functions apply the same rules as the arithmetic operators to combine an operand with more structure and one with less: short arguments may be treated as if replicated, unstructured ones as if they had the structure of the other operand.

The comparison operators apply to non-numeric data as well. In particular they apply to vectors whose elements are character strings, with the ordering corresponding to lexical ordering on the individual character strings. To compare two strings, the algorithm used is to find the first non-matching character position in the string; the smaller string is the one with the smaller character in that position (smaller in the sense of the ASCII character code). This includes the case that one string is shorter, since a string implicitly ends in the ASCII null or zero character. The same rule is used in sorting character string data; looking at examples of `sort` tends to make the rule clearer.

```
> sort(c("abc", "ab", "aba", "ab.", "ab|", "ab\n"))
[1] "ab"     "ab\n" "ab."    "aba"    "abc"    "ab|"
```

The two-character string sorts before any three characters starting with the same two characters. The newline character, "\n", and the dot sort before the alphabetic characters, but other special characters sort after.

Related to comparisons are logical "and" and "or" operations; in S, the operators `&` and `|`:

```
> out.of.range = x < 0 | x > 1000
```

The unary operator `!` performs logical negation:

```
> values = x[!is.na(x)]
```

As in other C-like languages, there are also operators `&&` and `||`, which expect single logical values as either operand. The operator `&&` returns a single `TRUE` if both its arguments evaluate to a single `TRUE`. The key point is that `&&` evaluates the right operand *only* if the left is `TRUE`. The operator `||` works analogously but evaluates to `TRUE` if either of its arguments evaluates to `TRUE`.

The `&&` and `||` operators in S really belong in the control discussion of section 4.2.4. Their behavior is much less S-like than any of the other functions discussed here, especially in that they don't deal with general objects, only single values.

Two logical summary functions, `all` and `any`, often provide the link between S's logical objects and the single values needed for control of computation.

```
if(any(y < 0) || any(x < 0))
    stop("none of the data should be negative")
```

The interpretation of these functions is usually obvious, but some special care is taken to handle missing values and objects of length 0. Here's a "formal" definition of function `all`: `all` returns `TRUE` if it knows that all of the values are `TRUE`, `FALSE` if it knows that any of the values is `FALSE`, and `NA` otherwise. Exchange "any" and "all" everywhere in that sentence and you get the corresponding definition for `any`: `any` returns `TRUE` if it knows that any of the values is `TRUE`, `FALSE` if it knows that all of the values are `FALSE`, and `NA` otherwise. A missing value (`NA`) could represent either `TRUE` or `FALSE`, so it will force a value of `NA` if the function needs to assert something about all the values, but not otherwise. When given a logical vector of length `0`, `all` returns `TRUE` and `any` returns `FALSE`.

Try out your intuition about those definitions against the following examples:

```
> x1 = c(T, NA); x2 = c(F, NA); x3 = c(T, F, NA)
> c(any(x1), any(x2), any(x3))
[1]  T NA  T
> c(all(x1), all(x2), all(x3))
[1] NA  F  F
```

A question that arises in S much more often than in other languages is: "Are two *objects* equal overall?" In other words, do the two objects agree in structure and content, so that we can treat them as identical? This question is meaningful in S, aside from a few subtle points, because objects in S are all self-describing and self-contained (part of the *Everything is an object* philosophy). S provides a function to test this strict requirement:

```
identical(x, y)
```

returns `TRUE` if the two objects are entirely identical and `FALSE` otherwise. It makes no allowance for fuzzy numerical results (hang on, we'll get to that) but otherwise understands any kind of S object. The function can be applied to numeric objects, S functions, hierarchical objects of any depth, or any other S objects. It works with the internal form of the objects and is substantially faster for large objects than operations on the lines of

```
all(x == y)
```

as well as being more general. Keep in mind that it is unforgiving. For example, two objects of class `named` are not identical if the names appear in different order, because the slots that contain the names won't be identical. Use `identical` when you need a solid guarantee that two objects are identical; it is quite fast and very hard to fool.

Asking whether two objects are equal in a more forgiving way is the purpose of a set of `all.equal` functions and methods. These have a different goal: they are intended to be used in testing, specifically in regression testing of S computations (see page 400). Here we want to be more aware that minor changes in how the computations are done or moving computations from one machine to another may produce small numeric differences or slightly different arrangements that represent the same data. Also, in a testing environment it may be essential to know something about *how* the two objects differ, rather than just that there was a difference, somewhere.

```
all.equal(target, current)
```

returns `TRUE` if it judges the two objects to be equal; otherwise, it returns a description of the differences found. The `all.equal` function includes methods specialized to particular classes of objects, to allow for known equivalences that might produce non-identical objects. Numeric objects are compared up to a small relative numeric difference (this is a thorny question in numerical analysis; the solution in `all.equal` is only intended to avoid minor differences in common situations). Classes of objects with special properties often have methods designed so the differences are interpreted in terms more familiar to the user of the objects than just the names of the slots.

```
> x = runif(10)
> all.equal(x, x + 1e-30)
[1] T
> all.equal(x, x + 1e-3)
[1] "Mean relative difference: 0.001693288"
> all.equal(named(a=1,b=2), named(d=1, e=2))
[1] "Names: 2 string mismatches"
> all.equal(list(1.,1.), list(1.,1.01))
[1] "Component 2: Mean relative difference: 0.01"
> all.equal(list(a=list(2.,1.), b="Other"),
+   list(a=list(2.,2.01), b="Another"))
[1] "Component a: Component 2: Mean relative difference: 1.01"
[2] "Component b: 1 string mismatches"
```

The methods for `all.equal` try to accumulate meaningful descriptions as they move down through the levels of the object. For objects with formal representations, but without specific methods for `all.equal`, the message will be interpreted in terms of slot names.

Comparison operations may also be related to *sorting* or *ordering*; S generally supports sorting of any data for which element-wise comparisons make sense. Functions `sort`, `order`, `rank`, and `sort.list` all either sort an object into ascending order or give some information about what sorting would do. The value of `sort.list(x)` is the permutation of the elements of `x` that would sort `x`. This allows ordering all sorts of things consistently with the order of the elements of `x`; for example, sorting the rows of a matrix.

The `sort` function differs from mathematical transformations in that it throws away any structure in its arguments. Although `sort(x)` has the same length as `x`, any spatial or temporal layout will have been destroyed, so `sort(x)` turns a structure into an ordinary vector.

4.4.2 Matching; Hash Tables

A set of S matching functions look up the values in one object in a table of possible values represented by another object. There are three kinds of matching: equality matching, matching of so-called "regular expressions", and matching with implicit string completion, called partial matching in S. All kinds of matching are defined for character string data, and equality matching is provided for other kinds of data as well. The computation

```
y = match(x, table)
```

returns an object with the same structure as x, but element i of y is the position in table of the value equal to element i of x.

```
> state.abb
 [1] "AL" "AK" "AZ" "AR" "CA" "CO" "CT" "DE" "FL"
[10] "GA" "HI" "ID" "IL" "IN" "IA" "KS" "KY" "LA"
[19] "ME" "MD" "MA" "MI" "MN" "MS" "MO" "MT" "NE"
[28] "NV" "NH" "NJ" "NM" "NY" "NC" "ND" "OH" "OK"
[37] "OR" "PA" "RI" "SC" "SD" "TN" "TX" "UT" "VT"
[46] "VA" "WA" "WV" "WI" "WY"
> match(c("AK", "WY"), state.abb)
[1]  2 50
```

Where elements of x are not found in table, the result has an NA element, although a third argument to match allows the user to specify something else (0 is often useful).

The table argument can also be an integer vector. With either string or integer data, match is much faster when x is large than directly comparing values with ==. S constructs a "hash" table from the values in table, in which each value in x can be looked up in constant time. If table is of class string, the table itself can be precomputed and kept in a slot of the object. Formally, you can also supply other kinds of numeric data to match, but be careful: the matching is to the exact values in table and takes no account of possible rounding error in numeric data.

Managing tables, especially tables of character strings, occurs often in computing with data. When the table is large or when matching will be done many times with the same table, you can make the computations faster by creating a special stringTable structure. This hashes all the strings in the table and stores enough information from the result to make matching against the table fast; in particular, once the stringTable exists the time required for matching should be essentially independent of the size of the table.

The stringTable is an optional slot in an object of class string. It may be created when the string is generated. If not, it can be added to the object later on. The stringTable is stored as S integer data, but the class stringTable interprets those integers in a special way. Because the table is large, this slot is not always computed when the string object is created. If not, stringTable(table) will be of length 0. You can add a string table to an existing string by the replacement operation:

```
stringTable(table) = TRUE
```

This will cause S to hash all the strings in table and store the resulting table as a slot in table. *If* you expect to do a lot of matching against the character strings in table, you will make the matching faster if table has a stringTable slot.

4.4.3 Regular Expressions

The other two kinds of matching only apply to strings. *Regular expression matching* interprets the strings in table in terms of special character patterns familiar from some text editors, such as vi or emacs, and from programming tools such as grep, awk, or perl. Regular expressions are a "little language" to describe classes of characters, positions in strings, and various other string patterns. For the details, see a document on one of the text editors or tools just mentioned. Here's a very brief refresher.

Regular expressions interpret certain characters as special in determining whether the regular expression string matches a target string:

```
^ $ [ ] . \ ( ) - +
```

The first two match the beginning and end of the target string. Square brackets enclosing some characters match any of the enclosed characters. The enclosed character can also be a character range; for example, "[a-z]" matches characters whose numeric code is at least that of "a" and at most that of "z". If the first character inside the square brackets is "^", the pattern matches any character *except* the remaining enclosed characters. The character "." matches any character. Characters other than the special patterns match themselves. A special character is made non-special by preceding it by "\". The character "+" after a pattern means one or more instances of the pattern. Parentheses group things together.

Regular expressions will win no prize for an elegant language. Programming complicated patterns with them is tricky and produces extremely obscure code. Still, regular expressions are widely used and quite well defined, at least in the sense that there is a standard definition, as part of the *Posix* standard for operating system interfaces. (S uses this standard definition of regular expressions.) Provided you keep to simple use of them, regular expressions can be a substantial help in manipulating strings.

Regular expressions occur three main ways in S:

1. when dealing with substrings; the `substring` function takes a regular expression to define what characters to extract or replace; for a discussion of substrings, see section 5.1.4.

2. as a logical-style subscript in extraction, where the elements matching the regular expression are selected;

3. through explicit matching via the `regMatch` and `regMatchPos` functions.

The second and third items are discussed here.

The expression `regMatch(x, table)` matches `x` to (usually) one regular expression in `table`.

```
> state.abb[regMatch(state.abb, "^A")]
[1] "AL" "AK" "AZ" "AR"
> state.abb[regMatch(state.abb, "K$")]
[1] "AK" "OK"
> state.abb[regMatch(state.abb, "^A|K$")]
[1] "AL" "AK" "AZ" "AR" "OK"
```

Since the single regular expression in the third example means "a string with either `"A"` at the beginning or `"K"` at the end", it extracts the union of the previous two expressions.

The function `regMatch(x, table)` matches analogously to the way that `match` matches ordinary strings, and returns a logical vector showing the matches. The function `regMatchPos` does the same matching, but returns a two-column matrix, with the first and last character position of the matches in the two columns. The number of rows in the matrix returned is the larger of the lengths of `x` and `table`. Where the regular expression did not match, both elements of the corresponding row will be set to `NA`.

It is possible to include more than one regular expression in `table` with either `regMatch` or `regMatchPos`. In this case the first expression is matched to the first element in `x`, the second expression to the second element, and so on. The shorter of the two objects, `x` and `table` is replicated sufficiently to match the longer. This use is much less common, but follows a paradigm used elsewhere in S. One application would be to match each row of a matrix to a different regular expression. Since matrices in S are stored column-by-column, a vector of regular expressions as long as the number of rows will be replicated correctly. (To match one regular expression to each of the columns of a matrix, you would need to do the replications directly or else transpose the matrix.)

Another use of the same idea exploits the S function `outer` to produce a matrix from matching each element of `x` to each regular expression of `table`.

```
> mm = outer(state.abb, c("^A", "^K"), regMatch)
> dim(mm)
[1] 50  2
> state.abb[mm[,1]]
[1] "AL" "AK" "AZ" "AR"
> state.abb[mm[,2]]
[1] "KS" "KY"
```

See the documentation of `outer` for the details of how this works; essentially, `outer` replicates its first two arguments into columns and rows of two matrices, and then calls the function supplied as the third argument (here `regMatch`) with the two matrices as arguments. Just the right interpretation for this example, as for a number of others. (By the way, for those familiar with the innards of C computations with regular expressions, be reassured that the implementation of `regMatch` only "compiles" each regular expression once, no matter how many times it occurs in `table`.)

The S function `grep`, named after the corresponding shell command, uses the same algorithm to return just a vector of the indices that match the pattern. It only works with one regular expression, but returns a more compact form if the pattern appears a few times in a large object.

Regular expressions can also be used with the operator `"["` in extracting and replacing parts of vectors and similar objects. In this case, the string must be made explicitly into an object with class `"regularExpression"`, by a call to the function of the same name:

```
files[ regularExpression("tex$") ]
```

This expression would extract all the elements of `files` containing strings ending in `"tex"`.

4.4.4 Partial Matching

Partial matching in S matches `x` to `table` in the same sense as `match`, but allows strings in `x` to only match the beginning of the corresponding string in `table`, so long as the match is unique.

Partial matching is used many places in S, most importantly in matching named arguments to functions; another way of describing partial matching is to say that it implicitly completes the strings, as happens in argument matching. Partial matching is also done when selecting subsets by name from named objects, from matrices or data frames with row and column labels, and by the "$" operator (but *not* when selecting slots or when using the `elNamed` function).

Partial matching is done explicitly by calling the function `pmatch`; its arguments and behavior look basically like those of `match`, but for each element of `x`, the match succeeds *either* if the element matches exactly some element in `table` *or* if the element matches the initial substring of just one element of `table` unambiguously.

In all cases, the rule is the same. In the call to `pmatch(x, table)`, the first step is to do exact matching, as in `match`. For each of the strings in `x` that fail exact matching, matching with completion is then tried. The match succeeds if the string uniquely matches the previously unmatched strings in `table`, but only one string in `x` can match any string in `table`. Two examples will illustrate:

```
> brandNames
[1] "Chevrolet"  "Volkswagen" "Eagle"      "Subaru"
[5] "Honda"
> pmatch(c("H", "V", "Hon"), brandNames)
[1]  5  2 NA
```

There are no exact matches here. The initial characters of the strings in `brandNames` are all unique, so the first two strings match, but when we get to the third the only partial match has been taken, so the match fails. You don't get any extra credit for matching more, but not all, the characters. Exact matches do count, however:

```
> pmatch(c("H", "V", "Honda"), brandNames)
[1] NA  2  5
```

In this case the first partial match fails because the only possibility has been taken by an exact match. The rules of `pmatch` are defined by the requirement that it behave the same way the S evaluator does in matching arguments, and the latter is fixed by long tradition. If the particular rules are acceptable, `pmatch` can be a useful tool.

One cautionary note: it's possible to do partial matching quite efficiently, but the standard S implementation of `pmatch` doesn't. The function is likely to be slow if there are many strings, especially if there are many duplicate initial substrings.

4.5 Extracting and Replacing Data

S expressions to extract pieces of an object use, most commonly, the `"["`, or "subset" operator. Other expressions use the `"[["` operator, which behaves like subsetting but extracts only a single element. Other operators and functions exist as well, such as `"$"`, `"@"`, and `el`.

Replacement expressions can use any of these on the left side of an assignment to replace the corresponding piece of the data. S is unique however, in having a fully general and extensible syntax for replacement expressions.

In this section we discuss the use of extraction and replacement expressions in S.

4.5.1 Extracting and Replacing Subsets

S uses expressions of the form `x[i]` to extract data from the object `x`, with the argument `i` indicating which elements of `x` we want. Such expressions are used throughout S. Both `x` and `i` can have many forms.

The general concept is that the value of the expression will be an object like `x`, but with data corresponding to the selection defined by `i`. In particular, the extracted object often has the same class as `x`. We often describe the computation as extracting a subset of `x` (it's not quite a subset in the mathematical sense, but never mind). The argument `i` is called the *index* defining the subset. The same form of expression can be used on the left side of an assignment operator too. In this case it means to *replace* the subset of `x` defined by the index with the values on the right of the assignment.

Some of the common forms for the index `i` are as follows.

1. If all the values in `i` are positive integers, then these are interpreted as indices within `x`, with `1` referring to the first element, and so on.

2. If all the values are negative integers, the selection is interpreted as being all of `x` *except* the indices defined by `-i`.

3. If `i` is a logical vector of the same length as `x`, it implies extracting or replacing all the elements of `x` corresponding to `TRUE` in `i`.

These interpretations are extended in various ways to make computations more convenient. `NA` values are allowed in the index (these generate corresponding `NA` values in extraction); `0` values are allowed, and ignored, in positive index vectors; extractions outside the range of the data generate `NA` values; logical index vectors are replicated if necessary to make them as long as the object.

Other classes of objects are interpreted as index arguments in special cases. For example, character string values are matched (by partial matching; see page 154) to the `names` attribute of a `named` object, and the resulting indices used to extract or replace elements of the data in the object.

```
> class(state.income)
[1] "named"
> state.income[c("Missis", "Alaska")]
 Mississippi Alaska
       10102  19100
```

Because `"Missis"` has a unique completion in the state names, it extracts the corresponding element.

The interpretation of index vectors for replacement differs in some necessary ways from extraction. Replacing an element beyond the length of the object extends the object to that length. Index values that don't refer to a legal position (`NA` and `0`) are ignored.

```
> temp
 [1]  11.9   3.4   2.9  -0.1  11.9   5.7   7.8  11.0   8.4
[10]   0.1
> length(temp)
[1] 10
> temp[c(10, 12, NA)] = 101:103
Warning messages:
  Replacement length not a multiple of number of elements
to replace in: temp[c(10, 12, NA)] = 101:103
> temp
 [1]   11.9    3.4    2.9   -0.1   11.9    5.7    7.8   11.0
 [9]    8.4  101.0     NA  102.0
> length(temp)
[1] 12
```

Element 12 was filled, and the length of `temp` set accordingly. Since element 11 was never set explicitly, it becomes `NA`.

The description so far is of extraction and replacement in vectors and other objects behaving like vectors. Extensions can be defined for other classes of objects, as they can with any S functions. One very special, but common case is that of multiple indexes used to extract or replace in matrices, multiway arrays, and other classes of objects with similar structure. In these objects, an element is associated with multiple index values; for example, the `[i, j]` element in row `i` and column `j` of a matrix.

A matrix or array object `x` can be indexed by expressions with as many index arguments as the length of `dim(x)`. For a matrix, the first index is applied to the rows and the second to the columns. Empty indexes imply all rows or all columns. Similarly for general multiway arrays, there can be as many arguments as dimensions. For example:

```
> dim(iris)
[1] 50  4  3
> iris[, 1, 1]
 [1] 5.1 4.9 4.7 4.6 5.0 5.4 4.6 5.0 4.4 4.9 5.4
[12] 4.8 4.8 4.3 5.8 5.7 5.4 5.1 5.7 5.1 5.4 5.1
[23] 4.6 5.1 4.8 5.0 5.0 5.2 5.2 4.7 4.8 5.4 5.2
[34] 5.5 4.9 5.0 5.5 4.9 4.4 5.1 5.0 4.5 4.4 5.0
[45] 5.1 4.8 5.1 4.6 5.3 5.0
```

Other specialized index objects are recognized for these objects as well. A 3-way array can be indexed by a matrix with three columns, in which the first column is interpreted as indexing the first dimension, and so on. The elements corresponding to each of the rows of the index are extracted or replaced.

This last example illustrates the case that the extracted object does *not* have the same class as the dataset. Since the extracted elements of the array won't generally make up a multiway array, they are returned as a vector. Similarly, matrix and array objects can be indexed by a single argument, just as if they were vectors. In this case too the extracted object will be a vector, not a matrix or array.

```
> iris[iris > 7]
 [1] 7.1 7.6 7.3 7.2 7.7 7.7 7.7 7.2 7.2 7.4 7.9
[12] 7.7
```

For replacement on the other hand, all forms of array indexing leave the object's class unchanged.

```
> trimIris = iris
> trimIris[trimIris > 7.5] = 7.5
> class(trimIris)
[1] "array"
```

The method used won't let you expand arrays by assigning outside their current dimensions (that's just laziness on the part of the implementers).

4.5.2 General Replacement Expressions

S has standard-looking assignment and replacement expressions similar to other languages:

```
sample1 = rnorm(n)
sample1[1:m] = log(sample1[1:m])
```

Compared to most languages, however, S has moderate extensions to assignment expressions, and very profound extensions to replacement expressions. Other assignment operators can be used (see page 137), the most important being `<-`. This operator must be used instead of `=` if you want to do an assignment inside an argument in a function call.

When we come to replacement expressions, the S syntax and its meaning are both very different and much more general than in many traditional languages. The syntactically legal replacement expression allows nearly anything on the left of the assignment; the set of *meaningful* replacement expressions is somewhat more restricted but still very general. Any function call whose first argument is a name is a legal left-side expression:

```
track1 @ x = xx
class(results) = "numeric"
dim(iris) = c(50, 4, 3)
```

Remember that infix operators are just functions, with their first argument appearing to the left of the operator, so the first example is a call to `"@"` with `track1` as the first argument.

Next, the name in the first argument can instead be any legal left-side expression, by a recursive definition. Instead of a name, a second call with a name as the first argument can be used, and again this name could be a third call, and so on.

```
tracklist[[i]] @ x = xx
class(el(output, 1)) = "numeric"
dim(bigResult$matrices[[3]]) = c(n,m)
```

The S evaluator will replace a nested replacement expression with a sequence of assignments and simple replacements operating on special intermediate objects. To see the expression, evaluate `replaceExpression(expr)`, with `expr` the unevaluated left side of the replacement expression.

The evaluator carries out the corresponding evaluation by unraveling the recursive definition, and storing the intermediate objects is specially named aliases. For example, to see how the expression `diag(el(mlist, i)) = dr` is evaluated:

```
> replaceExpression(Quote( diag(el(mlist,i)) = dr ))
{
  .A1 <- dr
  .A0 <- el(mlist, i)
  diag(.A0) <- .A1
  el(mlist, i) <- .A0
}
```

The individual replacement expressions in this expanded form are then interpreted in terms of the replacement functions, `"diag<-"` and `el<-`.

```
> replaceExpression(Quote( diag(.A0) <- .A1 ))
.A0 <- "diag<-"(.A0, .A1)
> replaceExpression(Quote( el(mlist, i) <- .A0 ))
mlist <- "el<-"(mlist, i, .A0)
```

Arbitrary replacement expressions in S reduce to a sequence of simple assignments and un-nested calls to replacement functions (with an object name as the first argument). *Don't* evaluate them yourself in this form, however: the evaluator takes advantage of the original replacement expression to economize on copies of the object being modified.

Since infix operators are just function calls with a different appearance, the expression

```
fit[[i]]$residuals[ wgts < .001 ] = 0
```

gets converted by the same interpretation, because we could write it in the form

```
"["("$"( "[["(fit, i), "residuals", wgts < .001 ) = 0
```

Converted to replacement functions, it has the form

```
{
  .A2 <- 0
  .A1 <- fit[[i]]
  .A0 <- .A1$residuals
  .A0[wgts < 0.001] <- .A2
  .A1$residuals <- .A0
  fit[[i]] <- .A1
}
```

This also illustrates that multiple nesting follows the same rule, with additional alias objects created.

4.6 Graphics

Graphical computations form a major part of the S programming environment, and an equally significant reason for its popularity with users. We will not cover graphical computations in detail here, but just indicate how plotting can be incorporated in programming efforts. Details of the specific plotting techniques available can be found in nearly all of the reference books in the Bibliography.

Rather than a single graphics library or system, S actually comes with several and with the potential for adding others. The multiplicity of possible approaches to graphics is another reason to concentrate here on the general principles that help incorporate graphics into S programming. Among current approaches are the following:

1. the traditional S graphics functions, such as `plot`, `lines`, etc., dating back to the earliest versions of S;

2. the `trellis` library, built from approach 1, but offering higher-level functions, more carefully designed displays, and special features for conditioning on time or other variables;

3. the S-Plus graphics enhancements, based in part on the *Axum* graphical software, and providing greater interactive capability, among other features;

4. an essentially unlimited potential for using graphics software that is *not* part of S, by exploiting one or another version of the ability to interface S to other software (an example of this approach being Java graphics toolkits).

The last approach, in fact, presents the most exciting direction for S graphics. The range of graphics software in modern computing is tremendous, and the potential for *interfaces* to S software equally striking. Interfaces described in this book, using `shell`, `pipe` and `fifo`, are examples. More advanced forms of distributed computing will make such interfaces more general and conventient as well. (At the time of writing, these are active research topics at Bell Labs and elsewhere.)

As for more traditional S graphics, we will settle here for a brief outline of the general approach and references for further information. The S-Plus manuals give detailed discussion of the S-Plus graphics. Chapter 3 of the book by Venables and Ripley, [10], deals with both traditional S graphics and `trellis`. Chapter 4 of the book by Becker, Chambers, and Wilks, [1], describing the earlier version of S graphics, is still valid for the current version of S.

Traditional S graphics uses a simple two-dimensional model, in which the current (rectangular) plotting surface is divided into a central rectangular *plot* region, surrounded by four *margins*. Graphical computations in the plot region are typically done in terms of co-ordinates determined by the data, and called *user* co-ordinates in S. The archetypical such graph is the x-y or scatter plot, produced by the S function call `plot(x, y)`. The two arguments are most basically two vectors of numeric data and a plot is produced using the first argument for horizontal position and the second for vertical. Many other functions add to such plots or provide alternatives to them, and all the functions are capable of extensive specialization, as we will discuss. S distinguishes functions intended to produce complete plots from those typically used to add to plots by the terms *high-level* and *low-level* plotting functions. Essentially, the former clear the display and recompute the user co-ordinates from the data in their arguments, while the latter do neither.

Graphics in the margins tend to produce supplementary information such as axes, overall labelling, or numeric labels to give the viewer the scale of the plot. Much of this information is produced automatically by functions such as `plot`, but the programmer can customize the graphics by direct use of the margins, typically with a slightly different co-ordinate system, with the appropriate x or y co-ordinate defining position along the axis and a co-ordinate in terms of the character height for position perpendicular to the axis.

Considerable control of graphics can be obtained by using optional arguments to both the high-level and low-level plotting functions. In particular the graphical system maintains for each device a set of *graphical parameters*, analogous to style parameters in typesetting languages such as LaTeX. These parameters can be queried and/or set persistently by calls to the function `par` and may also be set for the duration of a particular graphical function by specifying the name and value of the parameter as an optional argument to that function. Included are overall styles (such as those for determining axis labels), and specific details such as the symbols to use for scatter plots, colors, etc. See `?par` for the available parameters.

For some of the graphics functions available, see Table 3.10 on page 103.

4.7 Models and Advanced Numerical Methods

The S programming environment comes with a substantial collection of functions to perform numerical linear algebra and to fit models to data. A number of additional libraries have been written in S to extend the scope of model-fitting and related computations. Readers with special interest in modeling should see reference [3] for the description of statistical models in S; also, reference [10] covers numerical methods and the models library, as well as some of the extensions. In this section we will briefly introduce the essential functions and classes, with emphasis on the style of such computations in S.

Unlike some systems, S is a general computing environment for which models and linear algebra are natural applications, rather than a specialized system for such computations. As a result, some computations may not be as finely tuned in S, where general techniques are used, as they might be in a more specialized system. The compensation is that the numerical computations use general programming techniques that lend themselves to gradual refinement. S programmers are encouraged to specialize, modify, and extend numerical computations, and everything else as well. Once you understand the general approach, defining your own functions or other S software is a natural next step. The modifications can often include access to highly developed special techniques in other languages or systems.

4.7.1 Matrix Computations

S provides a class `matrix` and a variety of functions to operate on matrices. Most of the functions deal with numeric matrices, but the class is actually more general. It is an example of an S `structure`; that is, a vector object extended to define some structure among the elements. In the matrix case the structure is the ability to index the data by two subscripts, for the rows and columns of the matrix. As with all structures, the vector is a slot in the structure and any vector can in principle be in that slot. Other slots extend the vector suitably. For a matrix, these slots define dimensions of the matrix—the number of rows and columns—and optionally character string labels for the rows and columns.

Matrices can be manipulated by extracting or replacing portions of rows and columns, using the "[" operator with two indices: x[i,j] refers to the subset of x defined by applying the value of i to the row indices and the value of j to the column indices. In other words, i can have any of the forms for indexing a vector of length nrow(x) and j for indexing a vector of length ncol(x). Included in this is the notion that either i or j could be the empty expression, meaning all rows or all columns.

Similar expressions with the operator "[[" extract or replace a single element. There are various specialized manipulation functions as well: diag extracts or replaces the diagonal elements; t transposes the matrix.

Notice that the definition of matrices as a vector with added structure means that vector operations make sense as well. Using "[" with a single index implies operating on the vector of data in the matrix. Extracting data this way throws away the matrix structure, but replacements retain it. The computation of the one-way indices may need to take into account the way matrix structures are laid out in S (the data is stored a column at a time), but many computations can be written in terms of logical conditions:

```
x[ x <= 0 ] = 1e-7
```

The structure concept in S includes the notion that operations on structures retain the structure when that makes sense. For example, if x is a numeric matrix

```
x > 0
```

is a matrix with the same structure as x, but whose data slot contains the appropriate logical values, not numbers.

4.7.2 Numerical Linear Algebra

S provides a fairly wide selection of numerical computations on matrices, and both the language and the matrix class are designed to encourage other computations to be incorporated, for example by writing interfaces to algorithms in numerical libraries. Matrix decompositions, specifically the QR, singular-value, and eigen-value decompositions, are the principal building blocks. The functions qr, svd, and eigen provide these for numeric (i.e., double precision) matrices, in the third case restricted to symmetric square matrices. The Choleski decomposition of square positive-definite matrices is provided by the function chol.

Equation solving is provided by function solve, with the matrix inverse if solve is called with one argument. Some elementary approximation functions are provided for minimizing least-square or least-absolute-value residuals, but more elaborate fitting is done through the model-fitting computations described in section 4.7.3. Basic matrix multiplication is provided by the operators "%*%" and "%c%".

```
x %*% y
```

computes the matrix product of the two matrices x and y, treating vectors as matrices with one column. The "%c%" operator is a "cross-product" operator; that is, x%c%y returns the same result as t(x)%*%y.

The decompositions are returned in the form of objects containing separate matrix or vector components representing the parts of the decomposition. For example, svd(x) represents x as the product of two orthogonal matrices and a diagonal matrix: in S notation,

```
x == u %c% d %*% v
```

where conceptually, u and v are square orthonormal matrices and d is a rectangular diagonal matrix with the same shape as x. The object returned by svd has these three components, but d is just the nontrivial part, the diagonal vector. Options to svd can omit part or all of the two matrices in the decomposition. The eigenvalue and Choleski decompositions (functions eigen and chol) are similarly constructed in terms of their mathematical definitions. The QR (orthogonal-triangular) decomposition is constructed according to traditional numerical analytic ideas, with the result that the data are packed in an efficient but rather arcane form. In any case, the algorithm used is defined by a sequence of elementary reflection operations, so it makes more sense to think of the returned decomposition as a single object. There are a variety of related functions to perform various operations with the decomposition. See ?qr.fitted, for example.

4.7.3 Model-Fitting Functions

The models library provides a range of functions for fitting and analyzing numerical models for data. The library may be part of the standard search list for your installation of S; if not, you need to attach it by

```
library(models)
```

The model-fitting software is described in the S-Plus manuals and in several books on S. The full and original description is the *Statistical Models in S*, reference [3]; several chapters in reference [10] also discuss these models, and other statistical modelling software as well. We will give only a brief outline here; see the references for details.

Modeling in S takes data and a description of the model, applies a numerical fitting algorithm, and returns an object representing the fitted model. Methods exist to print, summarize, and plot these models. Typically, the user then may go on to repeat the process, either by changing the data or description before recalling the fitting function, or by using an updating method to modify the current model.

Models are described in two parts: the general kind of model being used (linear regression, nonlinear regression, tree-based models, etc.) is determined by the choice of fitting function (`lm`, `nls`, `tree`, etc.). The specific model form is defined by a `formula` argument to the fitting function. A formula is a stylized S expression; specifically, a call to the special operator `"~"`, which can be read as "is fitted to" in formulas. The left-side operand defines the variable (the data) to be approximated, the right-side operand the combination of variables to be used for fitting. So the formula

```
Fuel ~ Type
```

says to approximate the object `Fuel` by a model using the object `Type`.

The formula expression itself (the call to `"~"`) does no fitting, but just returns essentially the expression itself, as a `formula` object. It's up to the fitting function to interpret the formula as a fitting task. The `lm` function uses the formula to define a linear model that it then proceeds to fit by least-squares. The left side, or *response*, is an expression defining a numeric vector or matrix. The right side defines one or more predictor variables; by convention, different predictors are combined by `"+"`:

```
Fuel ~ log(Weight) + Type
```

This fits a linear model in two variables, `log(Weight)` and `Type`, plus an intercept, by the usual convention of linear models.

The variables appearing in the model look like ordinary S objects, and they can be just that. More commonly, though, they are defined in the context of some particular dataset. In this case, the user's call to the fitting function supplies a `data` argument. The fitting function attaches this object temporarily in order to compute the information needed to do the actual fit. Therefore, components of the object can appear as variables in the model formula. See section 5.3.4 for the general ideas here. The objects supplied in model fitting are often `data frames`; these are named lists whose elements are variables in the dataset to be used for fitting. Data frames are an old-style class, with no formal representation but with class `data.frame` and various methods for manipulating the objects. They appear to users to resemble two-way arrays, in that columns and rows can be addressed; for example,

```
x[ weights > 0, -1]
```

would return a data frame with all the variables (columns) except the first, and only the elements (rows) for each variable corresponding to `TRUE` in the logical expression `weights` > `0`.

For purposes of model-fitting, the variables in a data frame are expected to be numeric vectors, numeric matrices, or objects defining the levels of some factor (categorical variable). All the variables have to agree in the number of observations (that is, the lengths of the vectors and the number of rows of the matrices). The categorical variables work best if they have class `"string"`, or one of the extensions of string (see section 5.1.3). Specifically, the variables may have class `"string"`, `"stringFactor"`, or `"stringOrdered"`. The difference is that the levels for `string` objects may be recomputed when a subset of the object is taken, whereas levels of the other two classes are considered fixed until they are explicitly reset. Between `"stringFactor"` and `"stringOrdered"`, the distinction is that the levels of the latter are assumed to indicate an ordering (say, from small to large) and some model-fitting computations will fit these terms slightly differently.

Older versions of S used classes `"factor"` or `"ordered"` for categorical variables in models; these still work for most purposes, but `string`-based objects behave more consistently; see page 452.

4.8 Efficiency in Large Computations

Our goal in programming with data is to turn ideas into software, quickly and faithfully. Computations should be chosen because they express what we want to know about the data. Issues of how *efficient* the computations are, in the sense of the best use of machine resources to obtain a particular result, come later. But they will eventually intrude, because we will want to apply some of our ideas to larger datasets or implement ideas that intrinsically involve a great deal of computation. Some general understanding of the way S works and of the tools available for dealing with large problems will then be useful.

When starting on a project that is likely to stretch the computing resources (for example, because it involves bringing in large amounts of data), some good choices early on in organizing the data and the computations can make a big difference later.

This section presents a few general ideas and a number of specific hints. The hints mostly take the form of pointers to discussions elsewhere in the book. Of the general ideas, the most important is that of viewing basic computations in S as *whole-object* computations, conceptually acting on an object, or a major piece of an object, in one step. Grasping this view of S computations is often the single most important step in developing a good "style" of computing with S, for both understandable and efficient software.

4.8.1 The Whole-Object View

S has a unique model for computation; understanding some aspects of the model will help in using S effectively. Everything is an object in S, and all objects are dynamic and self-defining. S builds a functional computational model on these objects: nearly everything of importance in S happens by evaluating a call to a function.

S differs from other languages in that the "atomic" units underlying the objects and the computations are farther from the "machine". Even rather high-level languages, such as Java, usually start from things like single integers, numeric values, and characters. You have to move up several levels in programming to get to objects that correspond roughly to S objects, if such concepts even exist in the language.

S tries to make things easy for programming, especially for getting started. Dynamic, self-defining objects put more of the burden on the system rather than on the user to allocate space and keep track of data types. I think this is the right choice for most users.

A consequence, though, is that no S object gets as close to the hardware as a single integer or number, and no S computation can even conceptually be thought of as a single machine instruction on any ordinary computer. (The atomic operations in other languages are in fact often farther from single instructions than they seem as well, but the relative difference remains.)

Does this matter to the user/programmer? Usually not, because in many situations the user is unaware of machine efficiency anyway; the machine is often spending the bulk of its time idle, or on computations unrelated to the user's S tasks. But when designing computations to make reasonably efficient use of the machine *is* a concern, you will benefit from a viewpoint adjusted to computations in S.

The basic hint is simple to state: try to see your computations and the objects they return in *whole-object* terms. So, if you take two numeric objects and do some computations on them, think of this in terms of the whole objects, not the individual numeric values they contain.

For example, suppose we want to multiply the values in `y` by weights `w` when the weights are positive, but by some other function of the weights when the weights are negative, say `1/(1-w)`. If we thought of this in terms of individual values in `w`, we might be tempted to write a loop such as

```
for(i in 1:length(w))
  if(w[i] > 0) y[i] = y[i] * w[i]
  else y[i] = y[i]/(1-w[i])
```

This is perfectly legal and works fine; however, the number of individual S computations is several times the length of `w`. If the objects are large, the computation will be slow.

How should we think of this computation in whole-object terms? The key idea in the computation is the condition

```
w > 0
```

which controls what happens. The condition itself is an object, a logical vector. Like any logical vector, it can be used to extract or replace part of an object in a single S computation. From this idea it is a short step to the computation

```
pos = w > 0; neg = !pos
y[pos] = y[pos] * w[pos]
y[neg] = y[neg]/(1-w[neg])
```

The number of S function calls is now independent of the length of `w`. To an experienced S user, the computation also tends to look more natural as well, being expressed directly in terms of the data and the conditions on the data that define what we want to do.

A little more browsing (in this chapter) might have uncovered a function that expresses the idea of this computation even more directly: `ifelse`, taking a condition and two further objects as its arguments and returning values from the second or third according to the values in the condition (see page 134). Our computation can be carried out by:

```
y = y * ifelse(w > 0, w, 1/(1-w))
```

That expresses the computation about as directly as we could ask.

Here's another example.

```
for(i in 1:length(text)) {
    if(is.na(match(names(text)[i], c("x", "y", "z")))
      && substring(text[i], 1,3) == "##<")
      omit = c(omit, i)
}
```

The purpose of this code is to find all the elements of a named character vector such that the name does *not* belong to a reserved set of three special names and the character string element starts with the pattern `"##<`. The complicated test of the name against the reserved names is done on each string, where in fact what we're saying is that we only want to look at the non-matching elements. But in that case, the whole-object concept is to construct only the subset of non-matching elements, *as an object* and just compare the substring of that object.

```
omit = (1:length(text))[
  is.na(match(names(text), c("x", "y", "z")))]
omit = omit[substring(text[omit], 1, 3) == "##<"]
```

We constructed the indexes in `omit` as an object, starting with all the indices, `(1:length(text))`, pruning the reserved names, and then pruning again to retain only the matching substrings, all in whole-object terms. Here as in the previous example, the basic hint is to look over the computations and ask what they really mean in terms of the *objects* involved.

Should you wonder where this example came from, well, actually it was my own code, from a revision of the online documentation function. The actual example was a bit less easy to fix, but the whole-object principle had been ignored just as thoroughly. Clearly I'm not asserting that this approach is totally obvious from the start! It is, however, a useful step in making your code more efficient and at the same time clearer.

The whole-object concept arises again and again in S. See page 208 in section 5.1.5 for a discussion about how to think of tree structures in whole-object terms, and page 203 in section 5.1.3 for an example of creating a matrix with specified numerical values from a table of strings.

4.8.2 Techniques and Tools for Large Computations

The basic technique is classic: *Keep it Simple.* A long, complicated S expression or function, with many local variables, is likely to be harder to understand and also less efficient for large problems than a relatively small computation that combines calls to a few other functions to perform its essential tasks.

Similarly, in organizing data, S deals best with a fairly small number of simple pieces rather than a deeply nested organization of structures within structures. The organization of tree-structure data in section 5.1.5 is an example. Tree structures are conceptually made up of nodes connected to other nodes along with labels and perhaps auxiliary data attached to the nodes. The natural structure in S is to represent *all* the nodes by a few items: the connections, the labels, other node data. In contrast, an incremental view that works from a single node to other nodes is neither conceptually clean nor at all efficient. The whole-object view promoted in the previous section is the natural and efficient way to view most computations in S. The larger the object, the more such simplicity matters.

Another general technique to keep in mind is the *interface*. S encourages doing appropriate computations in other systems: an interface to the command shell leads to other programs that work in terms of lines of text; where that is a reasonable model for your computation and there is a tool or language—the PERL language, for example—that expresses the computation well, by all means use the interface. Similarly, C or Fortran subroutines can perform basic numeric or string computations and increase efficiency for large computations. After a numeric computation has been worked out interactively as an S function, there may be a natural translation of part of the computation into C or Fortran. The S function can then invoke the subroutine and handle the whole-object issues, such as organizing the data correctly. These interfaces are encouraged, as long as they fit naturally into the overall computations.

Such general techniques are best learned from practice and examples. Some examples are scattered throughout the book and in existing functions and classes. In addition to general ideas, there are a number of specific techniques and functions that can help with particular problems. Some are listed here, with references to more detailed discussions.

- If the whole session seems to be slowing down, with even simple computations taking much longer, after you have done some big tasks, it's possible that the memory being used for allocating S objects has become severely fragmented. You can call `reloadS` to start the session over from scratch, while preserving current session computations. You will see a banner as if you had just invoked S, but datasets in the session frame and other pieces of your previous session will have been retained. Dynamic memory will have been started from scratch, however, so *if* fragmentation was a problem, things should speed up. On a multiuser system, the slowdown could have many causes other than your own S session, so don't be surprised if this technique does not always help.

- If you need to do a *really* long computation, consider breaking it into a sequence of tasks. Running each task separately at the top level allows the evaluator to clear things out more thoroughly at each stage. Even aside from efficiency, running the computation as several tasks means that errors (including the computer going down) won't wipe out all your computations. Tasks can be set up using `setReader` and the general event management facilities (section 10.6).

- If you are building up a large object in stages, make heavy use of S replacement functions. S has a uniquely general way of replacing pieces of an object, and the evaluator deals with these specially, both existing replacement functions and your own new ones. Section 4.5 discusses replacement expressions and section 6.1.4 deals with writing replacement functions.

- If you want to match character strings in a large table, use the class `string` rather than `character` for the table, and make sure that the string object in which you're searching has its `stringTable` slot set. This matters most if you repeatedly match one or a few items against a big table. See page 151 for details.

- If you call a C or Fortran routine with large objects, you can minimize the preliminary computations needed to coerce and copy the objects by using the `CLASSES` and `COPY` arguments to the interface routines. See page 420 in section 11.3. Another useful tool, described in the same section, is the `unset` function, which tells S you no longer need a local object, typically when you are calling a C or Fortran subroutine. Both `COPY` and `unset` can economize on the extra copies S makes of objects involved in these calls.

- If you need to generate large amounts of atomic data (numbers, especially) outside S and then import the data, consider using the `readRaw` function (section 10.5.4). This maps binary data on a file into an object in S. For large amounts of data, time can be saved by not requiring the other program to format the data as text and then reading that data into S via `scan`. S will use memory mapping in the operating system to access large numeric blocks of data this way.

4.8.3 Iteration and the Apply Functions

Tradition among experienced S programmers has always been that loops (typically `for` loops) are intrinsically inefficient: expressing computations without loops has provided a measure of entry into the inner circle of S programming. To the extent that the loops can be replaced by a few calls to "atomic" computations, eliminating them will indeed provide substantial speedup, and can help conceptually by moving towards a whole-object view. The replacement is successful in a number of computations dealing with matrices and multiway arrays, by expressing an iteration as a computation on the whole array.

A family of functions generally referred to as *apply* functions provide alternatives to loops. The common notion of the functions is to apply a function call to all the subsets, defined in some appropriate sense, of an array, list, or other object. The results of all the calls are combined and returned as a single object. The apply functions are often useful because they simplify the expression of a computation; sometimes, but not always, they also have efficiency advantages (`rapply`, on page 174, is one example).

The original `apply` was defined for multiway arrays; since it has a different calling sequence than the other related functions, we can't quite make them methods for `apply`, but that's conceptually the notion. The simplest example of the family is `lapply`: this takes a list and calls a function iteratively with each element of the list as the argument to the function. Suppose we wanted to remove the names attribute from all elements of `x`:

```
lapply(x, function(obj){
      names(obj) = NULL
      obj})
```

This is a compact way to write the computation; providing the function explicitly may look a little odd at first, but has the advantage that nothing is hidden. The truth is that this computation is not significantly superior to the natural iterative version:

```
for(i in seq(along=x)) {
    names(x[[i]]) = NULL
}
```

No matter how efficiently `lapply` is implemented, the loop does not produce much overhead, and the inline replacement is able to save some storage. In general, the apply functions are useful and can improve the clarity of your code, but they are only occasionally essential for efficiency.

Two variants on `lapply` deserve mention. The function `sapply` works just like `lapply` but finishes up by trying to simplify the result. If each call returns an object of the same length, `sapply` returns a matrix with that many rows. If each call returns a *single* value, `sapply` returns a vector. Thus, for example,

```
sapply(x, class)
```

returns a vector containing the classes of each of the elements of `x`.

The newest member of the family is `rapply`, which applies a function *recursively* to all the chosen elements of a list, at any depth. The elements chosen must match specified classes. There are two typical ways to use the function; in the first, we want to compute some summary data on certain elements of the list. For example suppose we wanted to know the number of characters in each element of each character vector or string, anywhere in the list `x`. The result is given by:

```
rapply(x, nchar, classes = c("character", "string"))
```

As with `lapply` the first two arguments are a list and a function. The third argument restricts the application of the function *only* to elements that match the specified classes. For all elements of x that do match, the function `nchar` is called and `rapply` saves the result. For other elements, the result has a default value of `NULL`.

The difference from `lapply` is that for elements of `x` that are themselves lists or other recursive objects, the same computation is applied recursively. So in this case, we end up with all the character counts from all the character data at any level, plus a bunch of `NULL` objects. The last step of `rapply` is, by default, to call `unlist` on the results. This will merge all the character counts into one vector, throwing away the `NULL`s. For an example of `rapply`, see page 371 in section 9.5.

The other commonly useful way to use `rapply` is to modify a list: each element of the list to which the function is applied is *replaced* by the value of the call, while all other elements stay unmodified. Providing the argument `replace=T` to `rapply` signals this approach. For example, suppose we wanted to convert all the character string data in `x` to lower case (say, to make comparisons independent of case). The function `lowerCase` returns the result on a single vector. To replace the strings but leave the list otherwise unchanged:

```
xL = rapply(x, lowerCase, c("character", "string"), replace=T)
```

By the way, we could actually have given only the class `"character"` in the calls, because class `string` extends `character`, and so `nchar` and `lowerCase` would have been called after converting the strings to character vectors.

The `rapply` function is one case in which the elimination of loops *does* represent a substantial efficiency boost. To implement the recursive looping of `rapply` directly in S would often overwhelm the computations themselves, particularly when the classes supplied restrict the number of function calls to a small fraction of the number of nodes in the tree defined by the list object.

Finally the two apply functions that *don't* work on lists are the original function, `apply`, and the function `tapply`. The former takes an array, an argument, `MARGIN`, to specify a subset of the dimensions of the array, and a function. The function is applied to each combination of values for the specified margins. For example,

```
apply(x, 1, sort)
```

sorts each row of the matrix `x`. The function `tapply` takes an object, one or more indexing objects (implicitly these should each have levels), and a function. It calls the function on subsets of the object defined for each unique combination of levels of the indexing objects. The functions `apply` and `tapply` are useful in many applications. For details and examples, see the online documentation and the various books on S in the references.

4.9 The S Evaluation Model

You use S by giving S tasks to evaluate. The task is an S expression, usually involving one or more calls to S functions. S evaluates that task; more precisely, an S *evaluator* evaluates the task. For planning serious computations and for debugging problems that arise, some understanding of *how* the evaluation works may be useful.

4.9.1 The Evaluator as an S Function

The S evaluator is best understood as a hypothetical S function, say `Eval`, that behaves much like the actual function `eval`, except that the latter goes off right away to its implementation in C.

The arguments to `Eval` are `expr`, an S language object (a function call, an assignment operation, a `for` loop object, and so on) and `nFrame`, the index of the evaluation frame in which to evaluate the expression. The evaluation frames are elements of a conceptual list, `Frames`. The `Frames` list is initialized by the evaluator manager to be of length 1, containing an empty frame 1 (the task frame). Evaluating a top-level task corresponds to the call:

```
Eval(task, 1)
```

where `task` is the S object corresponding to the task.

The `Eval` function has methods for all the classes of S language objects. These methods generally include recursive calls to `Eval` to evaluate pieces of the objects (such as the arguments to a function call, the right side of an assignment, the condition in an `if` statement).

In addition to calling itself recursively, `Eval` invokes some utilities. Like `Eval`, these can be modeled as S functions that behave very like actual S functions:

- `Assign`, which evaluates a simple assignment, associating an object with a name in a particular evaluation frame;
- `New.frame`, which creates a new evaluation frame in the frames list;
- `Clear.frame`, which clears and removes an evaluation frame.

Actual functions `assign`, `new.fame`, and `clear.frame` provide similar facilities for the S programmer.

In addition, there are a variety of internal S objects, nearly all vectors parallel to `Frames`: for example, `Calls`, the function calls generating the frames; `Functions`, the corresponding function definitions; `Parents`, the integer indexes of the parent frames for each frame. These objects parallel the S functions such as `sys.calls` that return the corresponding objects (see Table 3.13 on page 106). As `Eval` proceeds, it extracts and sets elements of these objects.

4.9.2 Evaluating Function Calls

Since nearly all the evaluation consists of function calls, understanding how these work will be the most important step in understanding the evaluation model. The S evaluator begins with two objects: the definition of the function and the call to be evaluated. The evaluation of the call has either four or five main steps:

1. `New.frame` creates an evaluation frame for the call, as element N of the `Frames` list. The call and the function definition are stored in the corresponding elements of `Calls` and `Functions`.

2. The evaluator matches the *actual* arguments of the call to the *formal* argument names of the definition.

3. If methods have been defined for this function (i.e., if the function is known to be a *generic* function), the evaluator selects the best-matching method for the actual arguments.

4. `Eval(body, N)` evaluates the *body* of the function or of the selected method.

5. `Clear.frame` is called to clear frame N. The value of the body, moved by `Clear.frame` to frame `Parents[N]`, is the value of function call.

Let's take a fairly simple example to fix ideas. The matching and evaluation of arguments are the key steps; we will examine these in somewhat more detail in sections 4.9.3 and 4.9.4.

Consider the call to function `myscale` in the task:

```
> scaleRes = myscale(curfit$residuals)
```

and suppose `myscale` was defined as follows, with no methods specified.

```
"myscale" =
# rescale 'x', by default to the range '0,1'.  If supplied,
# 'location' is a central value for 'x'.
# After subtracting this off, the result is divided by 'scale'
function(x,
     location = min(x),
     scale = max(y))
{
    y = x - location
    y/scale
}
```

In this example, there is one actual argument expression, `curfit$residuals`. Argument matching will associate this with formal argument name `x`. The body of the function definition:

```
{
    y = x - location
    y/scale
}
```

is now evaluated. This is a braced list of sub-expressions, which are evaluated in turn.

The first sub-expression is an assignment; to evaluate that requires evaluating the right-hand side, the expression `x - location`. This in turn requires a call to the function `"-"` and eventually the evaluation of the two arguments `x` and `location` (more on that in section 4.9.4). The result is assigned as object `y` in the frame. Similarly, the next sub-expression generates a call to the function `"/"`. This completes evaluation of the braced list (its value is always that of the last subexpression evaluated). Since this in turn completes the evaluation of the body of the function, the value of `y/scale` is the value of the call. `Clear.frame` moves this value to the calling frame (frame 1 in this case) and clears frame 2, recovering its storage.

The assignment of `"scaleRes"` in frame 1 completes the evaluation of the task expression. While this is not part of function call evaluation, it is an important feature of the evaluation model, so let's consider it here. Since this is a top-level assignment, the evaluator notes it, and in addition to the immediate assignment in frame 1, copies the assignment to the working data when the task evaluation completes (without an error), conceptually by

```
assign("scaleRes", scaleRes, where = 1)
```

Because this conceptual `assign` only occurs when wrapping up an error-free task evaluation, it provides protection for top-level assignments; they are not committed to the working data if an error occurs in the task.

The specific features of the evaluation model for function calls are largely motivated by two basic concepts of S. Everything is an object, including unevaluated calls and the frame in which the evaluation takes place. S is a functional language, meaning in particular that the model is in terms of evaluating expressions and dealing with the resulting objects, rather than concepts such as addresses and pointers.

The objects and operations in the S evaluation model are nearly all accessible as tools in S itself. For example, `substitute(x)` returns the expression for formal argument `x` in the current call, and `sys.frame()` returns the local frame, as a named list.

4.9.3 Argument Matching

Actual argument expressions in a call can be named or unnamed. In the expression

```
myscale(curfit$residuals, scale = 1)
```

the first argument is unnamed, and the second has name `scale`. Named arguments look like assignments, and in a sense they are, but of a special kind. The name in the assignment must match one of the formal arguments of the function definition, and the assignment does not take place until the corresponding argument is evaluated.

The general rules for matching arguments in S function calls are as follows. Initially all arguments are unmatched. The evaluator then goes through the actual arguments in three passes, each time taking all the so-far-unmatched actual arguments in order and trying to match them to the unmatched formal argument names:

1. names for named arguments are tested for exact match to an unmatched formal argument;
2. names for any remaining named arguments are completed if possible by partial match to the unmatched formal argument names;
3. unnamed actual arguments are matched in order to any unmatched formal arguments.

There is a special argument name `...`, which is allowed to match arbitrarily many arguments. Since it messes up the semantic model somewhat, let's leave it for section 4.9.5.

Argument matching will fail if any of the *actual* arguments cannot be matched by the rules. On the other hand, any *formal* argument names can remain unmatched. The evaluator keeps track and acts accordingly when evaluating the argument in question. The programmer can test explicitly as well:

```
missing(scale)
```

returns `TRUE` if the formal argument `scale` failed to match any actual argument.

The rules are sufficient to handle most cases; the notion of named arguments is largely to allow S functions to have a few common arguments that are nearly always supplied, plus potentially many other arguments that are only used occasionally. By putting the common arguments first, the user can conveniently supply these unnamed, and then only have to remember the names (not the position) of the other arguments that happen to be needed in a particular example.

The function `objects`, for instance, returns the names of objects in a database.

```
> functionArgNames(objects)
[1] "where"   "frame"   "pattern" "meta"    "classes"
[6] "test"
```

Calls to `object` typically supply the argument `where`, specifying the database. The other arguments are all much less commonly used, usually only one or two of them in any call. For example, the `classes` argument restricts the classes of the objects selected and `meta` tells `objects` to look in a meta database. As an example of argument matching, consider:

```
objects("main", class = "classRepresentation", meta = "m")
```

The first pass matches argument `meta` exactly by name. The second matches the second actual argument, and completes it, as `classes`. The third pass then matches the remaining, unnamed argument to `where`.

When the call expression itself occurs in another function definition, the evaluator will in fact complete partially matched names in its copy of the function definition. Partial matching will then only occur on the first call. Since it is substantially less efficient than the other two steps for functions with many arguments, completing the names helps speed up function calls.

Partial matching, or argument name completion, is intended to make things a bit simpler by allowing users to specify just enough of the initial string in a name to make it unique, following the rules above. A few instances that are logically decipherable won't be caught by the rules above. For example, consider the function

```
> f = function(x1, x2) c(x1, x2)
```

The call `f(x2=2, x=1)` deciphers the two arguments correctly, but the call `f(1, x=2)` doesn't, because the unnamed argument is not matched until after the unsuccessful attempt to match the named argument. For most user-friendly choices of argument names, however, such problems are not likely to arise.

In obscure situations, name completion could also cause problems (for example, if you wanted to redefine `f` from one call to the next so as to change the argument names). If so, you can turn name completion off by setting the corresponding option `completions`:

```
options(completions=F)
```

But did you really want to deal in such obscure techniques?

4.9.4 Argument Evaluation

When the S evaluator prepares a function call for evaluation, it constructs the S expressions representing the actual arguments. The value of the argument is computed—that is, the expression for the argument is evaluated—only when it is explicitly needed. This strategy, referred to as *lazy evaluation*, permits flexible handling of missing arguments and computations depending on the expression for the argument, rather than its value. For example, a number of functions allow users to omit the quotation marks surrounding a name, supplying `xxx` instead of `"xxx"`. If the corresponding argument were just evaluated by ordinary rules, the value would be the object named `xxx`, if that existed. The following fragment of code allows users to be a little sloppy with the quotation marks:

```
function(what, ...) {
    if(is(substitute(what), "name"))
        what = deparse(substitute(what))
    .....
}
```

Whether this sort of sugar is good for the user might be debated, but there are examples in which it makes sense, such as cases where the function wants to operate on the names of objects, but the user's thinking is more naturally in terms of the objects themselves.

The most common use of lazy evaluation is just to examine an argument to see if it was supplied. Missing arguments can have default expressions, in the function definition. However, it may be more convenient or even necessary to take some more general action depending on whether an argument was supplied. The following code, for example, checks that only one of the arguments `where` and `frame` was supplied:

```
if(!missing(where) && !missing(frame))
    stop("meaningless to include both where and frame")
```

The expression `missing(where)` asks a question about the expression for the call, not about the value of `where`.

Computations dealing with missing arguments (other than default expressions) must, of course, be done before the evaluator needs the value of the argument. Otherwise, the user gets an error of the form:

```
> f()
Problem in f(): Argument "x" is missing, with no default
```

Avoiding these problems requires some understanding of when argument evaluation is required. Three situations cover nearly all examples:

1. Evaluation is required to pass an object through one of the interfaces to C or Fortran. For example, elementary arithmetic, logical computations, etc. obviously need the actual object. The interfaces `.Call`, `.C`, `.Fortran`, and (usually) `.Internal` all require that the expressions being passed to them be evaluated.

2. If the argument is needed to select a method for a generic function, it needs to be evaluated, to determine its class. For example, suppose the function `Arith` has two arguments and there are some methods whose signature involves both arguments, specified, say as:

   ```
   setMethod("Arith", c("track", "track"),
       ....
   )
   ```

   Any call to `Arith` will evaluate both arguments during method selection, but missing arguments are always allowed (see page 187).

3. An argument has to be evaluated if it is to be "copied" in an assignment in the function. If `x` is an argument, the expression

   ```
   oldX = x
   ```

   will force evaluation of `x`. But for what "copied" really means, see section 4.9.7.

Arguments will be evaluated when one of these conditions occurs either in the local function or in any function called from this one, once the argument is needed in that function. In other words, passing an argument down to another function does not require evaluation, since that call will use lazy evaluation also. Once the underlying argument needs evaluation, however, this will communicate back up to the current call and require evaluation of the argument here as well.

Arguments are evaluated by evaluating the corresponding expression and assigning the result in the local frame, associated with the name of the argument. If the actual argument appeared in the call, the expression is evaluated in the frame of the calling function. If the argument was omitted but there was a default expression, that expression is evaluated in the current function. Both these choices make sense, since the expression in the *call* refers to the objects in the calling function, while the *default* expression, being part of the function definition, refers to the formal arguments and local objects in this function.

For example, the default expressions for the arguments of `myscale` on page 177 use the argument `x` and the local object `y`. Default expressions need to know that any local objects will have been computed by the time the argument is evaluated, and can get into trouble if too many arguments are missing as well. The function below will get into trouble if called with no arguments.

```
function(x = max(y), y = min(x)) {
    ....
}
```

The actual S evaluator takes a few additional shortcuts that do not alter the semantics, but may cause the actual appearance of evaluation frames to look a little inconsistent with what we've said so far. Constant expressions for actual arguments are detected and assigned when the arguments are matched. Certain very special function calls are evaluated without forming a frame at all, namely, calls to functions whose whole body is a `.Internal` interface to C when the actual call is simple enough.

4.9.5 Arbitrarily Many Arguments

Computations sometimes need to examine or put together arbitrarily many objects: to form a list; to write out all the objects; to set any number of options or parameters. S provides a special argument name, `"..."`, for this purpose. When this name appears as a formal argument in a function definition, any number of actual arguments in a corresponding call will be matched to `"..."` in the evaluation frame. Then, if `"..."` appears as an argument in a function call in the body, the matching arguments will be passed down, exactly as they appeared, including any names.

The use of `"..."` is natural and useful in many applications, but it is nonstandard in terms of the general S semantics. In the rest of this section, we look into the mechanism and how to use it. Like all formal arguments, `"..."` is an object in the local evaluation frame, but it is a nonstandard object in several respects. The S evaluator will *only* accept `"..."` in the form mentioned above: as an unnamed argument in another function call. The result is always the substitution of exactly the set of arguments corresponding to `"..."` in the original call. This substitution is exceptional in S semantics and cannot be reduced to the more general model we have used so far.

Three issues arising from the special nature of `"..."` need discussing.

1. How can other arguments be included with `"..."` and matched in function definitions?

2. What corresponds to argument tests, such as `missing(x)`, for `"..."` arguments?

3. Suppose we do need to operate on the arguments in `"..."`, how can these be referenced?

S has specific rules for the first issue, and a special naming mechanism along with the use of substitution can handle the other requirements.

Other formal arguments can appear either before or after `"..."` in the formal argument list of a function. Arguments appearing before `"..."` are matched in the standard way, meaning that they get matched by position or by name, including name completion, as if neither `"..."` nor any arguments following it were included. The arguments after `"..."`, if any, will *only* be matched by name and then without name completion. These arguments should have some quite special meaning and unusual names, since they are to override arbitrary other arguments.

The `.C` and `.Fortran` functions for interfaces to C and Fortran routines provide an example.

```
> functionArgNames(.Fortran)
[1] "NAME"     "..."      "NAOK"     "COPY"     "CLASSES"
```

The first argument to these functions is the name of the routine. It is always supplied. The remaining actual arguments are passed down to the routine, in the order they appear, except for three special arguments that control the way in which all the arguments are interpreted: `NAOK` says whether missing values are allowed; `COPY` says which arguments need to be copied if they turn out to be precious; and `CLASSES` says what classes to which the routine would like S to coerce the various arguments (see section 11.2). A calling routine can use any other names for arguments. The names will be passed down to the interface (which, in this case just passes them back as names for the list returned from the computation). Normally, users supply `NAME` as the first, unnamed argument in the call to `.Fortran`. In contrast, `CLASSES` has to be named in full. Neither an unnamed argument nor an abbreviation such as `CL=` will match.

Because the evaluator applies substitution for "..." in any call, some of the usual tools for manipulating arguments in S don't work for this case. As an exercise, try using `missing(...)` in one of your functions. The effect when you call the function will likely be startling, until the effects of the evaluator's substitution are taken into account. See page 337 in section 8.2.3 for some techniques to deal with missing arguments in the presence of the "..." mechanism.

More general manipulation can use two basic ideas. First, a function call object containing the "..." arguments can be constructed from `substitute`:

```
dotsCall = substitute(list(...))
```

Then, for example, `nArgs(dotsCall)` is the number of arguments matching "..." in this call. The use of `list` above is purely arbitrary; any other name would do as well, but there *must* be a name. Don't try `substitute(...)`— what would that turn into?

While the "..." object is very special, S provides a naming convention to refer to the *individual* arguments matching "..." as ordinary objects. The name "...1" refers to the object obtained from evaluating the first argument matching "...", "..2" to the second, and so on. A function that wanted to concatenate all its arguments and store the result under the name supplied as its first argument could be written as follows.

```
keepAll = function(...) {
  value = c(...)
  assign(..1, value, frame=0)
  ..1
}
```

These objects are still a little special; for example, they do not exist in the frame unless you refer to them explicitly. And an error will occur if you refer to an element of "..." that was not supplied in the actual call. Our `keepAll` function, quite reasonably, can't be called with no arguments:

```
> keepAll()
Problem in keepAll(): Argument "..." is missing, with no
default
```

Otherwise, the objects can be used in the ordinary way.

4.9.6 Method Selection

When S evaluates a call to a generic function, it matches the actual arguments to the set of methods currently known for that function. These methods are combined from all the attached databases that have methods for this generic. When a call to the function first needs to be evaluated, the S evaluator builds a table of the available methods, keyed on the classes of those arguments used to specify the methods. If the set of methods changes (for example, if a database containing methods for this function is attached or detached), the table may need to be recomputed, but otherwise the table remains constant and is kept around by the evaluator, since recomputing it can be a fairly substantial job.

You can find which methods are currently available by calling `showMethods`. For example, the available methods for function `match` might be as follows.

```
> showMethods("match")
        Database            x         table
  [1,] "main"    "ANY"        "ANY"
  [2,] "main"    "numeric"    "numeric"
  [3,] "main"    "integer"    "integer"
  [4,] "main"    "character" "character"
  [5,] "main"    "complex"    "complex"
  [6,] "main"    "character" "string"
  [7,] "main"    "string"     "string"
  [8,] "main"    "integer"    "character"
  [9,] "main"    "factor"     "ANY"
 [10,] "main"    "category"   "ANY"
 [11,] "main"    "structure" "ANY"
```

In this case, we asked for all the methods for function `match`. The table has a first column with the name of the database where the method is stored, and as many other columns as the number of arguments involved in specifying methods. Each column is labelled with the name of the formal argument and the corresponding entry in each row is the class specified for that argument. Class `ANY` implies that this argument did not appear in the specification; in particular, a row of only `"ANY"` is the default method. The first method shown in the table is always the default method.

For `match`, the methods are all on library `main`. Most of the methods depend on both `x` and `table`; the methods for class `factor` and `category` were defined using only argument `x`, meaning implicitly that other arguments just had to extend class `ANY`, and any S class has that property.

When the S evaluator selects a method for a particular generic function, say `match`, it first evaluates as many arguments as are involved in the signatures for stored methods. The need to evaluate such arguments departs slightly from the lazy evaluation concept in S (as described in section 4.9.4):

the arguments involved in method selection need to be evaluated before the method itself can be known. (In this evaluation, however, missing arguments do not need to have defaults. The class "missing" can appear in signatures to match this case.)

If the actual argument classes correspond exactly to one of the stored methods, all is done. If not, and if the function has a non-empty *group generic* function, the evaluator looks for a method for the group generic exactly matching the signature.

```
> getGroup("match")
[1] ""
> getGroup("range")
[1] "Summary"
```

The group for `match` is empty, but if the call had been to function `range`, the evaluator would have looked for a method for function `Summary` for the same signature. For the details of group generic functions, see section 8.3.3.

If no method has been found so far, the evaluator tries next the signature that is *closest* to the current signature. S measures how far any object is from any class by a general rule, which we will describe below. The rule is not a distance in a strict mathematical sense; rather, it orders the distance from a particular class to other classes by the amount of change required to turn an object into an object of the other class. The evaluator uses this measurement to assess whether a given set of actual arguments should be transformed into the classes in a particular method signature. It then looks for a method, first for this function and then for a non-empty group generic, for the closest signature. The process is repeated until a method is found. Since the evaluator arranges that a default method always exists, the process must find a method eventually. Having chosen the closest such correspondence, the evaluator coerces the actual arguments correspondingly and evaluates the body of the selected method.

There are a number of functions in S that allow you to study the method selection process. The functions `selectMethodSignature` and `selectMethod` both take the name of a function and the signature corresponding to some hypothetical actual arguments. They return respectively the signature of the matching method and the definition of the method itself. If you only want to know whether the function *has* a matching method other than the default, you can call `hasMethod` with the same arguments. See the online documentation for details, and Table 3.20 on page 114 for all the relevant functions. These functions all contrast with functions `getMethod`, `existsMethod`, and `findMethod`, which only look for methods stored exactly under the signature supplied, without applying any extension relations.

S maintains for each class an ordered vector of all the classes the given class extends, in order from closest to farthest away. You can see this ordering as the value of the function `extends`, when you give it one class name as an argument. The same information can be obtained in more detail by calling `getClass` with the same argument.

```
> extends("integer")
[1] "integer" "vector"  "numeric"
> getClass("integer")

No slots; prototype of class "integer"

Extends:
Class "vector" by direct inclusion
Class "numeric" by explicit test and/or coercion
```

Any object has an essentially infinite distance from any class if the relation `is(x, Cl)` is false. Among valid `is` relations, S adds distance for indirection, for the need to test the relation, and for the need to coerce the object (page 319 in section 7.6 shows the details).

The evaluator marches through the candidate classes in order until it finds a matching method. When more than one argument is involved, as with function `match`, the evaluator fixes the candidate class for all but the last of these arguments, and looks for any matching method among the classes extended by the class of the last argument. If that fails, it moves to the next-best choice for the next-to-the-last argument and repeats the process. In effect, this makes matching the first argument more important than matching the second, and so on.

While the selection process is general, it's recommended that actual method selection keep to fairly simple situations. As the number of arguments involved and the variety of signatures increase, the user will find it increasingly difficult to understand which methods are being used, and why. The `match` example above is about as complicated as one would want.

Once the method has been selected by this indirect computation, the evaluator may be able to store the chosen method in its table. Then on subsequent calls with the same classes for the actual arguments, the method will be selected right away, as if there had been an explicit `setMethod` for this signature. For example, if the actual argument had class `integer` and there was a stored method for either class `vector` or `numeric`, the evaluator can store the method under `integer` as well.

There are two circumstances that foil this efficiency trick, however. First, any computation that alters the set of methods available may change the selection. In particular, attaching or detaching libraries that contain methods for a particular function causes the S evaluator to revise its method tables corresponding to that function.

Second, some extensions are *conditional*, true for some but not all objects from the class. If the actual object needed to be tested for the `is` relation, then another object from the same class might correspond to a different method. For example, if the actual argument `x` to `match` had class `matrix`, we don't know until we inspect the object itself whether it `is` a numeric object, or integer, or character, etc. So no single method for `matrix` could be stored in the table, if the only choices were for various vector classes. The longer selection process is needed each time.

In some cases, the programmer of the methods can partially overcome the inefficiency by defining an intermediate method. For a `matrix` argument, for example, an explicit method for class `structure` would *not* involve a test: a `matrix` object is always a structure. Typically, such a method will turn around and call the generic function again. The method for `match` is typical, in that it calls `match` again, using the vector part (the `.Data` slot) of the object.

```
> selectMethod("match", "structure")
function(x, table, nomatch = NA)
{
  x@.Data = match(x@.Data, table, nomatch)
  x
}
```

The main point is to get the sensible result for `structure` objects, but in addition this computation can be more efficient than having to repeat the full method lookup for each call.

4.9.7 Data Sharing

S is a functional, object-based language. In particular, this means that each function call is viewed as getting an *object* corresponding to each actual argument. What happens to these objects is just what you see in the definition of the function. There are generally no hidden side effects: if an object `x` in one evaluation frame is passed as an argument to another function, nothing going on in the called function can change `x` as a side effect.

An object in an evaluation frame is only changed when its name is the left of an assign expression. This effectively includes replacement expressions as well, because these are semantically equivalent to assignments of the value returned by a *replacement function* (see section 6.1.4). For example, the expression `diag(x) = v` is equivalent to (and evaluated as):

```
x = "diag<-"(x, v)
```

The function with name `"diag<-"` is defined to be the replacement function corresponding to `diag`.

There are a few functions that can play fast and loose with the evaluation model, in fact precisely the functions that can use the model explicitly: `eval` by evaluating an expression in another frame; and `assign` by doing an assignment in another frame. In very special situations, such cheating is valuable. The interactive browser functions cheat to allow users to test out computations on the fly. The result is an extremely powerful programming aid. Outside of these interactive situations, however, cheating on the S evaluation model is to be avoided. The model leads to code that is easier to understand and that behaves in a more predictable way than would be the case if things like pointers were used to modify objects at a distance, so let's avoid reintroducing pointers by the back door.

One potential disadvantage to the model is that we appear to be in danger of having many copies of objects, resulting in much wasted storage. In some cases the extra storage is indeed inevitable, a price required for the simplicity of the model. Often, however, a technique known in S as *data sharing* keeps unwanted storage duplication from taking place. This all happens automatically, but knowing how it works may help you understand the storage implications of some computation when applied to large objects.

The technique is simple in principle. Each object and each element of a recursive (list-like) object has an internal *reference count*. This counts the number of times the object has been assigned in some frame or database, either directly or implicitly as an element of some other object. In this discussion, *element* includes not only elements of lists and other similar objects, but also slots in objects defined with slots, as well as elements or slots arbitrarily many levels down in the objects.

When an object is passed as an argument in a function call, at the time that argument is evaluated the corresponding object is assigned in the new evaluation frame. Its reference count and that of all its elements are incremented by one accordingly, but no actual copy is made. When the function call completes, the evaluation frame is removed and all the objects assigned in it have their reference counts decremented. Whenever a reference count drops back to zero, the corresponding block of storage is freed.

Changing an object, for example by replacing one or more elements, causes some copying and alteration of reference counts, but generally just the necessary copying. A few examples will illustrate: this is essentially the "obvious" action, although some of the implementation details are not so obvious. Suppose we change one element of a vector by any computation, such as:

```
x[[i]] = value
## or
el(x, i) = value
## or
x$coef = value
```

If `x` came into the local frame as an argument, there may be other references to the same object. Therefore, we have to make a new object, like `x` but only local. Into this object we insert the new element `i`. Whether `x` is atomic (e.g., numeric) or a list, the memory cost of this is proportional to `length(x)`. In the case of a list, only the representation of the list itself has to be copied, not the elements. Also in the list case, `value` has its reference count incremented, and whatever object was in element `i` before has its reference count decremented.

Replacements typically occur in an iterative context. For example, here is a function that iterates an updating of elements of argument `x` until convergence. (The iteration and convergence are silly here, but the style applies to serious computations as well.)

```
"f1" =
# a silly function iterating until
# it converges to all 1's.
function(x, eps)
{
  repeat {
    if(sum(abs(x - 1)) < length(x) * eps)
      return(x)
    i = sample(length(x), 1)
    x[[i]] = sqrt(x[[i]])
  }
}
```

On the first time through the loop, `x` will likely be copied. It is now a local object with reference count 1. Any further replacements of elements in `x`, in this frame, will not need to do another copy.

Contrast iteration with recursion. Here is an implementation of the same computation, but recursively calling the function again until convergence is achieved.

```
"f2" =
function(x, eps)
{
  if(sum(abs(x - 1)) < length(x) * eps)
    return(x)
  i = sample(length(x), 1)
  x[[i]] = sqrt(x[[i]])
  f2(x, eps)
}
```

Each recursive call to `f2` generates another frame, with a shared reference to `x` in the calling frame. Each replacement of an element will have to copy `x`, to protect the caller's frame. The total memory growth will be the number of iterations times the length of `x`.

The recursion in this example can, of course, be trivially removed. More typically, the recursive form is the original, and natural, form. Converting to an iterative form is not something you should do unless there is a serious need for greater efficiency and it is clear that recursion is unacceptably bad. The `quicksort` example on page 69 is typical, in that recursion is essential to the concept:

```
quicksort =
function(x) {
  if(length(x) < 3) q2(x)
  else {
    fence = sample(x,1)
    c(quicksort(x[x < fence]),
      x[x == fence],
      quicksort(x[x > fence]))
  }
}
```

An iterative version would lose much of the elegance and clarity. How much storage would it save? An analysis of the memory growth in this algorithm would show it to be, on average, proportional to `n * log(n)`, where `n` is `length(x)`. If one were serious about implementing this algorithm in S, the memory cost is likely to be acceptable in most situations.

The general replacement function approach in S allows the evaluator to conserve on copies when replacements are evaluated. When the evaluator calls a replacement function, for example `"diag<-"`, it knows that the first argument is to be replaced by the function call's value. Therefore, the evaluator omits incrementing the reference count for the initial assignment of the first argument. Generally, this means that replacements can use replacements, to any depth, without making extra copies of the object. To make this work reliably, the evaluator has to sacrifice lazy evaluation of arguments in replacement functions, but only for actual arguments, meaning that the semantics of defaults and missing arguments is retained.

Data sharing and copying are relevant also in the interfaces to C and Fortran. The evaluator provides the routines called via these interfaces with pointers to the S objects: to the array of elements in the case of `.C` and `.Fortran`, to the objects themselves in the case of `.Call`. To start with, the S evaluator does not know whether the routines mean to write into the elements pointed to, so it assumes the worst and makes a copy, in the way it would if it expected a replacement operation to take place. If the programmer knows that certain arguments are "input only", and won't be overwritten, this information can be passed to the S evaluator, either through the `COPY` argument to the interface routine or by registering the particular routine through a call to `setInterface`. See page 420 in section 11.3 for more discussion.

Chapter 5

Objects, Databases, and Chapters

This chapter presents the techniques for dealing with objects in S. It discusses the classes of objects included with S, how the objects are organized into databases, and various techniques for creating and managing the objects. It complements Chapter 4, which presented fundamental computing techniques. Those were the actions in S; these are the objects acted on. The discussion here also complements that of Chapter 7, which talks about designing *new* classes of objects. The classes discussed here will be the building blocks for new classes.

5.1 Some Important Classes of Objects

All S objects have an explicit class. Information about classes and their relationships is stored in S libraries. The design of *new* classes is discussed in Chapter 7. This section presents some of the important classes provided with S itself.

Included are basic vectors that contain numbers, character strings, and other sorts of data. Included also are lists and similar classes that contain other S objects as their elements. On these basic classes, you can build arbitrarily complicated other classes.

5.1.1 Vector Classes: Fundamental S Data

Vectors are S objects containing n "elements" that can be indexed numerically. If `x` is any kind of vector, and `i` is a vector of positive integers, `x[i]` is also a vector, usually of the same class, containing the corresponding elements: `x[1]` contains only the first element of `x`, `x[c(2, 4)]` the second and fourth, and so on. The abstract definition of vector classes is essentially the ability to extract such subsets, and to get back an object of the same class. (The exceptions to getting back the same class are the `structure` classes, which add to vectors the notion of spatial or temporal structure. For structure classes, the extracted object loses the structural information.) The indexing expression `i` can also be a logical vector, selecting the subset of `x` corresponding to `TRUE` values in `i`. It can be a vector of negative integers, selecting all of `x` *except* the subset `x[-i]`.

There are additional operations defined for any vector. The expressions `x[[i]]` and `el(x,i)` both expect `i` to identify a *single* element of the vector. The first of the two is the one intended for general use; it can take all the forms of `i` used for extraction and methods exist for many S classes. The function `el` is intended for basic methods and *only* accepts single positive indices. The expression `length(x)` returns the number of elements in `x`.

The operators `"["` and `"[["` and the functions `el` and `length` can also appear on the left side of an assignment, with the interpretation that the corresponding subset, element, or property of the vector is replaced by the value on the right:

```
x[c(2, 4)] = c("Second", "Fourth")
length(x) = 4
```

The first replaces two elements of `x`, the second sets the length to 4. Replacing elements does not guarantee to leave `x` the same class. Generally, S will choose a class capable of representing both the current and the new elements, if these have different classes.

The virtual class `vector` glues all the specific vector classes together. A vector class is *defined* to be one that extends `vector`. To make such classes useful, methods need to be defined for the basic functions mentioned above. Once this is done, a new vector class can use the existing vector methods, which should only be defined in terms of the functions discussed here.

Section 7.5 deal with creating new vector classes; The present section presents the vector classes that are already defined in S. They include most of the lowest-level building blocks for organizing data. The most basic of S vectors contain numbers, logical values, character strings, raw bytes, or other S objects as their elements. All but the last of these are what we call *atomic* vectors in S. Atomic vectors are defined by the fact that a single element from such a vector is a vector of the same class as the whole vector. Vectors that contain other S objects as elements, rather than atomic data, are called *recursive*, in the sense that the elements of the vector can be any S class. In principle, gathering all the information in such an object involves looking at each element, then each element of that element, and so on, recursively. S has several classes of recursive vectors, differing in how the elements are interpreted. All of the recursive vectors are similar, though, in that their elements are each distinct S objects and the class of the elements need not be the same.

Of the atomic vectors, those containing numbers dominate the population of S objects in the world, and class `numeric` dominates those. Technically, it corresponds to double-precision floating-point data. For S users, that technicality usually doesn't matter; only when you need to understand issues of accuracy or when you communicate with subroutines in, say, C or Fortran does it affect your programming. There are other classes of numbers as well: `integer` for integers, `single` for single-precision floating-point, and `complex` for numbers in the complex plane (pairs of `numeric` values).

Logical values in S are `TRUE` and `FALSE`, as you would expect. The internal data type of these is the same as for class `integer`, for historical reasons and for consistent mapping to C and Fortran. Again, you don't care in most cases. If the idea of wasting all that space bothers you, the `raw` vector class gives you a way to deal directly with blocks of individual bytes, but the inefficiency is rarely important.

Character strings in S are represented as ordinary characters bounded by single or double quote characters. The class `string` represents a vector of such character strings, plus some optional additional information to deal with uses of strings in matching or to provide computations using the `levels`, the unique character strings meaningful for this object. Class `string` differs from the others in this section in that it is built up from basic classes in terms of slots, rather than being directly defined by its correspondence to a data type in C. If you need to operate directly with C code, you should use the class `character`, as discussed in 5.1.2. Section 5.1.3 discusses a variety of techniques for use with character strings.

The `raw` class holds n undigested bytes of data. You can manipulate these with the usual S functions for vectors, extracting and replacing subsets. They are not numbers, however; in particular, they cannot be `NA`. Raw objects are very useful when we need to store data from C or C++ structures. The programmer can register a particular structure by name and then generate or manipulate corresponding raw data in S. See section 11.5 for some of the special functions to deal with raw data.

Lists are the fundamental recursive class: a simple vector whose elements are arbitrary objects. In programming with lists, the distinction between a subset and a single element becomes important, more so than with atomic objects. If `x` is a list, the expression `x[i]` always returns a list, but the expressions `x[[i]]` and `el(x, i)` return whatever is in element `i`; by the definition of a list, it could be anything. In replacing subsets or elements, the distinction also applies. Consider the two expressions

```
x[i] = y; x[[i]] = z
```

In the first, `y` is expected to itself be a list, or something that will be coerced to one. The replacement will take elements from `y` and insert them in the appropriate places in `x`. The second expression takes the whole object `z` and makes it the single element `i` of the list; again, it is irrelevant whether `z` is a list or not.

The distinctions follow simply from the definition of the functions, but occasionally they require some care. For example, suppose `labels` is a vector of strings that we want to insert in both the first and third element of `x`. The expression that does it is:

```
x[c(1, 3)] = list(labels)
```

The function `list` on the right side may be surprising, but if it weren't there, the evaluator would try to interpret `labels` as a list, producing an object with each element containing one string. The best case would be that the length would be wrong and the evaluator would complain; the worst would be that `labels` happened to be of length two. In replacing subsets of lists, always remember that the right side will be interpreted as a list also.

Lists are the most common class of recursive objects, but not the only one. The narrow view of recursive objects is that they must be able to hold any object as any element, and that any subset is a valid object of the same class. The virtual class `recursive` corresponds to this concept. The only built-in classes extending this class are `list`, `expression`, and `"{"`. Class `expression` differs from `list` only in that the elements are expected (but not required) to be unevaluated expressions in the S language. The expression `eval(e)`, where `x` is of class `expression`, evaluates all the elements of `e` and, by convention, returns the value of the last element as the value of the whole expression. The class "{" represents braced sequences of S expressions; obviously, it's much more specialized, but technically it does satisfy the requirement: any subset of a braced list is a legal braced list.

The broader view of recursive objects includes all classes that can contain other objects as elements. For example, the objects that represent S function definitions contain other objects as elements, but in a specialized way. You can't extract an arbitrary subset of a function and expect the result to be a legal function. Similarly for objects representing other unevaluated S expressions, such as calls to functions. You should, if possible, *not* manipulate these objects directly with subset and element functions; instead, there are utility functions designed to extract and replace relevant pieces. For example, `functionBody` extracts or replaces the body of the function object; that is, the expression evaluated when the function is called. The collection of utility functions will inevitably leave some things out, and in this case you may need to figure out more basic manipulations, but try to use the utilities when you can, since they may provide more security and nearly always will make your own functions clearer.

The distinction between the narrow and broad view of recursive objects is between class `recursive` and the older function `is.recursive`. Like most such older functions (e.g., `is.matrix` or `is.numeric`), `is.recursive` takes an operational view, not a formal one. It defines objects that can be considered recursive in practice, in the sense that their internal representation allows objects to be inserted as elements. Whether `is(x, "recursive")` or `is.recursive(x)` corresponds to what you mean has to depend on the context. An example is shown in the appendix on page 431.

All vector classes in S must support functions to extract or replace numerically indexed subsets and the `length` function. Most support some additional functions. Comparison operations are usually defined, at least for equality and inequality. Comparisons that order the data, such as "<", are less universal—class `raw`, for example, does not support such operations.

Most vector classes supplied with S support the notion of missing data or `NA`. The best way to think of this is as a state that can be tested for each element of the vector, by calling the function `is.na`.

```
is.na(x)
```

returns an object like `x`, with `TRUE` wherever the corresponding element of `x` is missing. To set some elements of `x` to be missing, use `is.na` as a replacement function. On the right side of the assignment, put the elements you want to set to missing. To set all negative elements of `response` to be missing:

```
is.na(response) = (response < 0)
```

It is better to think of NA as a condition rather than a particular value; for example, in the case of floating-point data, S uses a standard concept of "not a number" that can be implemented in different ways depending on hardware. Not all vector classes support this concept. Class raw does not, since the byte patterns in raw data are deliberately not interpreted. For character strings, class string supports NA but class character does not. String objects are another example where you need to think of NA as a condition, not a value. No actual character string can encode NA.

5.1.2 Internal Representation of Vector Classes

Some of the fundamental vector classes in S have a special relation to data types in C or in Fortran. These classes may be called "compiled-in" in the sense that their representation is understood by the underlying code that implements S. More importantly, these are the *only* classes that correspond directly to arrays that might appear in C or Fortran routines not written explicitly for use with S. You need to know about these classes if you plan to write interfaces to subroutines in C or Fortran.

When S communicates with C using the .C interface function, each of these classes corresponds to a pointer to some C data type. The atomic compiled-in vector objects correspond to pointers to particular numeric data types or to C character strings.

All the compiled-in classes can be defined by a table that says what kind of data they are expected to hold, and how that data is represented in C. To understand how S manages data, imagine each basic vector as including a block of storage, such as a C program would represent by an array or pointer. Table 5.1 gives the correspondence of S class to C type; some of these classes also correspond to array types in Fortran. See section 11.2 for the Fortran correspondence, and the interface routines to both languages.

5.1.3 Character String Data

Character strings are used in four broad ways in programming with data: for display, for matching, for categorizing, and for string manipulation. S provides tools for all of these. This section describes the classes of objects designed to hold such data, in particular the class string, and tools for the first three applications. Section 5.1.4 discusses the interpretation and manipulation of substrings.

Class	C type	Data
numeric	double *	Numeric values.
logical	long *	Logicals: (TRUE, FALSE).
character	char **	Character strings.
integer	long *	Integer values.
single	float *	Single precision numeric values.
complex	complex *	Values from the complex field.
raw	char *	Unstructured raw data.
list	s_object **	Other S objects.

Table 5.1: *Basic Classes of S Vectors.*

String data used for display arise as labels, for example on rows or columns of printed tables, or on plots. Here is a random sample of 5 rows from a dataset used as an example for model-fitting.

```
> littleBit = fuel.frame[sample(60, 5),]
> littleBit
                     Weight Disp. Mileage     Fuel   Type
Chevrolet Caprice V8   3855   305      18 5.555556  Large
  Volkswagen Jetta 4   2330   109      26 3.846154  Small
    Eagle Premier V6   3145   180      22 4.545455 Medium
     Subaru Loyale 4   2295   109      25 4.000000  Small
Honda Civic CRX Si 4   2170    97      33 3.030303 Sporty
```

The labels on the rows, such as `"Honda Civic CRX Si 4"`, are in this case intended to be meaningful to the reader, and the software doesn't have to worry much about operating on the strings. Most of the time we just need to be sure the data is in the right place. The function `row.names` is the tool used to extract or set the row labels of a data frame:

```
> row.names(littleBit)
[1] "Chevrolet Caprice V8" "Volkswagen Jetta 4"
[3] "Eagle Premier V6"     "Subaru Loyale 4"
[5] "Honda Civic CRX Si 4"
```

Functions such as `names`, `dimnames`, and `slotNames` are other tools to extract or set labelling information in objects. Functions that do explicit display, such as `print` or `plot`, typically use methods that extract the appropriate labels based on the class of the object. These display functions may also have optional arguments to supply some of the labelling information.

Character strings are used for *matching* in many S computations. Those same row labels could be used to identify rows:

```
> fuel.frame["Honda Civic CRX Si 4", ]
                     Weight Disp. Mileage     Fuel   Type
Honda Civic CRX Si 4   2170    97      33 3.030303 Sporty
```

S provides several ways to use strings in matching—see section 4.4.2 for details about the computations. The best function for most purposes is `match`. Suppose we wanted to find out which rows of `fuel.frame` our call to `sample` selected:

```
> match(row.names(littleBit), row.names(fuel.frame))
[1] 52 13 42  9 18
```

The two arguments to `match` are some data in the first argument (character strings in this case) that we want to "look up" in the object supplied as the second argument. Objects of class `string` can arrange to make this operation faster for large tables; see the discussion of the `stringTable` slot on page 204.

Other S matching functions match strings with implicit completion of the first argument, which in S is called *partial matching.* In selecting rows of the data frame above, we actually only needed to give enough of the row label to uniquely identify the row. S will complete the implied string:

```
> littleBit["Honda",]
                     Weight Disp. Mileage     Fuel   Type
Honda Civic CRX Si 4   2170    97      33 3.030303 Sporty
```

Partial matching is also applied (and most importantly) in completing the names of arguments in function calls. It can be done explicitly by calling the function `pmatch`. See section 4.4.4 for how it works.

The third common form of matching is by *regular expression.* Those are dealt with when we talk about substrings in section 5.1.4.

The use of strings for matching expects the various character strings to be distinct in most cases, since they are often used to select some data items from a larger object. The third main application of character string data uses them in quite a different way: as strings to *categorize* various kinds of data. We expect the row names in a table to be unique: we would not like two rows to have the same name. The strings labelling the values of column `Type` in our example, however, are not unique. They represent some variable used to categorize the rows of the table. For objects of this sort, it is important to know what all the possible values are: S calls this the `levels` of the object.

```
> Type = littleBit[, "Type"]
> levels(Type)
[1] "Compact" "Large"   "Medium"  "Small"   "Sporty"
[6] "Van"
```

Notice that not all the levels appeared in the data. This may be important, if we want to do consistent computations with different subsets of a larger body of data. S allows the levels to be stored in the string object. Extracting a subset of that object will not recompute the levels.

Levels are frequently used to index another object consistently with the different character strings in a string object. The function `levelsIndex` accomplishes this: it returns an integer vector of the same length as the string object. The elements correspond to the levels of the corresponding elements of the string object. In our example, the first row has `"Large"` as the value of `Type`, corresponding to the second string in the levels. The second row corresponds to the fourth level, etc.

```
> levelsIndex(Type)
[1] 2 4 3 4 5
```

In model-fitting software, for example, the value of `levelsIndex(x)` would be used to choose rows of a matrix that parameterizes the levels of `x` for use in a linear model. If, say, `contr` is a matrix for parameterizing `Type`, then the matrix with `length(Type)` rows that represents `Type` in the model is:

```
contr[levelsIndex(Type), ]
```

Model-fitting aside, this is a very characteristic S computation, worth studying a bit if it is not obvious. The `levelsIndex` call returns a vector of the same length as `Type`; then that vector is itself used to select rows from `contr`. The matrix "subset" will often be much larger than `contr` itself, since the same level of `Type` may appear many times. The whole-object view of the computation encodes all the elements of `Type` in one simple expression.

Now we turn from the tools for dealing with character string data to the classes of objects to hold such data. The preferred class for character string data in S is `string`. This is not a compiled-in vector class; in fact, it has several slots that allow some extensions to primitive character string vectors, and adapt reasonably well to large string objects. An alternative class is the compiled-in vector, `character`. Class `string` formally extends `character`, which means that for most purposes requiring `character` you can still supply `string`. There is no extension the other way, however, so in some situations you need to generate string objects explicitly, perhaps by `as(x, "string")`.

String objects have the following four slots. Slot `text` is a `raw` vector holding the actual text. Slot `offsets` is a vector of offsets into `text`. Slots `stringTable` and `levelsTable` contain two optional tables for hashing, to find offsets from character strings and levels from offsets. The slots are all `integer` vectors except for `text`.

You should very rarely need to work with the slots directly, but knowing about them will help explain the situations in which `string` objects behave differently from `character`:

1. `string` objects can have missing values, which are equivalent to missing values in the `offsets` slot; `character` vectors contain pointers, for which S does not support the idea of `NA`.

2. If the `stringTable` slot is set up, it allows matching character string data in a `string` object without any preliminary computations. This means that the time spent in matching is essentially independent of the size of the `string` object that is acting as a table.

3. Very large `string` objects from a database are substantially more efficient to access than the corresponding `character` objects for most applications. (The main reason is that all the slots in a `string` object can be memory-mapped, while the pointers in `character` vectors can not.)

4. The notion of `levels` is built into `string` objects, which again makes them somewhat more efficient for applications that use levels, such as tabulation and model-fitting.

The one major counter-indication comes when interfacing to C subroutines. Because `character` objects are compiled-in vectors, they can be passed directly to such subroutines. String objects will be converted automatically to `character` objects when necessary, so specifying `character` for the desired class of an argument to a C subroutine includes `string` objects as well.

Two classes extend `string` for applications, such as model-fitting, where levels are important. Both are identical to `string` in their internal representation, but with some additional implications. The `stringFactor` class behaves like `string` except that a levels table is required, and considered to be independently defined. With class `string`, the levels table is a computational convenience that exists only to describe the levels present in the data. Subsetting a string will redefine the levels to those present in the subset. In contrast, the levels of a `stringFactor` object remain fixed under selection of subsets, until the user carries out some form of replacement that redefines the levels.

The `stringOrdered` class extends `stringFactor` with just the additional interpretation that the levels represent an ordering, which in turn causes some model-fitting software to choose some appropriate methods for fitting such data.

```
> levels(TestScore) = c("Low", "Medium", "High")
> TestScore = as(TestScore, "stringOrdered")
```

Levels, if they have been assigned, remain unchanged when you select a subset of the original object of a `stringFactor` or `stringOrdered` object. A subset, `TestScore[i]`, will always have the same levels as `TestScore`, assuming those levels had been set originally. If you want to reset the levels to only those appearing in the subset, use `findLevels` to find only those levels:

```
levels(x) = findLevels(x)
```

If there are no missing levels, `x` will be unchanged. If the levels were ordered, the new levels will appear in the same order as they did in the old levels; otherwise, they will appear in the order they occur in `x`.

5.1.4 Substrings and String Manipulation

General string manipulation computations interpret or modify substrings within character strings. The basic function for string manipulation in S is `substring`. In its primitive form, it indexes substrings within each of the character strings in an S object by referring to the first and last character positions. For example, to get the first three characters of each string in the object `files`:

```
substring(files, 1, 3)
```

The first and last positions are in principle arguments of the same length as the text object:

```
substring(files, nchar(files)-2, nchar(files))
```

This example extracts the last three characters in each string. If the vectors giving text, first or last are not of equal length, S replicates the arguments to match the longest. When the `last` argument is omitted, the end of the character string is assumed; a simpler version of the previous example is

```
substring(files, nchar(files)-2)
```

Similarly, omitting the `first` argument takes substrings starting in the first character position.

The same `substring` argument can be used to replace substrings; for example, to replace the last three characters in each string:

```
substring(files, nchar(files)-2) = "aux"
```

The substring computations try to make sense of most input. Last positions are pulled back to the length of the string. Substrings that are invalid are interpreted as empty; for example, missing values in the first or last positions, or first greater than last. Replacing an empty substring makes no change. For example, if we wanted to replace the last three characters only if the string had at least three characters, the following would do it, by inserting `NA` in the corresponding elements of the `first` argument.

```
substring(files,
  ifelse(nchar(files)>2, nchar(files)-2, NA) = "aux"
```

Substrings can be specified in other ways as well. The most useful alternative uses *regular expressions*. For a definition and some direct applications in S, see section 4.4.3. Besides direct use through the `regMatch` function, the two simplest applications of regular expressions are as subscripts and as an argument to `substring`. The `substring` function interprets any character strings in its second argument as regular expressions, and extracts or replaces the part of the text that matches the expression. For example, one more way to extract or replace the last three characters would be:

```
substring(files, "...$")
```

This is equivalent to the version that matches only strings of 3 or more characters, as we probably wanted, since each `'.'` has to match a separate character. More naturally, perhaps, suppose we wanted to retain only the characters before the first blank in the row names of our `littleBit` data frame:

```
> brandNames = substring(row.names(littleBit), "^[^ ]*")
> brandNames
[1] "Chevrolet"  "Volkswagen" "Eagle"      "Subaru"
[5] "Honda"
```

The first "^" in the regular expression means the start of the pattern, but the second "^", inside the square brackets, reverses the sense of the square brackets, so they now mean anything *but* the characters listed, the blank character in this case. (Elegance of design is not a characteristic of regular expressions, viewed as a language.)

The regular expression can appear in substring replacements as well. Suppose we want to replace all the substrings between characters "<" and ">" in the object Comments:

```
substring(Comments, "<[^>]*>") = "<...>"
```

As an exercise, and an example of the difficulty of programming with regular expressions, consider the difference between the regular expression used here and the perhaps more obvious "<.*>".

When used in S subscripts, regular expressions act as a logical expression. The subset extracted or replaced consists of those elements that match the regular expression. Since character strings can themselves be subscripts, the regular expressions have to be explicit. To get the elements of files ending in "tex":

```
files[ regularExpression("tex$") ]
```

Notice that we're not talking about substrings here: it is the elements of the files vector that are selected or replaced.

The function regularExpression returns an object of the same class. The unambiguous way to use regular expressions is by supplying objects of class regularExpression, but functions such as substring will interpret any character strings as regular expressions.

Regular expressions can also be used in two very basic functions, regMatch or regMatchPos to create selectors that can then be used elsewhere; see section 4.4.3. The logical vector returned by regMatch(x, table) shows which elements of x matched the expression in table; notice that regMatch only reports match, so it need not set unmatched elements to NA. The function regMatchPos does the same match, but returns a two-column matrix with the substring start and end positions of each successful matches in the corresponding row, or NA's if the match failed. The value of regMatch can be used in any S expressions, and in particular to express selections not easy to define in terms of the pattern alone. It's this combination of regular expressions with S logic that provides much of their value.

To get all the elements of `files` that do *not* end in `"tex"` is simple:

```
files[ !regMatch(files, "tex$") ]
```

The corresponding regular expression is a mess. Similarly, the numerical positions returned by `regMatchPos` can be modified to specify related substrings. In the replacement in `Comments`, suppose we wanted to leave the first and last two characters enclosed by `"<"` and `">"` and just shorten the rest of the substring to `"..."`.

```
pos = regMatchPos(Comments,  "<[^>]*>")
substring(Comments, pos[,1]+3, pos[,2] -3) = "..."
```

This computation also makes no change where we would not want any. If the expression does not match, the corresponding row of `pos` is `NA`, and so will be the values given to `substring`. Where there are fewer than five characters enclosed in the substring, the `first` argument will be numerically less than the `last`, and again the `substring` replace will do nothing.

5.1.5 Lists and Trees: Thinking about Recursive Objects

The ability of S objects to contain other S objects allows us to represent data of arbitrary complexity. S differs fundamentally from languages like C or Java, where individual values, pointers, or references are the basic ingredients in representing data. In S, whole objects are the elementary ingredients, and the natural approach to representing more complex data reflects this. This section presents some basic ideas and examples informally. Techniques for formalizing such notions by defining new classes are presented in Chapter 7.

Let's take an example. *Tree structures* are often a useful way to organize data about n entities, whether to relate biological species, to cluster observations on different individuals, or to represent a stepwise medical diagnosis. Each entity or *node* corresponds to some value and potentially to some number of other nodes, its *children*. In the common case of *binary* trees, each node has at most two children.

What would be the natural way to represent such data in S? The concept of nodes captures the essence of the tree structure: the information they contain and the relations among them define the tree. Saying this gives us the clue: we want to express structure and computations in terms of all the nodes at once. This expresses the "whole object" sense of trees in S, a way to talk about the data as a whole. S programmers sometimes refer to finding the whole object sense as "vectorizing" the structure.

Having grasped this concept, the task is easy. Let's imagine indexing things by nodes `1` to `n`, say. The fundamental structure of the tree is defined by the relations among the nodes: which nodes are the children of which other nodes? The tree structure is defined and can be computed with, once we represent this information. One approach would start by noticing that every node has a unique parent, except the top node. So an integer vector of length `n`, `parent` say, could give the index of each node's parent, with perhaps `0` for the top node. In addition to the structure, the tree is likely to associate some data with each node, say in a vector `value`. And we might want character string `labels` for each node, another object of length `n`.

This is far from the only way to look at the data. In the case of binary trees, for example, we might want to keep the children of the nodes in a two-column matrix, rather than the parents. These different approaches can be examined and played with, to see which matches our particular needs best. We may even choose to allow alternative forms, with functions to turn one form into another.

So, how might we have gone wrong in this situation? Inappropriate data structures often come from getting too fixated on the incremental details of algorithms that produce the object or operate on it. Suppose we started by thinking of trees as ways to partition the entities into subsets, and then to partition those subsets into other subsets, and so on. We might then think of a tree as a list of the `k` splits at the top level. The elements of this list then either do or don't get split again, so they represent either leaves of the tree or subtrees that get split again. So the list will contain sublists down to a depth equal to the maximum number of splits.

What's basically wrong with this approach is that we don't see the data as a whole. In order to operate in any way on the object, we have to write an iterative or recursive function; in the terminology of tree computations, each computation must *walk* the tree. Not getting beyond the underlying algorithm has led us astray.

Algorithms are important, of course, and structures related to trees have led to some of the most challenging algorithmic problems in computing. But S focuses on the user and the ideas, treating the algorithms as tools. Data structures should arise from what *you* and other users need from the data.

5.1.6 S Structures

We discuss now the class `structure` and some classes derived from it. These are particular recursive classes of objects, including many of the most commonly used classes. Intuitively, structures take a vector or other object containing n data elements and arrange the elements into some structure; for example, a matrix in S is some data arranged in a conceptual two-way layout. A multiway array generalizes this from two to any number of dimensions. A time-series is some data arranged at equally spaced points on a conceptual time scale.

The user's model for `array` is that the data values occupy cells in some hypothetical regular k-way layout. Arrays in S follow a convention that each of the k extents is indexed from 1 to its (current) maximum value. The vector of those maximum values is given by the function `dim`:

```
dim(x)
```

returns an integer vector containing k positive numbers. The data in the array is a vector of length

```
prod(dim(x))
```

Labels may be defined for the positions along each of the k extents. The function

```
dimnames(x)
```

then returns a list of length k, whose i-th element gives the labels (if any) for the i-th extent, if these properties have been specified. If no labels have been defined, `dimnames` can return `NULL`.

The `structure` class can be extended to define a class with the structure properties. It is a virtual class, but exceptionally for virtual classes it has a definition that includes one slot, `.Data` specified to have class `"vector"`. The structure concept is implemented by using the `.Data` slot as the vector part. Specific structure classes include `structure` and define additional slots to convey their own specific kind of structure. Arrays, for example, have `.Dim` and `.Dimnames` slots to implement the concepts in the `dim` and `dimnames` functions.

Earlier versions of S used structures for a number of informally defined classes, in addition to arrays and matrices. Time series objects included an attribute `tsp` to represent objects with data occurring at regularly spaced intervals in time or some other continuum. The software dealing with these is still available, largely unchanged from earlier versions of S. Categories and factors were both structures with levels defined as an attribute. These can still be used, although for new software, objects based on the `string` class are recommended instead. The model software described in reference [3] introduced a number of additional structures.

Section 7.1.7 discusses how to define new classes extending `structure`.

5.1.7 Raw Data

S provides a mechanism for including raw data in objects. By "raw" we mean some number of bytes of data that is not to be interpreted as one of the ordinary, built-in vectors such as numeric. Raw data can be passed around, included as a slot or component in other objects, and assigned to any S database. The contents will not be altered in the process.

Raw data is useful for introducing new data structures at the C language level into S objects. The `connection` class of objects, for example, uses raw data to pass around the internal information needed to identify and manipulate connections from the S session to files and other entities for input or output. These have to be manipulated at the C level, but the S connection objects identify the particular connection and contain all the information needed to handle it.

Raw data can include pointers and arbitrary C structures. The usual style of dealing with such data in S objects is to include identifiers that allow the software to validate the object. For example, pointers will not be valid over multiple S sessions, but an S object assigned to a permanent database will still hold the same data after the user quits one S session and starts another. Some code must check that the object corresponds to the current session: a common technique (used with `connections`) is to include the process id of the session in the raw data.

Raw data can be generated in four basic ways:

1. by reading the raw (binary) data from a file or other connection, using the function `readRaw`;

2. by using character strings that code bytes in either hex or ascii coding, supplied to the functions `rawFromHex` and `rawFromAscii`

3. by allocating a `raw` object and then filling it through a call to C code via the `.C` interface;

4. by calling an S-dependent C routine, usually through the `.Call` interface.

Details of how to use the third and fourth mechanism we leave to the discussion of programming in C (Chapter 11 and Appendix A). The reverse of the second mechanism is also available, via functions `rawToHex` and `rawToAscii`.

Raw data can naturally be thought of as a vector of byte data. The `raw` class itself is defined very leanly in this respect. You can do the fundamental vector operations on `raw` data—extracting and replacing portions of the data in terms of indices, computing and setting the length. But S puts no interpretation on the contents of the individual bytes; they don't have an intrinsic order, `NA`'s are not defined, and coercion to numeric or integer values is undefined. It's not that any of these additional properties is necessarily a bad idea, but rather that we prefer to see them defined in terms of other classes that extend or use the `raw` class explicitly. In this way, a variety of interpretations can be built on the basic notion.

5.2 Databases

The S *database* concept unifies all operations that group objects together by name. At any time during the S session, the S evaluation manager knows about some number of attached databases. Each database is essentially a lookup table in which some set of non-empty strings is each associated with an S object. The S programmer has access to the databases in a variety of ways; the function `database.attached` (see page 217) returns the object representing an attached database. Other S functions can test the existence, get, assign, or remove objects from the database, and can obtain the names of all the objects in the database. Besides the explicit functional access, the S evaluator will automatically search for objects by name and will assign objects in local databases (frames) when evaluating assignment expressions.

Databases strictly exist *only* when attached, and only in the evaluation manager. The databases can be attached by the programmer in a wide variety of ways, by interpreting something as a database. What the "something" is determines the type of the attached database. The most common case is to interpret an S chapter directory as a database; in addition, named lists and other S objects can be interpreted as databases, and S can create an empty table database unassociated with anything. The side effects of assignments and of detaching the database depend on the type; for example, assignments to directory databases write data to the file system, and detaching a modified object database can create an object to save the results.

Databases also differ in the purpose and duration for which they are used. The evaluator creates temporary databases, referred to as evaluation frames, during the evaluation of a task. These behave largely like other databases, but since it is their relation to function calls that really matters, we discuss them in the context of the evaluation model, in section 4.9.1. Other databases are attached explicitly, usually by the `attach` function. They may be added to the databases searched automatically by the evaluator, but can also be attached for explicit use only. They are usually attached for the duration of the session or until detached, but may also be attached temporarily, typically when their description appears in a call to `get`.

In the rest of this section, we describe various functions for examining and attaching databases, and for accessing the objects in them. We also discuss the computations by which the S evaluator finds objects corresponding to names and carries out assignment expressions.

5.2.1 Finding and Assigning Objects Automatically

The S database facilities are based on a few primitive functions to get and assign objects. Two of these operations happen automatically as part of evaluating S expressions: searching for the object corresponding to a name in an expression, and assigning an object whose name appears on the left of an assignment expression. These operations, and all the facilities for dealing with objects on databases, can be invoked explicitly, using the functions `get`, `assign`, `exists`, `remove`, and `objects`.

Let's consider the computations used by the evaluator to find an object referenced by name in an S expression. You may find it helpful to understand the rules, particularly if an object isn't found when you expected it to be.

When a name appears in evaluating an S expression, the evaluator *searches* for an object associated with that name, using a specific set of rules. Consider evaluating the function call:

```
log(x)
```

Once the evaluation needs the actual object associated with the name `x`, S looks for an object named `x` first in the local frame in which the expression is being evaluated. This includes, of course, the case that `x` is a formal argument in the current call; see section 4.9.4. S is a functional language, and good functional language programming says that nearly all ordinary names *should* refer to local objects. It is this locality of reference that tends to make functional languages relatively easy to understand and to use correctly.

However, S does not insist that all references be local. If `x` is not an object in the local frame, S looks next in the task frame (frame `1`), then in the evaluator frame (frame `0`), and then in the database search list (page 215). Frame 1 lasts until the current task completes; frame 0 as long as the evaluator exists, which is the duration of the S session in the unthreaded version of S described in this book. For how all this relates to the S evaluation model, see section 4.9.

Names are often *not* local when the they designate functions to be evaluated. In the expression

```
log(x)
```

the name `log` is looked up with the same search rules, but only in terms of function definitions. Non-function definitions are ignored. Since function name lookup occurs much more frequently than other non-local uses of names, the evaluator has some special tools to speed up the process. The name is looked up in the local frame, and frame 1 and frame 0; if a function definition is found, we're done. Otherwise, the name is then looked up in the database search list in the usual way. The lookup has an internal side effect of storing the definition in a special table, so that normally the actual definition only gets read in once (for the sake of efficiency). Also, the evaluator notes the existence of methods for generic functions, and sets up an internal table of these as well (see section 4.9.6 for how method selection works), again to speed up repeated use of the function.

Names appearing on the left of assignments are *only* assigned in the local frame. When the assignment appears in the body of a function, the local database is the evaluation frame, created by the S evaluator for the current call to that function. Section 4.9.1 discusses this kind of database. When the assignment was a top-level task, say an expression typed by the user, the object is still assigned in the local frame (frame 1 in this case), but when the task completes, the S evaluator commits that assignment to the working data, the database in the first position of the search list.

The same is true of replacement expressions, since these are the same semantically as a simple assignment of the value returned by the replacement function. Thus in all the following, `x` will be assigned locally:

```
x = y + 1
```

```
x[1:m] = x[m:1]
class(x) = "Matrix"
```

In the replacement cases, S requires that x be defined locally before the replacement takes place. (In the case of top-level tasks, "locally" is extended to include objects that are currently defined on the working data. These will be implicitly copied to frame 1; committing the replacement then overwrites the object in the working data.)

Both the search for an object and local assignment can be invoked explicitly as calls to the functions `get` and `assign`. If the calls specify neither the `frame` nor the `where` argument, they have the same semantics a referring to a name or assigning the name in the evaluator. The advantage of the functional forms is that the name can be computed and can be any non-empty string, not just what the parser thinks of as a name.

5.2.2 The Search List

At any time during the S session, the evaluator has a *database search list* of attached databases. When you start an S session, this typically is the session database (database 0), the working database (either the local directory where S was invoked or the user's home directory), and some prespecified set of S libraries. Evaluating calls to `attach` or `detach` can alter the search list.

The function `search` returns the character string names associated with the attached databases. Suppose that, when the session starts:

```
> search()
[1] "."    "models"     "main"
```

This is typical, and says that the current directory, ".", is the S working database, and that the two S libraries `main` and `models` have been attached on startup. A user can control the contents of the search list at startup by the techniques for customizing the session, such as by the use of calls to `attach` in a `.First` object (see section 4.1.1 for all the available techniques).

A database can be specified (in the `where` argument to `exists`, for example) in several ways. A positive number indicates position in the current search list. A character string name indicates the name associated with the database when it was attached, or equivalently the string that `search()` returns in the same position. The two expressions

```
exists("x", where=1); exists("x", where = ".")
```

are equivalent if the search list is what we showed above.

This notion of defining a position on the search list is a general one that recurs in many places: usually the corresponding argument is named `where`. S handles such arguments in a consistent way: any of the unambiguous ways to provide a `where` argument should work in each case. The two given so far are most convenient: either by numeric position in the list or by the character string corresponding, as seen by printing `search()`. Since S libraries are identified in the search list by the simple argument to `library` (`"main"`, `"models"`, etc.), this provides a convenient way to refer to the libraries without knowing where they come in the search list. Remember that the position of a database in the search list often changes when a `attach` or `detach` call occurs.

A third way to identify a database is by an object of class `"attached"`. This is the object returned by `library` and `attach`; if you forgot to save the object then, it is also the value of `database.attached(where)`. Such an object provides the most reliable way to give a `where` argument. It includes an `id` slot that essentially uniquely defines the database. Any programming that needs to refer to databases consistently throughout a session should always use the `attached` object.

Two variants on `search()` obtain the names of corresponding meta databases and the paths of chapter directories. Both are rather obscure, but here they are. The function `search` takes an optional argument `meta`; if it specifies `"methods"` or `"help"`, the returned value is the names of these databases, not the ordinary ones. About the only likely need for this is to determine whether there is a methods meta database, say, for a particular position in the search list. If there isn't, the corresponding name will be empty, so you can test by using the `nchar` function.

For example:

```
> tdb = attach()
> search()
[1] "."       "#12"    "models" "main"
> tdb
Database "#12"; type table; search position 2; status
"readwrite"
> nchar(search("methods"))
[1] 14  0 40 38
```

The table database attached in position 2 had no corresponding methods meta database.

By calling the function `searchPaths` rather than the function `search`, you get back the file system path to the chapter databases. Table and object databases are not directories, so they appear as empty strings in `searchPaths()`. The vector returned is useful if you need to work with the actual directories (e.g., in a shell command). See page 381 for an example.

5.2.3 Properties of Attached Databases

S provides a number of functions that return relevant properties of databases currently attached on the search list. All the functions have names beginning with `data.` and completed by something suggesting what they do: `database.position` returns the position of the database in the search list, for example. The functions all have a first argument that refers to the desired database. You can identify the desired database by any of the means described in section 5.2.2.

The function `database.attached` is in fact designed to go from some arbitrary way of referring to the database to a precise identifier. This returns an object of class `attached`. These objects are the absolutely unambiguous way to refer to an attached S database. For example, if we pick up the current position of a database and then attach or detach some chapters, that position may no longer be correct. But the `attached` object remains the unambiguous way to refer to the attached database (unless it is itself detached). Also, `database.attached` returns such an object for any attached database, including those that are not on the evaluator's search list. For example, chapters can be attached for explicit reference only, using the `purpose` argument in `attach` (see page 226). Such databases can be used with the `where` argument in calls to functions such as `get`, but are not on the search list. The object returned by `attach` is the same as that returned by `database.attached`; if you expect to refer to the database often, it's simplest to save the value of the call to `attach`. If you forget, call `database.attached`. Let's continue the running example begun on page 215.

```
> library(data)
> attach(fuel.frame)
> tdb = attach(name="Temp")
> search()
[1] "."      "Temp"   "#12"    "models" "data"
[6] "main"
> tdb
Database "Temp"; type table; search position 2; status
"readwrite"
```

```
> identical(tdb, database.attached(2))
[1] T
> database.attached(3)
Database "#12"; type object; search position 3; status
"readwrite"
> database.attached("data")
Database "data"; type directory; search position 5;
status "read"
```

The function `database.position` returns the position of the database in the search list, or `NA` if it's not attached.

```
> database.position("main")
[1] 6
> database.position("an excessively long name")
[1] NA
```

The position will be `NA` if the argument is not an attached database or if the database is attached but not in the search list. In the latter case, `database.attached` still makes perfect sense. If there is *no* attached database of the description given, `database.attached` returns `NULL`.

A second argument can be supplied to these functions to say that we want to deal with the meta databases in the chapter for `"methods"` or for `"help"`. These databases contain S objects used for special purposes in dealing with classes and methods or with online documentation. You shouldn't need to work with them directly, since it's much safer to rely on S functions that hide the database details. But if you *do* want to get at them, you specify the metadata wanted as a second argument. For example, here is another way to test whether the database attached in position 2 of the search list has an associated methods database:

```
identical(database.attached(2, "methods"), NULL)
```

since `database.attached` will return `NULL` if it fails to find the requested database.

Two more functions provide the `type` and read/write `status` of an attached database. The type is effectively the class of the database: S recognizes a number of types, and allows users to define additional database classes. S chapters have type `"directory"`; list and list-like objects when attached have type `"object"`; simple table databases have type `"table"`. In the example above:

```
> database.type("Temp")
[1] "table"
> database.type(3)
[1] "object"
```

The status of databases when attached is either "read" or "readwrite". In particular, library always attaches databases read-only. If changes have been made to a database, the status is "modified":

```
> database.status("."); database.status("main")
[1] "modified"
[1] "read"
```

Checking a table database for status "modified" before detaching it would warn us that changes would be lost unless the database was saved. If you have appropriate file system permissions, you can change the status of directory databases between "read" and "readwrite" by using database.status on the left of an assignment.

```
> cdb = attach("../clean")
> database.status(cdb) = "read"
> cdb
Database "../clean"; type directory; search position 2;
status "read"
> assign("xxx", 0, w=cdb)
Problem in assign("xxx", 0, w = cdb): database
    "../clean" is read-only
Debug ? (y|n): n
```

The status only applies to the specific database; if we wanted to prevent assignments also to the methods and help metadata, more calls would be needed. The function setDBStatus provides a more user-friendly interface for setting the status of an attached chapter.

5.2.4 The Objects in a Database

The functions described in this section manipulate the objects in databases or evaluation frames. They can return the names of all the objects, test for existence, get, remove, or assign objects in a database.

The function objects returns the vector of names in a database, the working data by default:

```
> objects()
```

```
 [1] ".Last.value"  "a"            "absRes"
 [4] "b"            "last.warning" "myGen"
 [7] "print2sided"  "rmat"         "view"
[10] "x"            "|res|"
```

It takes optional arguments, including `where`, `frame` and `meta` with the same meaning these arguments have for `get` and the other functions on page 221. In addition, it takes arguments to restrict the object names returned. For example, the `pattern=` argument provides a character string that `objects` interprets as a regular expression (in the style of the shell command `grep`; see section 4.4.3). Only names matching this expression will be returned. An argument `classes=` restricts the result to objects that have class equal to one of the strings supplied. An argument `test=` is expected to supply a function of one argument; this will be called with each otherwise eligible object as an argument and should return `T` or `F` to say whether to include this object in the returned names.

```
> objects(pattern = "^[ab]")
[1] "a"      "absRes" "b"
> objects(class=c("character", "string"))
[1] ".Last.value" "Tt"          "type"        "xx"
> objects(where = "testData",
+   test = function(obj)length(obj)>10000)
[1] "bigX" "bigY"
```

These arguments require some extra computation, increasing from `pattern` to `test`, the latter requiring S to actually read and use each of the objects in order to do the test. Nevertheless, `objects` called with these arguments is a relatively efficient way to find the objects you want before doing other computations.

A useful, related function is `getObjectClass`; it takes a vector of object names (which might be the result of a call to `objects`) and returns a string vector of the corresponding classes. If there is no object corresponding to some of the names (for example, if some other file was accidentally written in your chapter data directory), `getObjectClass` returns `NA`. So, if you were worried about such files being written, the following little function would reassure you.

```
"realObjects" =
# the names of the objects from database 'where', but make sure
# they are all really S objects, not just any old file.
function(where = 1, meta = 0, ...)
```

```
{
  what = objects(where = where, meta = meta, ...)
  what[!is.na(getObjectClass(what, where = where, meta = meta))]
}
```

The style of including "..." to be passed to `objects` lets the user supply arguments we don't care about (in this case, really only the `pattern=` argument).

The functions `exists`, `get`, and `find` provide tools to search the databases: all take an argument `name`, a character string specifying the name of the object to search for ("`absRes`", for example). The functions differ in what they return: `exists` returns `TRUE` if a matching object was found and `FALSE` otherwise; `get` returns the matching object and generates an error if there isn't one; `find` returns a vector of the names of all the databases on the search list that have a matching object.

These functions take other arguments to direct the search: `where` restricts the search to a single database, `frame` to a single evaluation frame. Searches can be made in the meta databases instead of the ordinary data by supplying the argument `meta`: `meta=1` says to search in the methods databases; `meta=2` to search in the help databases. There are a few other, more specialized arguments: see the online documentation. The function `find` does not have arguments `where` or `frame`.

The `where` argument can be anything that defines a chapter in the search list, as described in section 5.2.2—an integer, the name of the database or an `attached` object. In addition, a character string that is *not* a name from the search list is treated as a path name (interpreted as it would be as the argument to `attach`), so `where` can be used to look in chapters not currently attached. The effect is to attach the database temporarily for the duration of the call.

Instead of `where`, these functions can be given the argument `frame`; in this case the function looks in the evaluation frame specified. Two such frames have fixed definition: frame `0` is the session frame and frame `1` is the top-level frame for the current task. The other frames are usually best defined relative to the function in which the call to `exists` or `get` occurs, via functions such as `sys.parent` (see section 4.2.3).

If neither `frame` nor `where` arguments are given, the functions `exists` and `get` look in all the places that the S evaluator looks when resolving names: the current function call, frames 0 and 1, database 0, and the databases on the search list. The first match, in that order, determines the result.

There are functions specializing `get`, `exists` and `find` to important classes of objects. Corresponding to the special use of names in function calls, a call to `getFunction` looks for a function definition corresponding to a character string, and ignores non-functions. Similarly, `existsFunction` tells whether the named object exists as a function. The functions `getMethod` and `existsMethod` work similarly, but in this case you supply both the name of the function and the signature for the method you want (see section 8.1.3 for some details and alternatives). Documentation objects are also accessible by specialized functions, `getDoc` and `existsDoc`. The name in this case is the documentation topic; any topic will get the corresponding documentation object. See section 9.5.

The function `assign` relates to `get` and takes the same `frame`, `where` and `meta` arguments.

```
assign("fit", lm(Y ~ .), where = tdb)
```

This assigns the value of the `lm` call, with name `"fit"`, in the database `tdb`, an attached table database in the example on page 217. Via the `frame` argument, `assign` can also be used to do assignments in some other function's evaluation frame, but this is generally a bad idea since it breaks the S functional programming model. Exceptions are functions, such as `browser`, that are intended to supplement the user's interactive control over computations. Without either the `where` or `frame` argument, `assign` acts like an assignment expression

```
assign("absRes", abs(residuals(fit)))
```

This has the same effect as the ordinary assignment expression

```
absRes = abs(residuals(fit))
```

The assignment is done in the local evaluation frame (frame 1 if the assignment is evaluated at the top level of the task). The functional version of assignment can associate the object with any character string, not just those that are legal names in the S language:

```
assign("|res|", abs(residuals(fit)))
```

(Of course, you will need to use `get("|res|")` to refer to the object in an expression.)

The function `remove` removes objects from databases. It too takes the same three optional arguments as `get` or `assign`, but it can take a vector of several names if you want to remove all the corresponding objects at once:

```
remove(c("absRes", "fit"), where = 2)
```

The function `objects` with arguments `pattern`, etc., as described on page 219, often provides a suitable vector of names for `remove`. Via the `frame` argument, `remove`, like `assign`, can modify another function's evaluation frame, and it's an equally bad idea in most cases.

5.3 Attaching and Detaching Databases

New databases are attached to the search list by expressions such as:

```
cdb = attach("../clean")
```

The first argument, `what`, to `attach` defines the database to be attached. There are many possibilities. Depending on the arguments to `attach`, the effect may be to attach any of the following:

- an S chapter directory;
- a named object;
- user-defined database classes;
- table databases for temporary storage.

In particular, calling `attach` with `what` omitted,

```
tdb = attach()
```

creates and attaches to the search list an empty table database that persists throughout the session, unless detached.

The central concept to remember about S databases is that they fundamentally associate character string names with S objects. Given this, it is natural that S provides an open-ended way to extend databases from the usual library or chapter directories to arbitrary other ways to associate names and objects.

The call to `attach` usually has the side effect of including the new database in the search list, making it available for automatically finding S functions, methods, and other objects. The value returned by `attach` is an object that identifies the database. Optional arguments to `attach` control where it is attached in the search list and the purpose for attaching it. The function `library` is a specialized version of `attach` designed for attaching S system libraries, but not restricted to that use. When you are finished with a database, a call to `detach` removes it from the search list.

The rest of this section discusses details of the use of these functions and their side effects.

5.3.1 Attaching Chapters

If the first argument to `attach` is a character string, it is taken to be the name of a chapter directory. For example, to attach a chapter in subdirectory `"clean"` of the current directory:

```
> cleanDB = attach("clean")
> search()
[1] "."        "clean"   "models"  "main"
```

The value of the call to `attach`, an object of class `"attached"`, is the safest way to refer to the database in calls to such functions.

```
> objects(cleanDB)
[1] ".Last.value"  ".Random.seed" "TTT"
[4] "crash1"       "last.dump"    "x"
[7] "xmat"         "xmat2"
```

You can also supply the `where` argument as the the numerical position of the database in the search list or as the name. In our example, `objects(2)` or `objects("clean")` would have worked as well.

The argument to `detach` identifies the database in any of the ways that work for other functions such as `objects`:

```
> detach(cleanDb)
```

You can detach anything you attached, but S will prevent you from detaching the `main` library and possibly other databases it considers necessary to run at all.

The S function `library` works very much like `attach`, but is designed particularly to attach libraries defined in your version of S. This mainly amounts to `library` recognizing some short forms for such libraries; for example,

```
library(java)
```

is a short form for the subdirectory `"java"` under the S library directory. The library function has an optional argument, `help`. If you give the section as the value of that argument, some documentation for the particular library is given, instead of attaching it. With no arguments, `library` will print the available sections.

The `library` function also differs from `attach` in that it attaches databases read-only and has slightly different rules for the position in which the attached database appears.

The following example illustrates some combinations of attach/detach:

```
> search()
[1] "."        "models" "main"
> db1 = attach("./myfuns")
> library("java")
> search()
[1] "."        "./myfuns" "models"    "java"
[5] "main"
> detach(db1)
> detach("models")
> search()
[1] "."      "java" "main"
```

For the rules that determine where the new libraries appeared, and options to override them, see section 5.3.2.

The strings defining the chapter can contain shell variables, which will be expanded in finding the chapter directory but not in the name appearing in the search list.

```
> attach("$HOME/book")
> search()
[1] "."            "$HOME/book" "java"      "main"
```

If you have to use `attach` to attach an S library, you can do that by knowing that these are all subdirectories of `$SHOME/library`; library `emacs`, for example, is the directory `$SHOME/library/emacs`.

5.3.2 Optional Arguments for Attaching Databases

The call to `attach` has a number of other arguments that control what is done with the new database. You don't need to learn about these if you just want to attach a new database so its functions or other objects will be found automatically by name. But there are many other tools and tricks with S databases; this section describes some of them.

The argument `pos` controls the position of the new database in the search list. By default, `attach` puts the object in position `2`, and `library` puts it in front of "main". The position only matters if objects of the same name exist on more than one attached database: S uses a rule that takes the first acceptable version of an object that occurs in its search list. So, if shell exists on library main, which is initially in position 2, then attaching a new database with object shell hides the definition on main, which has now been moved to position 3. On the other hand, attaching the new database in position 3 or farther back would leave the old definition and hide the new one. If you know that the attached database has a different version of a standard S function and do (or don't) want this version to win out, the pos argument can ensure that. You probably want the new version, in most cases, so that the default position makes sense: library is designed so that libraries override the functions in main by default; attach is designed so that newly attached databases override anything but the working data by default.

The function library, for historical reasons, has an argument first as well, which behaves differently from pos. If supplied explicitly, it either puts the new database at the very end (first=F), or in position 2 (first=T). Generally, the pos argument is preferable.

One case in which pos makes a major difference is in the case pos=1. Now the new database becomes the working data, the place where all top-level assignments go. The previous working data now moves to position 2; its objects are still available (unless hidden by the new database), but no assignments will go there unless by an explicit call to assign. You're perfectly entitled to do this, although it's not recommended generally, because some objects in the working data help maintain continuity. For example, S arranges that random number generators in each evaluator can generate reproducible results, by storing the "seed" object in the working data. Attaching or detaching the working data will break this reproducibility.

Another useful argument to attach is purpose. This tells S what the new database should be used for; by default, it's intended as a "regular" database, to be inserted in the search list for S to look in whenever it is trying to find an object, either for data or as a function definition. One alternative is purpose="data". This means that the new database should be attached, but used only for explicit requests, *not* as part of the search list. Why would you want to do this? If you expect to get or assign objects from a database fairly often, but want to avoid confusion with objects of the same name on any of the databases on the search list, attaching the database in this way offers advantages of convenience and efficiency.

```
tData = attach("/home/testresults/newData", purpose="data")
```

Other values of purpose can be used as well; see ?attach.

5.3.3 Actions on Attaching and Detaching Databases

The attach and detach functions look for specially named objects, .on.attach and .on.detach, respectively, to provide optional actions just after a database has been attached and just before the database is detached.

For example, suppose chapter myModels had some functions that overrode the library models and you wanted users to be aware of this. Just define and include in chapter myModels something like the following function:

```
.on.attach = function(){
    message("Using a revised version of some models functions")
}
```

Another frequently valuable use of .on.attach is to initialize some code in C or Fortran when the chapter containing the code is dynamically linked to S. Similarly, .on.detach could call code to clean up before the chapter was detached. See section 11.1.2.

5.3.4 Attaching Objects

Any named list or similar object can be used as a database. The data frame objects used in S modeling are frequent examples:

```
> sdf = attach(ScottsDataFrame)
> objects(sdf)
 [1] "alp"    "avtens" "bw13"   "bw85"   "cdia"
 [6] "clado"  "dcore"  "ddt"    "del"    "ecc"
[11] "fid"    "fty"    "l1300"  "l1395"  "l850"
[16] "ldt"    "lth"    "ltm"    "mlen"   "na"
[21] "oval"   "pno"    "rod"    "twr"
> length(names(ScottsDataFrame))
[1] 24
> search()
[1] "."                  "#15"
[3] "clean"              "models"
[5] "main"
```

Why attach an object as a database? Usually, in order to write S expressions involving the named elements without having to extract them each time from the object. After the attach, we can write plot(dcore, alp) rather than

```
plot(ScottsDataFrame$dcore, ScottsDataFrame$alp)
```

There are some efficiency gains as well for multiple expressions involving large objects, though these should not be your dominant concern in most applications.

Notice that the attached database is different from whatever S object might have been involved in the call to attach. In our example, we just attached an object from some other database, but we could have done any kind of computation to produce the argument to `attach`. As always in S, it is the *object* that is the argument, not a pointer or reference to an object.

With databases, this means that assignments and other modifications to the attached database have no effect on the original object. In our example, no assignments or removals in database `"sdf"` alter the data in `ScottsDataFrame`. After some modifications, for example, we might want to compare some modified data with the original:

```
> sdfOrig = attach(ScottsDataFrame)
> plot(get("avtens", where = sdfOrig),
+       get("avtens", where = sdf))
```

The attached object, the value of `attached`, should be used in these cases to avoid confusion.

Directories and named lists are ways of defining databases that are built into S. In addition, programmers can define new classes that interpret database operations any way the programmer wishes. The argument to `attach` can also be any object from a database class, a class that extends the virtual class `database`. Roughly, any definition is allowed as long as some fundamental methods are implemented for the class, such as reading and writing objects to the database.

Finally, you don't need to give S any initial data at all to create a new database:

```
newDb = attach()
```

will generate a new database and attach it to the current session. Such databases are used as a place to store named objects during the S session. Any number of different computations can share information this way, so long as they also share the `newDb` object that identifies the attached database.

5.3.5 Attaching Databases Temporarily

Databases in any form can appear in arguments to functions such as `get` and `assign`, as the `where=` argument. If the argument refers to a database currently attached, the interpretation is straightforward: get or assign the data in the corresponding database. In the case of assign, this means that the function call will have a persistent side effect of changing the contents of some chapter or other database. Such side effects are perfectly legal, but as a matter of keeping your S programming easy to understand and debug, you should use them sparingly and only when you have a clear purpose in mind. S is a functional-style language, where nearly everything happens as a function call, and nearly all function calls *only* need to be understood in terms of the object returned.

What happens if the database in the where= argument is not currently attached? S interprets this as a request to attach the database temporarily, just for the scope of the current function call.

```
> stdNames = get("standardCountyNames",
+  where = "/usr/local/S_data")
```

Assuming "/usr/local/S_data" is not currently attached, S attaches it for the duration of the call to get and then detaches it again. You can also attach databases temporarily in an explicit fashion, which might make sense if you want to do several gets or assigns, but all during the same function call.

```
tDb = attach("/usr/local/S_data", purpose="temp")
for(which in possibleNames)
    if(match(thisName, get(which, where=tDb), 0))
     {ok = TRUE; break}
```

The advantage of temporary databases is that they disappear without any action on your part, rather than lying around for the rest of the session cluttering things up.

The corresponding disadvantage is that you have to attach them again each time. In extreme cases, the extra computation required can slow things down noticeably. In the example above, suppose we had written the code as:

```
for(which in possibleNames)
    if(match(thisName, get(which, where="/usr/local/S_data"), 0))
     {ok = TRUE; break}
```

Then the get call will attach and detach the database each time. This is more computation than the get operation itself, though you're unlikely to notice, unless possibleNames is very long. The sensible solution is usually to attach the database for purpose `"data"` if you expect to use it frequently, or to attach it with `purpose="temp"` in a function call that will do *all* the interaction you expect to do with that database. An attached database of purpose `"data"` is not used in the standard search for objects, and so imposes relatively little penalty on the rest of your S computations; in particular, it can never cause you to get an object from that database when you didn't mean to.

A few possible uses of a temporary database do not make sense at all. For example, attaching an *object* as a temporary database for the purpose of `assign` ends up doing nothing. As we emphasized, assigning to an attached database defined from a list-like object does *not* reach back to change the original object on another database; instead, it develops a new object especially for the database. So a temporary version of such a database will just disappear if you don't explicitly save it, and you can't explain how to save it if it's only accessed in the call to `assign`. S will try to warn you if you seem to be doing something meaningless of this form.

5.4 Chapters

The S chapter is the basic unit for organizing collections of S objects (functions, datasets, classes, methods, code written in C or other languages for use with S, and so on). A basic principle of organization, then, is to make a chapter to contain all your work on a particular project. To get S going at all, you should have created a chapter at some time. If not (perhaps because you started in an earlier version of S), you really should do so, to avoid missing out on some of the tools that S creates in the chapter.

The prime chapter for a user is recommended to be a subdirectory named `S` under your login directory. This directory is recognized as special when S starts up: if the current directory is not an S chapter, S will look for this subdirectory as the working database for the session. You can also use your login directory itself for the same purpose, but this is largely a relic of earlier versions of S and is not recommended.

As you expand your use of S, you will likely want to create further chapter directories to keep different projects from conflicting with each other. Chapters are especially important for substantial programming projects. You can put all the relevant source code in S and, perhaps, in other languages as well, into one chapter directory, and organize it all by the `make` and other tools described in Chapter 6. Since the S documentation is implemented in

S objects, the same tools will manage the documentation.

Even for projects that are primarily *using* S as opposed to programming in S, chapters provide the key organizational element. This section discusses tools for managing the chapter, independent of whether it contains data, software, or both. Section 6.2 presents techniques specifically related to the programming aspects of the organization.

5.4.1 Creating a Chapter

From a shell, the S shell command

```
CHAPTER
```

creates an S chapter in the current directory, provided you have write permission in that directory. You can run `CHAPTER` and other S shell commands in one of two ways. Either precede the command with the shell command you use to run S, for example,

```
Splus CHAPTER
```

or else define the shell variable `SHOME` to be where S resides and include the directory `$SHOME/cmd` in your shell's path. (See page 406 in Chapter 11 for details.) The `CHAPTER` command has two main effects:

1. It creates a subdirectory `.Data` to hold S objects (plus two further subdirectories underneath to hold metadata for methods and documentation).

2. It generates a `makefile` including rules to dump and restore the chapter, and optionally to compile subroutines to be linked with S.

The command will not overwrite an existing S chapter, and in particular will not harm any data or overwrite an existing `makefile`; it will append some S rules to a `makefile` that does not look like an S `makefile`.

You can control what goes into the makefile, or whether one is created at all, by arguments to the `CHAPTER` command; see `?CHAPTER`. For most applications you should not need to worry in detail about what the command does in terms of specific files and subdirectories; you should think of the whole directory as a chapter and leave S to organize subdirectories and files. Operate on the chapter directory as much as possible through the utilities in S. Of course, you can and should create S objects or software in other languages inside the chapter directory. The whole of the chapter then represents a project in programming with data. Later in this section we will discuss how to define such files as part of the chapter.

5.4.2 Dumping, Moving, and Rebooting a Chapter

After working on a chapter for some time, you may want to make an archival record of all the results. This section describes the standard procedure to create a portable dump of everything in an S chapter, as well as how to ship the results around and reboot the chapter somewhere else, not necessarily in the same computing system. This is the recommended procedure for:

- distributing your chapter to other users;
- rebooting your chapter after changing computer or operating system version;
- creating a permanent, readable archive of the chapter;
- cleaning up a large or messy chapter.

This procedure works because it creates files that can represent any S objects, to the full accuracy of the local computations, but does so in text files that are not themselves dependent on the local system, except in the sense that numeric values in the objects themselves may reflect local computational accuracy.

To start the process off, run the S function `dumpChapter`.

```
> dumpChapter()
A list of the files defining the chapter "." was written to
"./DUMP_FILES"
```

The file `DUMP_FILES` contains the names of all the files you need to archive the chapter. All those files will now be ready. You can make an archive using shell tools such as `tar` or `cpio`. For example,

```
tar cf myChapter.tar `cat DUMP_FILES`
```

will create a `tar` archive file, `myChapter.tar` that can be shipped to another computer to regenerate the chapter there. On the new machine you would invoke `tar` again to recreate the files. Or,

```
cpio -pmd $HOME/newChapter < DUMP_FILES
```

would recreate the files in directory `newChapter` under my login directory. You can also use the files in a version management system to keep track of revisions; this is most important for chapters that are primarily for programming, rather than data (see page 259 for some examples).

Suppose, by whatever sequence of operations, we have now moved all the files to another directory, perhaps on a different machine. This directory should start out empty; now we want to regenerate the chapter there. There are three steps, each an S shell command to run in the new directory. As always with S shell commands, you need to precede them with, say `Splus`, or have `$SHOME/cmd` in your shell's path.

1. `CHAPTER`
2. `make boot`
3. (if there were C or other compiled files) `make S.so`

The first turns the directory into an S chapter:

```
$ CHAPTER
Creating data directory for chapter
```

The second step is to boot the contents of the chapter:

```
$ make boot
S datasets booted
```

S will not reboot into a chapter that already contains any S objects. You must create a clean chapter first. Basically, this policy was adopted so the result of the boot could be asserted to be identical to the chapter originally dumped.

If you have C, C++, or Fortran code shipped with the chapter, you will want to take a third step:

```
make S.so
```

This creates the shared object, `S.so`, that is linked automatically when the new chapter is attached. You will redo this step each time you make changes to the source code in the chapter.

That's all you need to know for dumping and moving a chapter. There are a few other facts about the dump process that may be useful.

The files `all.Sdata`, `meta.Sdata`, and `help.Sdata` contain symbolic dumps of the objects in the original chapter. They are in fact special, a little different from what you might get with the function `data.dump` in other contexts. They should *only* be used to boot the new copy of the chapter. The symbolic dump format in general is the way to ship S objects around between computers or to different languages (see section 5.5.2). The key properties of the format in this context are that it preserves all the numeric information in the original objects, that it can represent any S object, and that it contains only standard, printable characters.

Symbolic dump files, on the other hand, are useless for humans trying to read anything but the simplest S objects. For this reason, the dump also produces a file `all.S` containing an ordinary dump of any functions or language objects in the original chapter. This file is not strictly needed, and in fact is not used in booting the chapter. It's there so you can have a readable definition of the functions in the chapter to supplement the symbolic dumps.

5.5 Dumping and Restoring: Moving S Objects Around

This section discusses techniques for moving arbitrary collections of S objects, perhaps editing them in the process, perhaps restoring them in a different environment.

When you need to move a few objects from one S chapter to another, on the same machine, you can do that in S just with the functions `get`, `assign`, and `remove`. The following little function, for example, takes the names of some `objects`, assigns each object on `to`, and removes it from database `from`:

```
"moveSome" =
# move each object named in ‘objects’;
# Get them in database ‘from’, assign in ‘to’.
# If ‘remove’ is ‘T’, remove each after it's re-assigned.
function(objects, from, to, remove = T)
{
  for(what in objects) {
    assign(what, get(what, where = from), where = to)
    if(remove)
      remove(what, where = from)
  }
}
```

As an exercise, you might want to make `moveSome` friendlier about an object not being found.

Two other applications for moving objects around need more specialized computations. We may want to examine and possibly edit the data in the objects, then recreate them from the edited versions. Or we may want to move S objects around in a well-defined form that does *not* rely on how S stores objects internally on a particular machine.

For human viewing and editing, the function `dput` and its relatives create general but (fairly) readable text output from S objects. For moving objects between machines or between S and other systems, the function `data.dump` and the format that supports it provide the key tools. Moving whole chapters was discussed in section 5.4.2.

5.5.1 Deparsing and Dumping for Editing

The function `dump` takes the character string names of any number of objects and optionally a connection (often a file name) to which it should dump the objects. The file so produced is suitable to be parsed and evaluated by the standard S evaluator, by a call to `source`, for example.

Our ability to dump objects to a file for editing just relies on a way to deparse the object; that is, to turn the object back into some lines of text that look more or less like the input to the parser. This computation is performed by the function `dput`.

```
> dput(myscale)
function(x)
{
    (x - min(x))/(max(x) - min(x))
}
```

The function `dump` effectively uses `dput` in a loop:

```
> dump("myscale")
[1] "myscale.S"
```

The call to `dump` begins by writing `"myscale ="` to the file, followed by the output of `dput(myscale)`. If we don't tell `dump` what file to use, it either constructs a file name, as here, from the name of the single object to dump, or uses the default file name `"dumpdata"` if there are several objects.

We can give `dump` a second argument, `file`, specifying any writable connection. We can build up the output for editing in various ways (maybe combining other computations to put things on the file besides the dumped object definitions).

Understanding the general idea of `dump` will help to use it more generally. It writes one or more S assignment expressions to the file, followed by some expression that will recreate the corresponding object when the file is presented as input to the S standard evaluator or to the source function. A very schematic version of dump would be

```
function(object, file) {
    for(what in object) {
        cat(what, "=", file=file)
        dput(get(what), file)
    }
}
```

(assuming that file is a connection open for writing). The essential requirement is that file can be used as source for S tasks, say as the argument to source.

There is a nice generality here that exploits S classes and methods. While dput writes *function* objects in the form the S parser can turn directly back into functions, in general any S expression that evaluates to the original object is fine. Methods exist for dput designed to produce reasonably readable and editable output for particular classes of objects. Anyone can write additional methods, which will then be used by a call to dump when dumping objects from the corresponding classes.

The generality comes in that dump doesn't need to know how to dump different kinds of objects, and the knowledge built into dput is just in the form of methods for individual classes. The designer of a new class can, if it makes sense, provide a method to help users edit objects from that class; setting that method for dput will make it available to anyone using dump or other similar utilities to create editable versions of objects.

The default method for a class with slots is to generate a call to the extended version of the new function, with arguments specifying the values of the individual slots (these will be printed by calling dput recursively for each slot). Class track is defined on page 281 in section 7.1.1. An object of class track has two numeric slots, x and y:

```
> class(trx)
[1] "track"
> dput(trx)
new("track",
x = c(1., 1.25, 1.5, 1.75, 2., 2.25, 2.5, 2.75, 3.)
, y = c(-0.22, -0.02, -0.88, 1.32, 0.05, -2.37, 1.48, -0.49, 0.67)
)
```

This is not too bad for editing. We might want to create a specific method, perhaps to align the corresponding values in the two slots (see page 338 in section 8.2.4 for an example using this class).

5.5.2 The Symbolic Dump Format

S supports a portable, symbolic format for dumping and restoring arbitrary S objects. The call:

```
> data.dump("myscale")
[1] "myscale.Sdata"
```

produces a file containing a symbolic representation of the object `myscale`, completely general and portable (only containing text) but not intended for humans to read—just look at one to be convinced. A call to `data.restore` to read the file generated by `data.dump` would restore the dumped objects on any system running modern S. The contents would be identical, up to inevitable differences due to different word length or floating-point hardware. Section 5.5.3 gives details.

While symbolic dump files are hard for humans to read, they are much simpler to describe formally, and therefore much better for other programs to read. Therefore, they also form the basis for a way to communicate S objects to other software systems. The format is used in several ways in S itself; in particular, for dumping and restoring S chapters (section 5.4.2) or individual objects (section 5.5.3). Its potential usefulness extends beyond this, however, and so this section describes the format itself explicitly.

The format has three characteristics that make it generally useful. It is *simple* and therefore software to read or write it can be written for a variety of programming languages, so long as they can represent similar objects to those being transmitted. It is *general*, so that extending its applications does not require extending the format. It is *portable* in that it represents arbitrary data as ordinary text, so that data can be exchanged across hardware and operating system boundaries.

Every S object is represented in this format as three lines of header information followed by the contents of the object. The three header lines are the *class* of the object, the *mode* of the object, and the number of *elements* in the object. Following this the format contains each of these elements (which may themselves be S objects, represented in the same format), with new lines separating successive elements. Since new lines play this special role, any newline characters occurring within the data itself are escaped in the usual way (`"\n"`).

The terms *class*, *mode*, and *element* have specific meanings in defining the format. The class is object's class, represented as a character string. The mode is a string defining how the individual elements are to be represented; it is sufficient to have the fixed modes for the various kinds of atomic data S understands, plus one or more modes indicating that the elements are themselves S objects. The atomic modes are `"numeric"`, `"integer"`, `"logical"`, `"single"`, `"character"`, `"raw"`, and `"complex"`. For the atomic modes, the elements have a predefined form (e.g., numbers, character strings, raw bytes) and each of these has a defined representation in the format.

The representation for atomic vectors is largely the obvious one. Numbers are printed in integer or floating-point formats, with the format for output chosen long enough to represent all the information in the data. At the C level, S constructs format strings suitable for use with `fprintf` and related C routines. The field widths in the format strings are set to be sufficient to capture full accuracy in output. See page 417 in Chapter 11 for a brief discussion of the C formats.

Mode `"raw"` needs somewhat more consideration. Raw data is represented exactly, by hex coding. Since S accepts arbitrary bytes in raw data, there can be no guarantee that what the raw data represents is *itself* portable. If you are aiming at full portability of dumped data, you should restrict the use of `"raw"` data to cases where the information coded in the bytes is not machine-dependent. In contrast, using `"raw"` to represent C structures or other internal data that contain pointers will not work if portability is needed.

There are a fixed set of atomic modes (in S generally, as well as in this symbolic format). Everything else either has length zero or has elements that are themselves S objects. For a class that has slots, the elements are the contents of those slots, in the order the slots appear in the representation. A class that uses an atomic object as its prototype but no slots will have a class string that is different from its mode, but the contents of the object will be correctly dumped.

The complete format is then defined by defining the representation of the atomic vector classes and handling everything else recursively. This means that the format does not restrict the set of recursive modes (modes for objects that contain other objects as elements). S has traditionally used a variety of such modes, in particular for representing the S language itself. With classes available, this would not be needed nowadays, but the format accommodates it for back-compatibility.

When using the symbolic format for communicating with other languages you may not need this generality, and could restrict all list-like objects to one mode, `"list"`.

The actual representation of an object's class is not automatically included in dumps of data, so the two applications exchanging objects from specially defined classes need to believe they have a consistent definition of each class exchanged. For all classes, with or without slots, the symbolic format guarantees that the contents of objects will be consistently transmitted, but leaves the application responsible for interpreting the contents consistently with the definition of the classes.

As an example, the symbolic format for an object of class `sequence` is shown below. Class `sequence` is represented by a vector of three integer values `c(1, 120, 12)`. Its symbolic format would be:

```
sequence
integer
3
1
120
12
```

The representation of the data elements will be determined by the data mode field. In this case, mode `integer` determines that the three data elements will be dumped or read as integers.

As a second example, here's the dump for a small object of class `track`; this class has two numeric slots. The dump format uses newlines to separate everything, but let's cheat and imagine it could use commas as well, so we won't produce a huge piece of empty page. With this convention, here's a dump of `trx[1:3]`, the first three points in the `trx` object on page 236.

```
track, list, 2, numeric, numeric, 3, 1, 1.25, 1.5, numeric,
numeric, 3, -0.22, -0.02, -0.88
```

The class `"track"` is followed by `"list"`, the number of elements (slots in this case), then by the dump for two objects each of class and mode `"numeric"` with three elements.

5.5.3 Using the Symbolic Dump Format

The format described in section 5.5.2 allows S objects to be written and read in a general, complete format that is portable across machines and operating systems. Also, the files are lines of ordinary text and so are suitable for various utilities, in particular, for source code and version control systems. S provides tools at three levels to use the symbolic dumps. At the chapter level, `dumpChapter` dumps a complete chapter, for shipping around, version control, etc. (see section 5.4.2). The other two levels, lists of objects and single objects, are described in this section.

The `dump` and `restore` functions are aimed at dumping one or more S objects in a form that is (usually) complete, but relatively easy for human editing. If we don't care about human editing, but rather want to ensure that we dump *all* the information in a concise, well-defined format, the functions `data.dump` and `data.restore` are what we need. The file written by `data.dump` contains any number of objects in symbolic dump format, along with the names of the objects. When this file is read by `data.restore`, all the corresponding objects are assigned in a database on the target machine. As with `dump`, the first argument to `data.dump` is a vector of the names of the objects.

A call to `objects`, using some of its optional arguments, is often a good way to generate the first argument to `data.dump`. To dump all the function objects in chapter `"myfuns"` under the current directory:

```
> db1 = attach("./myfuns")
> data.dump(objects(db1, class="function"), "myfun.Sdata",
+   where = db1)
```

Now the file `"myfun.Sdata"`, which is just ordinary text, can be compressed, copied, mailed, put on the web, or shipped around however we want. To restore the objects, perhaps on another machine:

```
> data.restore("myfun.Sdata")
```

This will copy to the working data all the objects named in `"myfuns"`.

Functions `dataGet` and `dataPut` read and write individual objects in the symbolic dump format. Unlike `get` and `assign`, they read from and write to files, or more generally S connections, not databases.

Suppose we want to write an interface between S and some Java application, such as a graphical user interface. An interface of this form could provide users with "non-programming" access to some S functions we have written. The symbolic dump format is an excellent way to implement such an interface, because both S and Java are capable of writing and reading such data.

On the S side, we need to be able to read data written by the Java application and to write data that Java will read. Suppose we have set up two S connections, `fromJava` and `toJava` to handle the communications; e.g.,

```
> fromJava = fifo("./fromJava")
> toJava = fifo("./toJava")
```

Then to write an object, say `result`, in symbolic dump format:

```
> dataPut(result, toJava)
```

and to read something the Java application wrote on `toJava`:

```
> nextTask = dataGet(fromJava)
```

The Java application will require some software to generate the symbolic dump from its side. This can be very simple, for example a character vector containing one or more lines of input as a user would type them to the S evaluator. If the user interface wants to generate an S task to find the names of all the objects on the working data, it might write the following information to the `fromJava` file.

```
character
character
1
objects(where=1)
```

The S side of the interface would read this, recreate the character data, and arrange for this to be parsed and evaluated, with the output written back to the `toJava` file. The Java application will need to read data in a similar format and to display the output to the user.

Actual implementation of the interface requires managing *events*; for example, the S process needs to respond when input is waiting on the `fromJava` file. In the example, we used a `fifo` connection specifically because these are suitable for event management. A call to the function `setReader` will arrange for a specified function to be called whenever input is available on the fifo. The reader function in this case will likely call `dataGet`, arrange for parsing and evaluation, and return the output to Java. Section 10.6 discusses how to set up readers and manage events from S.

Chapter 6

Creating Functions

This chapter presents techniques for writing, modifying, debugging, and documenting S functions. Because of the "everything is an object" nature of S, programming techniques operate on objects—there is no barrier between S computations on data objects and computations on functions and other parts of the language. Getting used to this duality and the freedom it provides is the most important step to powerful programming in the language.

6.1 S Functions and Expressions

Section 1.5 introduced some useful tools for developing S functions. We will examine them again here, along with some additional techniques, but this time with the focus on a slightly more serious, large-scale programming project. As programming goes from casually creating a few tools to an organized project, developing some general habits and strategy will pay off. A variety of tools in S can help. Key to the philosophy of S is that the tools are not special, unchangeable operations but simply examples of computing with S objects, when the objects are functions or other language objects, rather than numerical data. The S functions and database tools described in earlier chapters will turn out to underly the programming process as well.

An S function is an S object having class `"function"`. That statement is the clue to programming with S functions. All the tools we will discuss start from some simple computations that S knows how to do with function objects.

6.1.1 Creating and Editing Functions

How should one get the first version of a new function? The overall goal, as always, is to turn the ideas we have into software, quickly and faithfully. That suggests creating a function object quickly, a function that won't likely do all we want but will be a starting point. In one approach, try to create a function in a line or two of typing, dump out that version, edit it and source the edited version back in to start trying it out on some examples. It is rarely a good idea to write out a long function away from running S: that style is more suited to a compiled, offline programming language. Occasionally you may have to work that way, if you have a very complicated new idea, but programming in S is usually more productive and more fun if you can turn the idea quickly into a function and then gradually refine the idea after trying it out.

Three ways to get started apply frequently.

1. If there is an S function that does something related, but the idea is to change how it works or its arguments, either copy over the existing function, or start the new function out as just a call to the existing one. When we come to discuss creating new methods and documentation, this idea is even easier: S will provide us with the current definition of the method for free.

2. If there is no current version of this computation, but you know what arguments you want, just create an empty or extremely trivial version of the function in a one-line expression.

3. If you've just been doing some related computations directly, as tasks typed to S, use the `function=` argument to the `history` function to grab those tasks.

Here are some examples. The S function `all.equal` compares two objects, allowing for minor differences. It returns `TRUE` or some character strings describing the differences found. Suppose we want a function using the same tests as `all.equal` but only returning `TRUE` or `FALSE`. The function `identical` works that way, but it insists on every bit being equal. The new function, let's call it `equalTest`, should behave like `identical` but use the testing and special methods of `all.equal`. The least typing to get started is:

```
equalTest = all.equal
```

Now the new function has the same arguments as `all.equal`, which is probably what we want, but the whole body of the function is likely going to be thrown away. Alternately, we could start with a call to `all.equal`:

```
equalTest = function(target, current) all.equal(target, current)
```

With either initialization of the function:

```
> dump("equalTest")
[1] "equalTest.S"
```

The function definition is on file `"equalTest.S"`, ready for editing. There is in fact a simple one-liner that does what we want; we'll show it on page 248.

For another example, suppose we want a function to scale some data to a given maximum and minimum (an application would be to scale the data in order to add it to an existing plot). The arguments are going to be the data, say `x`, and the target minimum and maximum, say `xrange`. If that's as far as we've got in thinking about the computation, fine, let's create an empty function. It's of no use yet, so we immediately dump it out for editing.

```
> reScale = function(x, xrange){}
> dump("reScale")
[1] "rescale.S"
```

We can edit `"rescale.S"` to contain the few lines of S computation to do what we want.

With the same example, but in the third situation, we might just have spent some time doing this rescaling on a particular set of data. We can use the `history` function to grab the relevant lines and dump them out for us as a function.

```
> history(max=20, function="reScale")
Function object reScale defined
and saved on file reScale.S
```

The `max=20` can be just a rough estimate; in any case, we'll have to edit the dumped file.

The tradeoff between the two ways of creating `reScale` is between argument list and function body. The trivial version has the right arguments (probably) but we have to compose the body from scratch. The dumped history has no arguments, but the expressions in the body are at least a start on how to do the computations.

Let's pursue the approach using `history` in this example. Suppose that we had an object `grid` in the working data. This contains some grid values used to plot horizontal lines on an earlier plot. We would like to plot a similar distribution of grid lines on the current plot but now we have a different plot, say

```
plot(Weight, Fuel)
```

The idea is to scale `grid` to the range of the data on the y axis, `Fuel`. Working interactively, we figured out the computations. The following sequence of tasks did it (comments added later):

```
> gmax = max(grid); gmin = min(grid)
> ymax = max(Fuel); ymin = min(Fuel)
> xxx = grid - gmin ## starts at 0.
> xxx = xxx/(gmax - gmin) ## 0 to 1
> xxx = xxx*(ymax - ymin)
> xxx = xxx + ymin ## ymin to ymax
> abline(h=xxx)
```

The last line generates the grid on the plot.

Having done this work, we would like to preserve the computations as a function. Using `history` to dump the expressions, we start out with the file `"reScale.S"` containing:

```
reScale =
# function created from
# history(function = "rescale")
function()
{
  gmax = max(grid)
  gmin = min(grid)
  ymax = max(Fuel)
  ymin = min(Fuel)
  xxx = grid - gmin
  xxx = xxx/(gmax - gmin)
  xxx = xxx * (ymax - ymin)
  xxx = xxx + ymin
  abline(h = xxx)
}
```

The next step will be to edit the dumped file. We need to decide what the arguments should be called, edit the file to use those arguments instead of the actual objects, and of course return the computed object, rather than assign it or plot it. In this example, we just change `grid` to `x`, `Fuel` to `xrange`, and modify the last two expressions a bit.

One more step, strongly encouraged even at this first stage: edit the comments at the top of the function to say what it's doing and provide online documentation. With these changes, we get the following contents of the file:

```
reScale =
# the data in 'x', rescaled to the range of 'xrange'
function(x, xrange)
{
  gmax = max(x)
  gmin = min(x)
  ymax = max(xrange)
  ymin = min(xrange)
  xxx = x - gmin
  xxx = xxx/(gmax - gmin)
  xxx = xxx * (ymax - ymin)
  xxx + ymin
}
```

A call to `source("rescale.S")` will parse revised definition and assign it to `reScale` To test it, since we just did the computations on `grid` and `fuel`, we can verify that example with the new function, as well as checking the range.

```
> all.equal(xxx, reScale(grid, Fuel))
[1] T
> all.equal(range(xxx),
+ range(reScale(grid, Fuel)))
[1] T
```

We can verify the new self-documentation as well.

```
> ?reScale
Title:
        ##<Function reScale >
Usage:
        reScale(x, xrange)
Arguments:
  x:    argument, no default.
```

```
    xrange: argument, no default.

 Description:
      the data in `x', rescaled to the range of `xrange'
```

Your preferred user interface may have shortcuts for dumping and sourcing (in s-mode for emacs, for example, the dump and source operations are bound to key strokes and operate directly from editor buffers).

The example prompts a few comments on style in programming with data. There are many fixes one could make to the body of `reScale` to make the source more compact and elegant: those multiple assigns to `xxx` were natural when working interactively, but could be collapsed into a couple of expressions in the function. In fact, `xxx` isn't even needed; we didn't want to overwrite `grid` in the working data, but in the function `xxx` can just be edited to `x`. The argument is a local object in the function so we may as well overwrite it.

It's a matter of taste when to do this sort of fixing up; if you can't stand the clutter, do it right away. But as a strategy, trying out the ideas before making the function pretty or efficient is usually a better investment. If the basic idea turns out not to be what you want, better to find that out quickly.

In this example, you might have been tempted to assume that the range was computed outside `reScale` and supplied as arguments `ymin` and `ymax`. Nothing wrong with that, but notice that the function as written is more general, with only a trivial difference in the amount of computing done. If the range *was* computed outside and passed in as the `xrange` argument, redoing the `max` and `min` calculations is irrelevant. With the more general assumption, we get a more flexible function, setting the range of one object to the range of another. The choice between the two approaches is not reducible to a rule or an algorithm, it's a matter of style and philosophy. Practice in designing functions and thinking about applications often leads to a style of computing with data that imposes fewer assumptions or restrictions.

Now, about the function `equalTest`. The dumped version just called `all.equal`, but we want to return `FALSE` whenever `all.equal` finds a discrepancy, rather than the character vector describing the differences, which is what `all.equal` returns. The one-line expression to compute this is:

```
identical(all.equal(target, current), TRUE)
```

Convince yourself that this works, either by examining what it does, or trying it out.

The use of `identical` may look like a gimmick, but in fact it expresses another basic idea. When testing the result of a computation, there may be an extreme or default case in which we know exactly what will have been computed: `TRUE` in this case, in other cases perhaps `NULL` or a zero-length vector. The meaningful, reliable, and precise way to test this is with `identical`. Whether or not this one-line solution is obvious, the effort of creating the function could be well worthwhile, if it records the solution for other users. The edited source file might contain:

```
"equalTest" =
# test using the method of 'all.equal', but always
# return 'TRUE' or 'FALSE'.
function(target, current)
  identical(all.equal(target, current), TRUE)
```

Once again documenting comments were added to the source.

6.1.2 Dealing with Optional Arguments

The functions you write will often be more flexible if they allow users to supply optional arguments, for example to change the way the computations are done or to supply information in different forms.

There are two basic techniques for handling optional arguments: default expressions in the function definition, and testing directly by calling `missing` with the name of the argument. A third technique that is sometimes useful is to compute the number of arguments, `nargs()`. Since this is often useful when dealing with `"..."`, it can be left for section 6.1.3.

The recommended strategy is to use default expressions when the default computation is simple and `missing` for anything more complicated, for example when more than one argument can be missing. In simple cases, the default values make what's happening easier to see in reading the function. But complicated default expressions can produce errors, particularly if they refer to other arguments. The two techniques are not mutually exclusive: `missing(x)` will correctly report that argument `x` was not included in the call, whether or not there is a default expression for `x`.

Default expressions are particularly common for labelling information and for numerical parameters in computations. In both cases, the motivation is that *something* is needed, but the user may or may not care. The solution is to provide a reasonable choice, but make the object an argument so that users can override.

The function `zapsmall` takes some data and sets to zero any elements that are more than a factor of `10^digits` smaller than the maximum. (The purpose is to ensure that the data will print nicely, without being forced to a fancy format to represent some relatively small values.) A slightly simplified version of `zapsmall` is as follows.

```
"zapsmall" =
# reduce to zero any value smaller than '10^-digits * max'
function(x, digits = .Options$digits)
{
  m = max(abs(x), na.rm=T)
  digits = max(digits - log10(m), 0)
  round(x, digits)
}
```

The values in `x` will print out to `digits` significance either as they would have before, or as 0 if they were relatively small. The default for argument `digits` is the `digits` option, a reasonable choice since this is also the default precision used in printing. Values set by the `options` function are often good defaults. Users have a two-level way to specify such values: temporarily through the argument to the function or persistently through a call to `options`.

Another good source of default values is the object `.Machine`. This is a `named` object created to hold the local machine characteristics. Its components are rather more exotic than the options values, but they provide a portable way to relate default values to the smallest and largest reasonable numbers. The named components come in three flavors: `"double.`*thing*`"`, `"single.`*thing*`"`, and `"integer.`*thing*`"`. These correspond to double-precision (i.e., class `numeric`), single-precision (class `single`), and integer (class `integer`) values. Actually the only integer value is `integer.max`, the largest representable integer.

A typical use of the values would be for a numerical tolerance, say for some iterative computation. The component `double.eps` is the relative precision of numeric computation, the smallest value $\epsilon > 0$, such that $1 + \epsilon > 1$. For a really precise computation, you might insist on accuracy to nearly that tolerance; otherwise, some heuristic depending on it could be applied. All sorts of other heuristics are practiced, none with any compelling argument except in those rare cases where the desired accuracy can be quantified clearly. One heuristic used in the `models` library is to say that, although the computations are double-precision, accuracy to single-precision is about as much as meaningful. So, for example, the tolerance for linear dependency in the routine that fits linear models is given as follows:

```
"lm.fit.qr" =
function(x, y, singular.ok = F,
   tolerance = .Machine$single.eps, qr = F)
{
  ....
```

Use of the square-root of .Machine$double.eps is also common; the two values are often about the same size. The key technique, however, is not the specific value but submerging machine differences through the .Machine object.

The missing test for an argument is needed if the function uses a different computation based on which arguments were supplied. The abline function used to plot horizontal lines on page 246 is called in three different ways: with argument h= or argument v= to plot horizontal or vertical lines, or with one or two arguments giving the equation of the line. Depending on the arguments, it computes suitable line segments to plot. The function then needs tests roughly of the style below:

```
abline =
function(a, b, v, h) {
    if(!missing(v)) {
    ## compute vertical grid
      ....
    }
    if(!missing(h)) {
    ## compute horizontal grid
      ....
    }
    if(!(missing(a) && missing(b))) {
    ## use equation for line
      ....
    }
}
```

Similar examples are the functions get and exists, which allow arguments where or frame. Both can be missing, or one (but not two) can be provided.

Sometimes an optional argument to one function is then used, if provided, to define an optional argument to another function. In relatively simple examples, missing handles this fairly well.

```
"again" =
# re-evaluate the last task, or the last one
# matching 'pattern'.
```

```
function(pattern)
{
  if(missing(pattern)) history(max = 1)
  else
    history(pattern, max = 1)
}
```

The function again, shown here slightly simplified, reruns a single task from the history file, and passes its optional pattern argument to history. This approach quickly becomes a mess, if there are several optional arguments to pass down. An alternative involves using "..." as an argument, and we deal with it in section 6.1.3.

Default expressions are evaluated in the frame of the function called, and so can involve local objects in that frame, including other formal arguments to the function:

```
"identify" =
function(x, y, labels = seq(along=x), n = length(x),
  ....
```

Both labels and n have defaults based on the argument x. With such default expressions, you need to be sure that the computation won't die because too many arguments are missing: identify requires argument x, for example.

It is *almost* true that default argument expressions are never logically needed, in that the default for n above could be written:

```
if(missing(n)) n = length(x)
```

The difference is the question of *when* the default is computed. Defaults in the argument list are computed when the corresponding argument is first needed. In some functions, it may be hard to know just when this is. For such cases, you may prefer to put the default assignment in the argument list, so the computation is done automatically in the right place.

6.1.3 Programming with Arbitrarily Many Arguments

In a variety of situations, you may want your function to take some arguments in a situation where the names and number of arguments is not fixed. You may want to pass some arguments down to another function, but not want to examine each of those arguments individually. Or, you may allow the user of your function to supply some named components for a list or similar object, with the names of the components entirely up to the user.

For these, and for other less obvious applications, the special argument "..." may be the best tool. When this name appears as a formal argument in a function definition, any number of actual arguments in a corresponding call will be matched to "..." in the evaluation frame. Then, if "..." appears as an argument in a function call in the body, the matching arguments will be passed down, exactly as they appeared, including any names.

For an example, `colorDevice` is a little function to set up a graphics device with a prechosen color map:

```
"colorDevice" =
# set up the graphics device given in 'deviceName' with
# a color map.  Pass '...' as parameters to the device function.
function(deviceName, ...)
{
  device = getFunction(deviceName)
  switch(deviceName,
  x11 = {
    shell("S x11Colors")
    device(...)
  },
  postscript =
    device(colors = myPs.colors, ...),
  ## default
  device(...)
  )
}
```

The function takes a string giving the particular device and, optionally, any number of arguments to the particular device function.

The `colorDevice` function needs to have an arbitrary argument list, since it wants to pass arguments to whatever device is chosen, and does not want to have all possible arguments to each device as explicit formal arguments in `colorDevice`. A call to `colorDevice` might look like this:

```
colorDevice("postscript", "report.ps", w = 4.5, h = 4.5)
```

The function retrieves the `postscript` function and stores it locally as `device`. The call that actually takes place to `device` is equivalent to:

```
postscript("report.ps", w = 4.5, h = 4.5)
```

These arguments are matched to the definition of `postscript` by the standard S rules.

The same "..." mechanism is helpful if there are several arguments to be passed down to another function, but we need to pass them down *only* if the user provided them. The example of `again` on page 251 used an explicit test:

```
if(missing(pattern)) history(max = 1)
else
  history(pattern, max = 1)
```

But if there had been other arguments supplied to `again` that needed the same treatment, we end up with a maze of tests and an exponential number of different cases.

If the arguments to be passed down are absorbed into "...", only one call is needed. Suppose we want a utility, `allFunctions`, that returns the names of all the function objects on some database or frame. The function `objects` will do this for us, but we need to provide `objects` with the argument `where` or `frame` to say where to look, and either of these could be missing. A simple implementation is then to use "..." for all the arguments to be passed down.

```
allFunctions = function(...)objects(class = "function", ...)
```

The user of `allFunctions` sees it as a function that has all the arguments of `objects`, except maybe `class`. In fact, the documentation of `allFunctions` should present the function that way, perhaps with an explanation that it formally uses "..." as the mechanism.

The use of "..." is not without some problems, or at least complexities. One is that we seem to have lost along the way the ability to deal individually with the arguments. This is not entirely true, but some extra care is needed. The test `missing(x)` no longer works, but there are three alternatives. In simple cases, if we only want to know if all of the "..." arguments are missing, we can just count the arguments. In the `colorDevice` example, if we wanted to check whether any options had been supplied in "...", the following test would work:

```
if(nargs() > 1)
{
  ## some ... arguments supplied
   ....
```

S supplies the function `hasArg` to check for argument names in either "..." or other arguments. If `x` is a standard argument, `hasArg(x)` returns `TRUE` if `missing(x)` is `FALSE`. Otherwise, `hasArg(x)` returns `TRUE` if "..." is an argument and the actual call has a named argument of the form "x=". For an example of the use of hasArg, see page 337 in section 8.2.3.

Using hasArg will detect arguments reliably *only* if the "..." argument is known to be supplied with named arguments, as for example when the "..." is passed on to options or par to provide named parameters. Otherwise, "..." may itself contain unnamed arguments. If the arguments are to be passed to objects, say, then the argument where might be supplied with no name as the first argument, in which case hasArg cannot detect it. If we do know how "..." is to be used, a third technique will handle all possibilities. S supplies tools to do the argument matching explicitly; for example, match.call returns a call object with the correct argument matches and all arguments explicitly named. The argument where will be missing in the call to objects if and only if "where" is not one of the names in the corresponding match.call. In allFunctions, for example, missing(where) can be computed as follows:

```
makeCall = substitute(objects(...))
makeCall = match.call(objects, makeCall)
is.na(match("where", names(makeCall)))
```

The first line generates the S call object, the second line converts this to the same call, but with all the arguments named. If the S evaluator could match argument where in the call to objects, then makeCall will have an element named "where". The third expression is TRUE exactly when the argument where would be missing from the call.

For more on "...", you may need to read section 4.9.5 for a discussion of it in the context of the S evaluation model, including the formal rules for matching "..." in calls and an explanation of what happens in the use of substitute above.

6.1.4 Writing Replacement Functions

Assignment expressions in S can have general expressions on the left of the assignment operator:

```
length(data) = 1
x[i, ] = -x[i, ]
substring(messages, "^") = "!!! "
fit$residuals[ wgts < .001 ] = 0.
```

The generality arises from a simple interpretation of assignment operations in which the left-side expression is a function call, with the name of an object as the first argument. If f is any function, an expression of the form

```
f(x) = y
```

is evaluated as if it were the expression

```
x = "f<-"(x, value = y)
```

The function `"f<-"` is called the *replacement function* corresponding to f, and by convention it has the name f, with `"<-"` pasted on the end. So the replacement `length(data)=1` is equivalent to the expression

```
data = "length<-"(data, value = 1)
```

The replacement function has the same arguments as f, plus a final argument named (again by convention) `value`.

The generality of replacement expressions follows from this single convention. As always, infix operators such as `"["` and `"$"` are really just functions called with a different syntax. The first argument to the replacement function can itself be any function call, rather than a name, if the resulting nested replacement can be unraveled into a sequence of replacements. The S evaluator performs the unraveling and does the nested replacement using local alias objects for intermediate results (see section 4.2.5).

Writing replacement functions essentially just requires remembering the conventions. The function must have the same arguments as the corresponding ordinary function, plus the final argument `value`. When the function is called, the evaluator always supplies this argument, named, with the actual argument being the right side of the original assignment expression. The most common programming error in writing replacement functions is to forget to return the revised version of the *whole* object as the value of the function. The point is not just to modify the object locally, but to return the revised object, with which the evaluator will replace the previous assignment.

As an example, the expression

```
diag(x) = 1
```

sets the diagonal elements of matrix `x` to the right side of the assignment. Here is the corresponding replacement function from the main S library:

```
"diag<-" =
function(x, value)
{
    if(!is.matrix(x))
        stop("Can only replace the diagonal of a matrix")
```

```
        i = seq(min(dim(x)))
        if(length(value) != 1 && length(value) != length(i))
            stop("Replacement diagonal of wrong length")
        x[cbind(i, i)] = value
        x
    }
```

Because replacement functions are just ordinary functions, they can make use of all the S programming tools for functions. Debugging tools such as `trace` work identically, when given the name of the replacement function; e.g, `trace("diag<-", browser)`. Replacement functions can have methods based on the class of any of their arguments, including `value`. The requirement in the `diag` replacement function that `x` be a matrix could be handled by defining a method for `"matrix"` and not supplying a default method. For more on replacement methods, see page 340.

We said that replacement functions should agree with the ordinary function in arguments, but the requirement is for the sanity of users, not something enforced by S. You should adhere to it: no user will be pleased to find that a different argument name is needed when a function appears on the left side of an assignment. The one useful exception is that there need not be *any* ordinary function at all corresponding to a replacement function. For example, suppose we want to write an expression

```
trim(x) = alpha
```

to trim all values of `x` beyond a limit `alpha` according to some criterion. So long as the replacement function is defined suitably, this replacement is perfectly legitimate, even if there is no definition for the function `trim`. Once again, the only special feature is that S redefines a replacement expression as an assignment of the value of the corresponding replacement function.

6.2 Organizing Your S Software

If you plan to develop S functions for any substantial project, you will want to decide how to organize that software so you can manage and understand it. Even a project that starts out as a small effort can grow into a more serious one. Most of us have wished we had organized projects early on, rather than trying to put some order into the chaos later.

This section discusses the programming cycle, techniques for testing software, and organizational tools at the chapter level.

6.2.1 The Programming Cycle

The approach to programming with data described in this book is object-based. The data, functions, documentation, classes, and methods are all maintained persistently in S chapters, as S objects. The programming cycle is the cycle of operations we go through to modify some of these objects. New functions, classes, or methods may be defined, or existing ones modified. The results of the programming need to be tested. Documentation will be created or revised.

The model for the programming cycle proposed here starts and ends with the S objects. When we want to make changes, we dump the relevant objects in a suitable form to edit, edit them, source them back to create new and revised objects, and test the results. We can also insert in the cycle steps to archive or distribute the software, usually at the level of the chapter (section 6.2.2).

For revising objects—editing function definitions, for example—S provides a family of `dump` functions. They all take some S object(s) and dump them to a file (more generally, an S connection) in a text format. You edit the files to make whatever changes are needed, then use a corresponding S `source` function to go from the file to the (revised) S objects.

We've shown this operation for functions in section 6.1.1 and elsewhere. The general call to `dump` takes any number of object names. The objects can be arbitrary, not just functions, although large amounts of data may not be too appealing to edit, unless the dump uses a method designed for the particular class of data (see page 337). Whatever you dump, the dumped file contains a sequence of S tasks; evaluating those tasks after editing the file will modify the chapter corresponding to the editing changes. A call to `source` is the standard way to get the edited file back, creating the revised S functions or other objects.

For classes, methods, and documentation, an analogous dump-edit-source cycle applies. The function `dumpClass` dumps the definition of an S class (see section 7.1.1); `dumpMethod` dumps a method and `dumpMethods` dumps the definition of a generic function and its various methods (see section 8.1.2). For all of these, the function `source` can be applied to the edited file to get the revised software back into S.

The call to `source` parses and evaluates the contents of the file in a single task. If there are any errors, none of the objects will be restored. For very large source files, we may prefer to evaluate each task on the file separately, for example by diverting the standard evaluator's input to the file or connection:

```
sink("dumpdata", "input")
```

At the end of the task containing the call to `sink`, the evaluator will start reading tasks from the file `dumpdata`. Now each complete expression on the file will be evaluated as a separate task. Errors will not halt the evaluation and any successfully evaluated expressions will restore the corresponding objects.

A similar pair of functions, `dumpDoc` and `sourceDoc`, serve to create and get back files containing S online documentation, once you decide to move beyond the self-documentation provided by comments in the functions. Section 9.3 provides details; notice that the documentation case differs a little, in that the files are dumped as documents in the SGML language. The `sourceDoc` function has to be used, rather than `source` or similar methods. It converts the file to S source and then evaluates the result.

Nice user interfaces can hide much of the detail in dumping and getting back the various S objects. Menus or hot keys in graphical user interfaces can save you a good deal of typing. With luck, your version of S comes with such tools (my own programming environment of preference is `emacs`, with its customized environments for S and other languages). As far as organization is concerned, though, it's important to keep in mind that we're treating the S objects as the master copy. The cycle is dump, edit, source back into S.

6.2.2 Organizing the Chapter

S software comes in two fundamental units: the *object* and the *chapter*. Everything is an object, so programming with data is programming with objects: designing them, creating them, and computing with them. The objects that S maintains for you as permanent data or software are, in turn, organized into chapters: directories inside which the objects and other information are organized. By now, you will have created one or more chapters for your projects. It's usually a good idea to see each project as a chapter; that way, there is no confusion about where the software for the project resides. S creates the chapter within some directory in your file system, typically in response to the `CHAPTER` shell command.

The definition of the chapter is kept in several files within the chapter. To get an up-to-date dump of the chapter from inside S, use the function `dumpChapter`. This function is called with the name of an S chapter, but it should normally be run on the working data, when running S in the chapter's own directory (so there is no question where all the files will appear).

```
> dumpChapter()
A list of the files defining the chapter "." was written to
"./DUMP_FILES"
```

The call to dumpChapter writes a number of files in the chapter directory, containing *all* the S objects belonging to the chapter, plus the other information necessary to recreate the chapter. (You can do the dump from a shell command as well; see the documentation of dumpChapter.)

One of the files written is DUMP_FILES, a list of all the files needed to define the chapter. Your next step depends on why you dumped the chapter. If you want to copy the chapter to another directory or another computer's file system, you can invoke next a shell tool to create a single file or to copy all the files. Some examples of the use of such tools are either of the following shell commands (run from the chapter directory where the dump was done):

```
tar cf myChapter.tar `cat DUMP_FILES`
cpio <DUMP_FILES -o > myChapter.cpio
```

The archive file would then be moved to another directory (maybe on another machine), and the chapter restored there. See page 232 in section 5.4 for the sequence of steps to restore the chapter. Copying and restoring a chapter isn't a bad idea, even on the same system, if you have done a great deal of revision and have a large number of functions. The restored version is somewhat more compact and efficient at accessing the objects in the chapter.

Another useful application of dumpChapter is to keep versions of your chapter. After running dumpChapter, you can then commit or check in the revised files, providing a timeline of major changes in all the material of the chapter. This is a somewhat unusual application of version control, but in fact makes considerable sense. By definition, the files such as all.Sdata provide a complete definition of the S objects in the chapter at any time, something that is not easy to do otherwise. You need to archive DUMP_FILES and all the files it refers to.

All the dumped files are text files (that's why they can be shipped around to different computer systems), and can usually be handled by version control tools, such as cvs, rcs, or sccs. The cvs system suits S particularly well, since it operates on directory hierarchies. Once you have installed the files in DUMP_FILES as the cvs definition of the S chapter, you can commit the changes to the chapter by calling dumpChapter and then running the shell command

```
cvs commit .
```

in the chapter directory. By restoring DUMP_FILES from a particular date, and then all the files it refers to from the same date, you can get a snapshot of the entire chapter's source and S objects at that time.

For other version control systems, you will need to commit each of the files explicitly; for example, in rcs, assuming you made an RCS subdirectory in this chapter:

```
ci -u `cat DUMP_FILES` -m"fixed all the data-reading bugs"
```

For other options in dumping and restoring chapters, see section 5.4.2.

The combination of the various S dump functions with editing, file management, and version control tools will give you a consistent programming environment to manage even fairly large systems. (The S software itself at Bell Labs is managed essentially this way.) If you would rather, though, you can manage the software in a number of other ways. You might want to keep a set of source files for all your functions, classes, etc., and regard these source files, rather than the S objects, as the master copy. This mostly means that you will not use the dump functions, except for dumping and moving the chapter, and for getting initial versions of documentation files. I would slightly urge the S-object-based approach instead, mostly because more than one source file might define the same object, particularly as you get into classes and methods.

The key strategy, though, is to settle on a single approach to managing the software and stick to that consistently, until you consciously decide to change it.

6.2.3 Tools for Testing

One stage in the programming cycle described in section 6.2.1 remains to be described: testing the modified software to see if it works. As your programming project becomes more serious, testing will become more important, and usually less fun. This section describes some tools that can help.

No really large, general-purpose collection of software can be expected to work perfectly. Even if it did, its perfection can almost certainly never be proved. The many ways in which the software can be used, and the definition of what it should then do, make the testing problem essentially impossible in any complete sense. What *can* be done is to build some tests along with the application itself, so that some of the functionality continues to be tested as the software evolves.

The general approach suggested in this section is to organize the tests as S expressions, specifically as *assertions*. We define an assertion to be an S expression that is expected to evaluate to `TRUE`. As your software evolves, you can build up a collection of assertions; testing them after making changes to your chapter will help catch problems and provide some confidence in the software.

Some functions provide basic tools to make things more convenient, and the everything-is-an-object nature of programming in S helps organize the testing process. Once you have a collection of assertions, the function `do.test` will arrange to evaluate them. Assertions that succeed will evaluate silently; if an assertion fails, `do.test` will report the expression and the value.

In building up the assertions themselves, you will need some tools that test objects. Two basic functions are `identical` and `all.equal`. Both take two arguments and return `TRUE` if they judge the corresponding objects to be the same. For testing, you want to provide them with either two ways to compute the same result or else a target object and a computation asserted to produce that object.

The two functions differ in the criteria they use, and what they return.

`identical`: The two objects must be exactly equal in all respects; if not, `identical` returns `FALSE`.

`all.equal`: The two objects are expected to be identical up to small differences that might be considered irrelevant, and up to differences that are considered unimportant for this class of object. If the objects are not judged equal, `all.equal` returns one or more character strings describing the nature of the differences found.

Use `all.equal` for complicated computations including numerical results that might be slightly affected by moving from one machine to another, and also for objects that can be "equivalent" even though not equal in every byte. Use `identical` if the assertion says that the computation should produce exactly this result, and that anything else must be a mistake.

As you might expect, `all.equal` was written and designed specifically for testing software. If the objects you're computing are at all complicated, you probably want to use `all.equal` at the top level of the assertion, just because it will provide some clues as to what went wrong. If you want to make sure that x and y are identical, but still get the description from all.equal if they are not, combine the two functions in the form

```
identical(x, y) || c("not identical", all.equal(x, y))
```

Even if all.equal is fooled, you will still get a report.

One of the trickier aspects of testing is to verify that your software correctly detects problems. The function error.check takes an expression as argument, but expects the evaluation of that expression to end in an error. By using the error-handling facilities in S, error.check intercepts that error and returns TRUE. If no error occurs, error.check returns FALSE. The error message received is printed.

To make the test more reliable and also to keep it silent when the asserted error occurs, give error.check as a second argument the error message expected. The usual technique is to run error.check first with one argument, then copy the printed error message to provide a second argument.

```
> error.check(sqrt(letters))
error.check got: "Problem in x^0.5: Non-numeric first operand"
[1] T
> error.check(sqrt(letters),
+ "Problem in x^0.5: Non-numeric first operand")
[1] T
```

The revised call is the one to include in your tests.

In devising assertions, remember that you want to learn as much as possible when one fails. Where in normal programming you would naturally "and" together conditions which all have to hold, the best approach here is to generate a list of the subsidiary assertions. If a computation was expected to define a non-virtual class "trackMatrix" that extended class "tracks", the assertion

```
!isVirtualClass("trackMatrix) && extends("trackMatrix", "tracks")
```

is less useful than the logically equivalent assertion

```
all.equal(list(F,T),
  isVirtualClass("trackMatrix), extends("trackMatrix", "tracks"))
```

If the point of the individual tests needed emphasizing, you could use named instead of list, giving the elements suggestive names.

Organizing the test expressions can be done in various ways. The argument to do.test can be a connection or file name, in which case do.test will parse the file and treat each complete expression in the result as an assertion. Alternately, you can give `do.test` an `expression` vector, in which case it will evaluate each element of the argument. Keeping the tests as expression objects in the chapter will ensure that they are dumped, moved, and restored along with the rest of your chapter.

6.3 Debugging and Error-Handling

This section presents some techniques to figure out what happens when problems occur, both problems that cause S to stop evaluation and those that just show up as wrong answers. The functions `recover` and `browser` conduct an interactive dialog with the user from inside an evaluation frame. The language and evaluation in this dialog are essentially like those in the standard reader interface users use to provide tasks to S, with the exceptions that computations are done locally in the evaluation frame and that some additional facilities are provided.

The `browser` function is intended to be inserted in a function being debugged by calling `trace`; the `recover` function is normally called as part of the error action defined by the user. Sections 6.3.2 and 6.3.3 discuss how to specify tracing and error options.

The browsing functions are powerful tools, because they provide the full functionality of the language in the context of an ongoing computation. Developing facility with them and using them frequently will help to make your programming more effective.

6.3.1 Browsing in the Evaluator

The `recover` and `browser` functions are S functions that conduct an interactive dialog with the user, when the evaluator is running interactively. They parse input from the user as an S expression and evaluate that expression in an evaluation frame, the *browsing* frame. The expressions are shown automatically as with the standard evaluator for S tasks. Assignments are interpreted in the browsing frame, so the interaction can alter the contents of the frame. Neither function is designed to be used in a non-interactive context.

The `recover` and `browser` functions are largely identical in their interaction, but designed for different contexts. The difference in their behavior is largely in the interpretation of a few special commands.

The steps in the interaction with the `recover` function are

1. `recover` asks whether the user wants to debug this problem.

2. If yes, `recover` picks the first nontrivial evaluation frame, working up the nested evaluations currently in progress. In this frame, if any, the user is offered an interactive mini-session.

3. When the user is finished debugging, `q` will quit `recover`, returning to the regular interface; `dump` will dump the evaluation frames (see page 266); and `go` will retry the expression, presumably after the user has changed some data in the local frame.

Before beginning interaction with the user, `recover` sets up special definitions in frame 1 for functions `up`, `down`, `where`, `dump`, `q`, `c`, and `go`. The definitions interpret the first two functions to change the browsing frame, the next to show the traceback of calls, and the last four to exit `recover` as in step 3 above—`c` is a synonym for `q`. The special evaluator in `recover` also interprets the name of these functions, with no parentheses, as a short form for a call to the function with no arguments, *provided* there is no local object in the browsing frame with the same name. The operator `"?"`, with no arguments, prints a summary of available shortform commands, and a list of the local objects.

The `browser` interface uses the same special functions, except that `go` is redundant, and `dump` is not defined. There is no preliminary dialog and interaction always begins with the browser frame being the frame from which `browser` was called.

The following example illustrates the automatic error recovery.

```
> xxx = reScale(fuelData$Type, fuelData$Fuel)
Problem in max(x): Numeric summary undefined for mode
"character"

Debug ? (y|n): y
Browsing in frame of max(x)
Local Variables: .Generic, .Signature, na.rm, x

R> where
*3: max(x) from 2
2: reScale(fuelData$Type) from 1
1:  from 1
R> up
Browsing in frame of reScale(fuelData$Type)
Local Variables: x
```

```
R(reScale)> ?
Type any expression. Special commands:
'up', 'down' for navigation between frames.
'where' # where are we in the function calls?
'dump'  # dump frames, end this task
'q'     # end this task, no dump
'go'    # retry the expression, with corrections made
Browsing in frame of reScale(fuelData$Type)
Local Variables: x

R(reScale)> class(x)
[1] "string"
R(reScale)> q
>
```

When we had seen enough, we used `q` to exit from the browser. Alternately, we could make some changes in the browsing frame at an appropriate level, and try again.

```
R> up
Browsing in frame of reScale(fuelData$Type, fue....
Local Variables: x

R(reScale)> class(x)
[1] "string"
R(reScale)> x = levelsIndex(x)
R(reScale)> class(x)
[1] "integer"
R(reScale)> go
>
```

The assignment changed `x` in the frame of the call to `reScale`; the `go` function then restarted computation for that function call.

Remember that the special commands are just shortcuts: typing `where` is interpreted as `where()`. These functions have temporary special definitions. In some applications, there may be a local variable with the same name as the special command; you can still use the function, but not the shortcut. If you were browsing in a call to `get`, `where` is an argument, so you need to type `where()` to get the functional form. In addition some of the special functions have useful optional arguments: `up(2)` or `down(2)`, for example, says to move up or down by 2 levels of call.

If the solution to your debugging problem eludes you, you may want to come back and browse again. Exit from `recover` by typing the keyword `dump`.

The evaluation frames you have been browsing in will be saved in the object `last.dump` on the working database. When you want to look at them again,

```
debugger()
```

will invoke a browser on the dumped list of frames. The slight difference is that these are no longer active evaluation frames, but rather images of those frames produced by the dump. You cannot try to re-evaluate the expression with command `go`, but otherwise you should be able to debug much as you would have at the time the error occurred. The function `debugger` is also useful with the error option `dump.frames` discussed in section 6.3.3.

6.3.2 Tracing and Interactive Browsing

Interactive browsing can be used before an error, by tracing S functions. The mechanism is a general one, typical of S programming tools in that it works by operating on S functions as objects. The function `trace` creates a new version of a specified S function by inserting some additional expressions into the current definition and storing the new version in the session frame. The usual application inserts a call to `browser` at one or more places in the function:

```
trace(reScale, browser)
```

This modifies the function `reScale` to include the expression specified by the second argument to `trace`. If this argument is the name of a function, as here, `trace` expects this to be a function it can call, with no arguments, from the body of `reScale`. Alternatively, the argument can be any S expression, which `trace` will insert, as is, into `reScale`. The expression usually needs to be quoted, if we don't want it evaluated by `trace` itself. For example, to print the class of the argument on each call:

```
trace(reScale, Quote(print(class(x))))
```

By default, `trace` puts the new expression at the beginning of the call to `reScale`; by supplying the argument as `exit=`, an expression can be inserted just before `reScale` exits in addition to or instead of at the beginning.

```
trace(reScale, Quote(print(class(x))), exit = browser)
```

The argument `at=` to `trace` allows expressions to be inserted at other places as well; see the detailed documentation of `trace`.

The side effect of `trace` is to create a version of the modified function on the session database. To stop tracing function `reScale`, call `untrace`:

```
untrace(reScale)
```

Called with no arguments, `untrace` stops all tracing. It is important to untrace a function before modifying its definition, since some tools may not be smart enough to realize that the version of `reScale` on the session frame is only for debugging!

If we are debugging a particular method for a generic function, rather than an ordinary function such as `reScale`, the same tracing technique can be used. Instead of `trace`, call `traceMethod`, whose first two arguments are a generic function name and a class signature, the same as the first two arguments to `setMethod`. The effect of `traceMethod` is just like that of `trace`, except that it creates a temporary version of the particular method, rather than of an ordinary function. The function `untraceMethod` corresponds to `untrace` to turn off tracing for methods for particular signatures. If you want to turn off tracing for *all* methods, though, you can just call `untrace` with no arguments.

6.3.3 The Error Option

When the S evaluator encounters an error or interrupt (a signal), it looks for an error action; that is, for some S function to call or expression to evaluate. This section discusses what such actions should do, and how to set your preferences for error actions. In section 6.3.4 we go into some techniques for adding error actions temporarily during computation and for distinguishing particular signals.

The `recover` function is the usual interactive error action in S. To change this, use the `error=` argument to the `options` function. The argument should be a function, an expression (typically as an argument to `Quote` to prevent its being evaluated), or `NULL` to indicate that we don't want any error handling. For example, the default error action is actually a call to `recover` if the current task is being run interactively, but a call to `dump.calls` otherwise. The corresponding call to `options` would be:

```
options(error = Quote(
    if(interactive()) recover()
    else dump.calls()
  )
 )
```

The `dump.calls` function saves the traceback of the function calls in effect when the error occurred. Another useful choice for the error option is `dump.frames`. It dumps all the local evaluation frames as well, suitable for examining later with `debugger`.

The `options` function interprets its arguments as named elements, which it inserts in the object `.Options`, and stores. When the evaluator needs to know the value of an option, it looks in that object. By default, `options` stores the revised options in the session frame (frame 0), meaning that they remain current for the rest of the session unless changed. The special argument `FRAME=` instructs `options` to store the new options in that frame, rather than frame 0.

If you are setting an option temporarily, it's a good practice to store the new option in frame 1, the task frame, and then to arrange to restore the old options settings when the current function has completed, as in the following example:

```
oldError = options(error = NULL, FRAME=1)
on.exit(options(oldError))
```

The first line turns off any error action for this function call, while the `on.exit` call ensures that the old error option is restored. More detailed control of the error action can be achieved, however, by the techniques in section 6.3.4; these allow specifying actions in addition to the error option, and actions depend on the particular signal received.

The error action itself can be any expression or, more conveniently for testing it out, it can be a function. The function will be called with no arguments when an error occurs. As an example, suppose we expect that some problems may occur, and would prefer to just collect the information about them for future inspection. Setting the error action to `dump.frames` will only leave us with the last dump, so let's design a function that uses `dump.frames` but saves multiple dumps. From looking at the online documentation for `dump.frames`, we observe that it takes an optional argument for the name of the object to hold the dump. Using these, we write a function `save.frames` that stores successive dumps as `dump.1`, `dump.2`, etc. on the working data.

```
"saveFrames" =
function()
{
  nSave = if(exists("nSavedFrames", where = 1))
    get("nSavedFrames", where = 1) else 0
  nSave = nSave + 1
  what = paste("dump", nSave, sep=".")
  dump.frames(what, where = 1)
```

```
    assign("nSavedFrames", nSave, where = 1)
}
```

To examine one of these dumps, one could use `debugger(dump.1)`, for example. The call

```
options(error = saveFrames)
```

sets this as the error action for the current session.

6.3.4 Additional Control of Errors and Interrupts

Two other mechanisms are provided to specify error actions in addition to the option specified by `options(error=`*action*`)`. These are intended to provide added, temporary actions and to provide actions that apply only to specified kinds of errors or signals.

To introduce the actions, it's best to look at what actually happens when an error or interrupt occurs. The internal, C-based code for the S evaluator always does the same thing: it evaluates a call to the function `doErrorAction`, giving it as arguments the message defining what happened and a numeric error code. If the call resulted from an error in S evaluation (for example, a call to `stop`), the code is `1000`, by convention. If the call was the result of a signal to the process, the code is the standard Unix code for that signal. An interrupt signal has code 2, for example.

The S function `doErrorAction` now tries to figure out what action to take by looking for the following:

1. an object named `".onError.`*n*`"`, where *n* is the numeric code in the call to `doErrorAction`;

2. an object named `".onError"`;

3. the element named `"error"` in the object `.Options`.

Each of these objects is interpreted, if found, as an error action and evaluated, in the order above. If the object is a function, it will be called with no arguments; if the object is `NULL`, nothing happens. Otherwise, the object is expected to be some (`Quoted`) S expression, which will be evaluated. All the evaluations, both the call to `doErrorAction` and the error actions themselves, are done in the frame of the function call that was current when the error occurred. If no non-`NULL` action at all was found, `doErrorAction` will print the error message but no other error action will be taken.

There is a slight, subtle difference between specifying a function and an expression as the error action. The expression could involve objects in the local frame where the error occurred, but the function will be evaluated in its own frame, called from the current frame. Of course, using local objects only makes sense if you know which frame the error occurred in or arranged that the error action only applied during that call.

The search for an error action uses the standard rules for finding objects in S. This gives us a chance to get detailed control over error actions. The search rules are defined in section 5.2.1; for the present purpose the relevant rules are that S looks first in the local frame, then in frame 1 (the task frame), and then in frame 0 (the evaluator, or session, frame). By storing objects in these frames, we can obtain control over the error actions for the scope desired.

To establish a supplemental error action only for the current frame, assign `".onError"` locally:

```
.onError = Quote(remove(thisName, w=1))
```

This has the effect of executing the call to `remove`, using a locally defined object name. However, this action will not take place during any calls *from* the current frame, since those calls won't find the local `.onError`. Therefore, a slightly different way of doing the same thing would be to put the error action into frame 1. In this case we can't refer directly to a local object, so we need to construct the error action with the name in it. The `substitute` function is good for such constructions.

```
assign(".onError", frame=1,
  substitute({remove(NAME, w=1); recover()},
    list(NAME = thisName)))
```

Frame 1 is the right place to put the error action if we want to ensure that any error in the current task causes this action to be taken. A search for `.onError` from any call in the current task will include frame 1, but the frame is cleared when the task completes.

If we only wanted the error action to be used until the current call completes, we can remove the object from frame 1 with a call to `on.exit`:

```
on.exit(remove(".onError", frame=1))
```

Even friendlier would be to look in frame 1 first, save any previous `.onError` found there, and restore that object on exiting. I'll leave that as a programming exercise.

All these approaches can be restricted only to certain signals by defining the "`.onError.`*n*" object. It's possible, for example, to communicate between processes by sending a signal from one process to another. Signal 16, for example, is a so-called "user signal". If we knew that our S process had process id `12345`, for example, the shell command

```
kill -16 12345
```

would send that signal to the S process. An object `.onError.16` would provide an error action for such signals, but would be ignored for any other errors or signals.

If a signal occurs *between* tasks, when the evaluator is not active, the S manager still calls `doErrorAction`, but `doErrorAction` only looks for `.onError.`*n* corresponding to the signal received. Invoking a general browser when there is no task makes little sense, but an action designed to respond to a particular signal will likely behave much the same at any time.

Finally, a couple of general comments on strategy:

- As with much of S programming, error handling is designed to define nearly everything in S. Aside from the call to `doErrorAction` itself, everything can be changed in S. You can even write your own version of `doErrorAction`, given the courage.

- Signals provide a mechanism for interprocess communication, but not a very precise or controllable mechanism. If you want to share data between two processes or pass tasks to S from some other process, the better mechanism is usually to write the data to a `fifo`, for example, and establish an S reader to look for input waiting on the fifo. Actions on signals are still of some value, though, in that very little programming outside of S is needed. If you expect your S process to lie around doing nothing much until a signal wakes it up, a signal-specific error action could work reasonably well and be easier to implement.

6.4 Programming the User Interface

This section takes a brief look at designing special-purpose programming interfaces for users. User interfaces can lead to some very substantial projects indeed, but in this section we will just take a small nibble out of the topic. Even so, the techniques to be discussed can help you to make your software more convenient and useful.

User interface programming involves two basic ingredients: communication and control. We need to communicate with the user, presenting information and getting back instructions. Based on the communication we then need to control what S does, usually arranging to evaluate tasks, similar to the standard S task evaluation that takes place in any S session. In other words, we are creating our own variations on the standard parse-and-evaluate approach to S tasks. Some basic tools, and the availability of the S evaluation model in S itself, allow such variations.

6.4.1 Functions for Parsing and Evaluating

Parsing is the operation of taking input from a connection, interpreting that input as S expressions and returning the S objects that represent the expressions. Evaluation takes an S object representing an expression and returns its evaluated form.

There are two functions to parse input according to the grammar of the S language, `parse` and `parseSome`. They deal with exactly the same language, but differ in their approach to getting input. The function `parse` reads input from the connection it's given until it reads at least one complete S expression. If the first line read is not a complete expression, `parse` will continue to read; if it reaches an end-of-file in the middle of an expression, it generates an error. The approach of `parseSome` is roughly the opposite: it reads one line or perhaps more if asked, but never uses more input than required for one complete expression. In most cases, `parseSome` should be used if you want to control interaction with the user, while `parse` is more suited to non-interactive parsing, say of source files.

The two functions differ also in what they return. The value of `parseSome` is an object with class `"parse"`. It has slots `expr` for the parsed expression, `text` for the input text, and `message` for an error message if there was a syntax error in the input. You use `parseSome` in the context of some interaction with the user; so long as the input so far is partial, you keep recalling `parseSome` with the previous output now as an argument. (In an interactive situation, presumably some prompt or other sign has been given to the user that more input is needed.) Once the input is complete, you can evaluate the `expr` slot of the value returned. If a syntax error occurs in the input, no S error is generated. Instead, `parseSome` captures the error message and returns it as the `message` slot in the value. Since this is an interactive dialog, the function calling `parseSome` should inform the user of the error, and provide an opportunity to edit the previous text or otherwise recover.

The value returned by `parse`, on the other hand, is always of class `expression`. This class holds a vector of zero or more S expressions, each of which is suitable for evaluation. An important technical detail in using such objects is that the evaluator evaluates all the elements of an `expression` object, but only returns the value of the last; otherwise, there would be no way to distinguish, for example, a list of n evaluated expressions from a single evaluated expression returning a list of length n.

Let's turn now from parsing to evaluation. When an ordinary S session starts up, input presented on `stdin()`, the input connection of the corresponding process, is handed to a standard reader, as described in section 10.6. The standard reader does a little more than just evaluate expressions. It arranges to copy assignments in the top-level expression to the current working data when the task is complete. For tasks other than assignments, the value of the expression is assigned to `.Last.value` (so you can get it back if you realize you *should* have assigned it). Non-assigned values are normally displayed automatically, by calling `show(.Last.value)`, unless the task itself has suppressed automatic display by assigning `.Auto.print=FALSE` in frame 1. (This is what a call to `invisible` does.)

This whole convention can be invoked explicitly as `standardEval(expr)`, where `expr` is the object representing the task to be done. So, for example, if you want to construct the expression by some path other than the usual parsing of input, you can ensure that evaluation will still have the standard semantics of an S task, by passing the expression to `standardEval`.

On the other hand, if you do want to modify the evaluation semantics in some way, you should use the lower-level evaluation function `eval`. This only does evaluation.

6.4.2 Generating Messages and Errors

When you decide that the user should be informed of some situation during the evaluation of a function call, four related functions are available: `message`, `warning`, `stop`, and `terminate`. They all take any number of arguments, which they concatenate into a single string (with no additional separators inserted) and present to the user. The functions differ in how bad they consider the situation to be.

A call to `message` just writes the string to the user, specifically to the `stderr()` connection; there is no implication that anything is wrong. A call to `warning` records the string as a warning message. What happens next depends on the corresponding option `"warn"`. This is an integer option with values indicating how seriously the user wants to take warnings. The default level, `0`, causes warnings to be accumulated until the end of the task. At this point, the evaluation manager will print the warnings, if there are not too many, or else will notify the user who can call the function `warnings` to look at them. With `options(warn=1)` warnings are printed immediately; with `warn=2` warnings are not allowed, and are turned into errors by the evaluator. Conversely, with `warn= -1`, warnings are ignored and thrown away.

The function `stop` suspends the evaluation of the current function call and starts the evaluator on its error action, as described in section 6.3.4. Usually the task will be over at this point, but functions can call `restart` if they want to be called after an error. If the expression

```
restart(TRUE)
```

is evaluated in a call, and an error subsequently occurs, the evaluator will then *restart* the call. What this means precisely is that the evaluator will evaluate the body of the function again, but without any re-initialization of the frame. Whatever arguments and local variables have been assigned will retain their values as of the error.

To make use of `restart`, the function needs to be prepared to test whether this is a restart. Anything will do, but be careful that the test can't itself generate an error. One simple mechanism would be the following:

```
if(exists("restartSet", frame = sys.nframe()) {
    ## do whatever we're supposed to on restart
    .....
}
else {
    restartSet = T
    restart(T)
}
```

The `restart` mechanism is used by `recover` and `error.check`, among others; look at those functions for examples. As you might imagine, programming with `restart` is risky, in the sense that errors in a function that has called `restart` are very likely to generate infinite loops of errors. Make heavy use of `trace`, `browser`, and other programming tools. You are also likely to need the following expression if you do manage to get into a browser:

```
restart(0); stop()
```

This turns `restart` off and ends the task.

Finally on the escalation of seriousness comes the function `terminate`. As its name implies, it terminates the S session, although it will try to honor the `.Last` action. This function is normally reserved for situations in which the evaluator believes that the memory for S objects has been corrupted, or some other truly unrecoverable catastrophe has occurred. For ordinary interactive programming, it's hard to imagine any situation in your functions that could be that bad. However, if you are using S for some more specialized, non-interactive purpose, you may want to ensure that no more tasks are evaluated if, for example, there is a danger of corrupting a database. In such situations, `terminate` will do it: the session will end without starting any further tasks.

6.4.3 Communicating with the User

A number of functions deal explicitly with displaying some information for the user and retrieving a response. Beyond these, of course, you can construct any combination of function calls to display information and read responses. The packaged communication functions have the advantage of being simpler to use in most cases, and of extending from the plain, text-only versions we show here to graphical user interfaces of various sorts.

The functions `menu` and `dialog` both involve presenting the user with choices among possible responses. The visual format is different, and they extend in different ways to graphical interfaces, but the end result is essentially the same. The user gets a set of choices, represented by a character vector of length `n`, and the response is meant to match one of the choices. The functions return a single integer on the range `1` to `n`.

```
> colors = c("red", "green", "blue")
> menu(colors, header = "Color for boundary?")
Color for boundary?
1: red
2: green
3: blue
Selection: g
[1] 2
> dialog("Color for boundary?", colors)
Color for boundary? (red|green|blue): g
[1] 2
```

There are some minor differences between the two functions, other than the presentation; for example, `dialog` requires a question as its first argument;

the corresponding argument to `menu` is optional. Other differences apply to a particular user interface format. In the text form shown here, `menu` takes integers as an alternative to strings in the response, so that 2 typed by the user also selects the second item, and `dialog` allows for a default choice. More important in choosing one or the other is that they will often map into quite different graphical interface tools if you extend your use to that form. Each function has an argument referring to the connection or device on which the choice is presented. The argument is named `graphics` for `menu` and `connection` for `dialog` for reasons of back compatibility. But they are used in exactly the same way: if an object is supplied as this argument for which a `menu` or `dialog` method is defined, S will use that method. In this section we only consider ordinary text interfaces, however.

As an example, let's construct a function to do a slightly more extended communication with the user than `menu` or `dialog`. In the process we will apply some of the useful basic tools for communication.

Suppose we want to prompt the user to select two colors at one time. Or, to put it generally, we want to give the users some choices, and allow multiple, possibly duplicated selections. This extends slightly beyond what `dialog` or `menu` does: we need to read several selections from the user and match all of them against the choices. Looking at the implementation of `dialog`, for example, is a good way to start. This function is fairly complicated, with lots of checking for problems. But after some study, the key portions seem to be the following:

```
function(question, answers, deflt = 1)
{
    ....
    cat(question, " (", paste(answers, collapse = "|"),
        "): ", sep = "", file = stderr(), fill = 2)
    resp = readLines(stdin(), 1)
    ....
    pmatch(resp, answers)
}
```

The call to `cat` communicates the choices to the user, `readLines` reads back one line of response, and `pmatch` matches this to the choices. Our application needs to read several strings back, and we don't want to require each to be on a separate line. Reading in free form is something that the function `scan` does well, so let's use it instead of `readLines` to read any number of strings.

The selection of the response by `dialog` only requires the user to supply enough of the text to make a unique choice among the alternatives. S implicitly supplies name completion. The algorithm used, *partial matching*, is directly available as the function `pmatch`. This takes two character vectors and matches the strings in the first to those in the second. There can be any number of strings in the first argument.

The call to `pmatch` looks fine, except that we will discover right away that `pmatch` does not allow duplicate responses unless we tell it to. So, perhaps not our first attempt, but the second or third might look something like the following:

```
pickSeveral =
function(header, choices)
{
  cat(header, "\nChoices:",
      paste(choices, collapse=", "),
      file=stderr(), fill = T)
  strings = scan(what = "")
  pmatch(strings, choices, duplicate=T)
}
```

In the example where we want the user to pick colors, our function might contain the call:

```
cols = pickSeveral("Please pick colors for boundary, interior.",
   colors)
```

and the user's interaction might go as follows:

```
Please pick colors for boundary, interior.
Choices: red, green, blue
1: r g
3:
 ....
```

with the interaction setting `cols` in our function to `c(1, 2)`.

As always, the emphasis here has been on getting something quickly that does what we need. Now that our role includes designing software for others to use, there is more incentive to go back and make the interaction more convenient for the user. The `pickSeveral` function could be improved in a number of ways. Like `dialog` and `menu`, it should detect mismatched input, which will produce `NA` in the match to `choices`. When the calling program needs only a known number of responses, it should stop reading input as soon as it gets that many responses. Try these as an exercise (the implementation of `dialog` offers some clues).

Chapter 7

Creating Classes

Classes are the fundamental organizing principle for data, just as functions are for computations. Most substantial projects will involve designing new classes of objects. This chapter presents techniques for creating new classes and modifying the relations between classes. Section 7.1 covers the essential tasks of defining and modifying a class definition; the remaining sections discuss other specific needs. See Chapter 5 for the use of existing classes of objects and the general ideas underlying data in S. See also section 8.2 for the other key part of creating classes: defining the methods that make them useful.

7.1 Specifying a Class

S uses the concept of *classes* of objects to organize its treatment of data. Classes can be treated informally, but the real power lies in the mechanisms for explicitly defining classes and their relations to other classes. Specifying a new class defines the *representation* of the class; that is, what kind of data an object from the class contains. The representation is most often defined by *slots*: a new object is specified to have named slots, each of which is an object of another specified class. Each class also has a *prototype*, the object from which a new member of the class is created. By default, the prototype for a class with slots is a list whose elements are the default objects of the class for the corresponding slots. Prototypes may be specified explicitly; in particular, classes can be usefully defined with no slots but with a nontrivial prototype.

Defining a new class gives you immediately two important tools. You can define methods for S functions that will be applied automatically when the argument in a call to that function matches the new class. You can generate and use objects from the class, knowing from the definition the kind of data the object contains.

Beyond the standard definition of classes, S provides a number of other tools. Classes may be *virtual* if no objects are actually to be generated from the class. Virtual classes are a powerful technique: they capture properties that otherwise distinct classes may have in common.

A class in S may *extend* another class, by adding some information to that of the other class. Extensions are usually implicit in the representation, but may be set explicitly. The relation may be conditional, with a defined method to test the condition. The relation may also require an explicit coercion. Extension is a strong statement about the two classes and should be used only when it really makes sense. A similar relation between the two classes can be defined with an `as` relation, with the difference that programmers have to coerce the object to the class explicitly. Section 7.2 discusses extensions and coercion.

The definition of a class, including all `extends` relations, is stored in the metadata of an S database. A variety of S utilities examine the definitions and use them for method selection and other computations.

7.1.1 Defining and Editing a New Class

A class of S objects is defined by a call to the function `setClass`. This function, like most S functions named "set*Something*", has the side effect of setting some information in an S database that, in this case, defines the new class. You can create a new class by calling `setClass`; after that, the function `dumpClass` will dump out the current version of the class to a file. After you edit the file to make any modifications wanted to the class, running `source` on the file will recreate the class with the modified definition.

The procedure for managing a class that you've created is much like that for managing a function, but slightly more complicated for several reasons. First, the actual object stored in your chapter to define the class is created by the S evaluator, using more information than just the call to `setClass`, since nearly all classes depend on other classes for their definition. You don't assign this object directly (at least, I hope not); instead, each modification of the class involves calling `setClass`, and perhaps some other functions as well. Dumping the class via `dumpClass` writes all the relevant calls to the output file. A second reason that class management is more complicated is that other S objects *depend* on the class definition; namely, any object created from the class after it is defined. Changing the class definition raises some issues about interpreting objects created from the earlier definition. For this reason, S provides *version control* as an option for classes. It's better not to worry about that just yet, while we're starting in with class definition. Section 7.4 will deal with updating and version definition.

The first argument to `setClass` is the character string name of the new class. Other arguments provide the representation of the class and/or a prototype. To define the class `track`:

```
setClass("track",
  representation(x = "numeric", y = "numeric"))
```

Evaluating this expression creates class `"track"` in the working data; in particular, methods can now be defined with `"track"` in the signature. The `representation` argument specifies the structure of objects from the new class. In this case, the new class has two slots, both numeric. *Any* object belonging to this class is guaranteed to have these slots and the slots will be of class `numeric`.

A third argument to `setClass`, `prototype`, specifies the contents of a newly created object from the class. When it is omitted, the prototype is a list of the default objects from the corresponding classes. The prototype for `"track"` is a list of two numeric objects, each of length 0. If we didn't want this as a prototype, we could supply another object, typically a list of length two whose elements could be interpreted as valid objects of class `numeric`. You can provide a prototype but *no* slots: this is a useful technique to define a simple class that uses objects in a special way. For example,

```
setClass("sequence", prototype=numeric(3))
```

says that objects of class `"sequence"` will be created from a numeric vector of length 3. But we want to treat these objects specially, and *not* as a numeric vector.

The arguments to `setClass` other than `name` are all optional. If only the name is provided, the new class is a *virtual* class:

```
setClass("database")
```

Objects are not going to be created from this class, but it can appear in signatures for methods. Virtual classes are a powerful and useful concept; they are discussed in section 7.1.5 on page 292. More usually, however, the class object specification includes either the *representation* of the class or a *prototype* of what an object from the class should be, or both.

There are a few additional arguments to `setClass` that are less frequently used. The information defining the class is stored internally in the S chapter to which the `setClass` applies, the working data by default. The `where=` argument to `setClass` can specify any chapter, if you are allowed to assign objects there. The `validity=` argument supplies a function for validity checking, discussed in section 7.1.6.

The call to `setClass` exists for its side effect of storing the definition of the new class for S to use in future computations. The `setClass` call does return an object, though, of class `"CLASS"`.

```
> Track = setClass("track",
+ representation(x="numeric", y="numeric"))
> Track
[1] "track"
> class(Track)
[1] "CLASS"
```

The `CLASS` object looks like a character string and, like that character string, can be supplied as an argument where a class is wanted, such as when specifying methods or `is` relations. For most purposes, we can ignore the distinction between the class object and the the class name—the character string supplied to `setClass`. Very occasionally, the context may not make it clear whether a class name or some other use of a character string is wanted; in such cases, the class object must be supplied, but an expression of the form

```
as("track", "CLASS")
```

will do the job.

7.1.2 Representations and Extensions

The *representation* of a class defines how objects belonging to the class are constructed. The function `getClass` returns the representation of an existing class, while the function `representation` constructs a new class representation in terms of named *slots*, each of which is defined to be an object of some other class.

```
representation(nodeData = "vector", parents = "integer")
```

The value of the call to either `getClass` or `representation` is an object of class `classRepresentation`, which will be stored as part of the metadata definition of the new class (see section 7.6). A `show` method for this class describes the slots and the extensions of the class.

```
> getClass("file")

Slots:
 description  pars        mode
 "character" "raw" "character"

Nonstandard prototype for slots description, pars, mode

Extends:
Class "connection" by direct inclusion
```

Optional arguments to `getClass` let you get back only the information from the metadata object, perhaps only on a particular library. See page 305 and the detailed documentation.

The representation for a new class can *extend* an existing class; that is, the new class has all the properties of the old class, and can be used wherever an object from the old class is required. If the new class has the same representation as the old, just give the name of the extended class as the second argument to `setClass`:

```
setClass("regularExpression", "string")
```

The class `regularExpression` extends the class `string`. All the methods for the old class apply to an object from the new class. By extending `string`, we imply that class `regularExpression` has additional properties, likely to be expressed as methods that apply to `regularExpression` but *not* to `string`. We'll examine on page 285 whether this is, in fact, what we *want*.

A new class will often extend an existing class *and* have some additional data as well, expressed as slots not present in the old class. The extended class then appears as an *unnamed* argument in the call to `representation`. For example:

```
setClass("array",
    representation("structure",
         .Dim = "integer", .Dimnames = "list or NULL"))
```

The class array extends the class structure (it has all the slots in the representation of structure), plus two explicitly named slots. The technique of extending one class to another is very powerful but needs to be used carefully. Discussions of this important (and difficult) design issue recur throughout this and other chapters.

The simpler and more general forms are in fact doing the same thing, constructing the representation of the old class, maybe with some added slots, and using it as the representation for the new class. The definition

```
setClass("matrix", "array")
```

is only syntactic sugar for

```
setClass("matrix", representation("array"))
```

The extended class does not need to have any slots in its representation:

```
setClass("measured",
  representation("numeric", fuzz = "numeric"))
```

The implementation is slightly different in this case; section 7.1.7 discusses the details.

Because the matrix class extends class array, wherever an object of class array is needed, S will automatically treat an object of class matrix as equivalent. This does not depend on the representations being identical: if the new class has additional slots, S will extract the portion of the object corresponding to the old class. In fact, a class can extend another class without their representations being related at all. The extended class may be a *virtual* class, for which the representation need not be defined at all. For example, the various vector classes such as numeric all extend class vector. There is no implication for their representation, but the relation is valuable because it allows us to write methods for class vector that apply to all its extensions.

Even for non-virtual classes, new classes can be extensions without any relation between their representations. In this case, the new class is defined with an explicit method for coercing an object from the new class to the old class. For example, class string extends the elementary vector class character even though their representations are completely different. A method is defined, by a call to setIs, to turn string objects into character objects (see section 7.2).

When a class extends another class, a call to as on the left of a replacement is often a convenient way to insert the extended class's contribution to the object—it can be thought of as shorthand for copying all the slots from the included object into the new object. If `z` is from a class that extends `matrix`, and `x` is a matrix,

```
as(z, "matrix") = x
```

will insert into `z` all the matrix slots from `x`.

Such `extends` relations are an extremely important and powerful programming technique. They economize, sometimes dramatically, on the effort of writing methods for new classes. There is no need to rewrite a method for general multiway arrays to make it apply to matrices: the `extends` relation extends the method automatically. Defining new classes as extensions can also help to keep the definition of the existing class from becoming too complicated. When objects from a class may have optional structure, consider whether it might be cleaner to define a new class for the objects with the structure.

As with most powerful programming techniques, though, class extension needs to make sense when it is used. The assertion is a powerful one: for example, we are saying that *whatever* we want to do with `character` objects, we can do that with `string` objects equally meaningfully. In that example the assertion is pretty obvious: `string` objects are designed to be an alternative way of representing character string data, with some additional properties and computational advantages. For a more ambiguous example, consider two alternative definitions of the class `regularExpression`:

```
setClass("regularExpression", "string") # or ...
setClass("regularExpression", prototype = string())
```

We want to define regular expressions to be strings in which some characters have special meaning when we want to match character-string patterns (see section 4.4.3).

The first definition causes all string computations to apply automatically to `regularExpression`; with the second, either special methods must be applied for string-like operations (for example, `substring` extraction or replacement), or the `regularExpression` object must be turned into a string by a call to `as`. The second definition is less convenient, but safer. Regular expressions treat some characters specially; therefore, some operations on character strings could produce invalid regular expressions. There is no magic formula: in such cases we are trading off convenience against unintended results. If we wanted the convenience of extensions, specifying a validity method is often an important safeguard.

One particularly controversial aspect of class extensions concerns *multiple* extensions. S allows you to define a class to extend more than one existing class. All that's needed is to supply more than one unnamed argument to `representation`. Multiple extension is a technique that should be used only when you are very sure it makes sense. There are a variety of possible problems. Intuitively, the behavior of an object from the new class may be ambiguous, since we have asserted that an object from it can be regarded as belonging to either of two different classes, neither of these extending the other. Different object-oriented programming systems take different approaches to this issue.

S allows multiple extensions, through both direct inclusion and the use of `setIs`. Direct inclusion in restricted to unambiguous cases. Slot names must be unique and only one non-virtual class without slots can be extended directly. But unless your new class clearly merges the properties of two other, largely unrelated kinds of object, simplicity and clarity usually suggest avoiding multiple inheritance.

Each slot in a representation should be a previously defined class, but it is legal to give any name as a slot class. If the class is not currently defined, S will generate a `NULL` object in the corresponding prototype, and will issue a warning from `setClass`. In particular, there is no absolute prohibition against an object having as a slot another object of the same class. In most applications, such *recursive* class definitions are not a good idea. The conceptual model for S classes is to build up information from simple to more complex. Including a class in itself suggests an incremental view of the object, rather than the whole-object perspective that works best.

For example, one could define a *tree* structure in a variety of ways. We looked at trees informally on page 208 in section 5.1.5. The essential idea is that the tree consists of some number of nodes, each node having a single parent node and some node-specific data. A natural representation for general tree objects could be something like

```
setClass("generalTree",
    representation(nodeData = "vector", parents = "integer"))
```

The `parents` slot defines the parent node of each node as an index (with 0 for the top or root node) and `nodeData` is a parallel vector of the node data. Both slots are required to have length n, the number of nodes in the tree.

However, if you learned about tree structures while programming in other languages, you might be tempted to define a tree structure recursively, in terms of the node data and the children. In a binary tree, the representation might be

```
## a bad example!
setClass("binaryTree",
    representation(nodeData = "vector",
        left = "binaryTree", right = "binaryTree"))
```

Here `left` and `right` are the two sub-trees. This could be sensible in other situations, but it's a *very* bad idea here. The whole-object view of the tree is essentially impossible; all computations have to be expressed recursively, working on one node and then recursively on the children. One additional, insidious problem would be to distinguish nodes that had either zero or one subtree, not two. We would need some concept of a `NULL` tree; perhaps by introducing the notion of a "vector of nodes" of variable length. But all this comes from not thinking in whole-object terms, so let's abandon it.

Suppose we did want to specialize the `generalTree` class to binary trees? One solution would be to stick with the `left` and `right` slots, but treat them as indices, like the `parent` slot in `generalTree` but being either `0` or the index of the corresponding child node. A slightly cleaner alternative is to simply use the representation for general trees unchanged:

```
setClass("binaryTree", "generalTree")
```

Not all general trees are binary trees, so this solution requires writing a validity method for class `binaryTree`; see section 7.1.6. If we can do that sensibly, we get the advantage of using general tree methods on binary trees, only having to write binary tree methods when there is a reason to do so.

7.1.3 Prototypes; New Objects

Once you define a new class, you and users of your software need to be able to generate objects from the class. The usual and best approach is to write a generating function, which we will discuss in section 7.3. Generating functions allow your users to concentrate on the meaning of the class, rather than the details of its implementation. You can also provide more defaults and better checking of the data provided.

For simple classes, some adequate tools are provided automatically by S from the class definition, including the class prototype and the general function `new`, both of which we discuss in the present section. These tools can be used directly when you're first designing a class and will then likely become the basis for writing generating functions.

All classes have a prototype, stored in the metadata defining the class. If the class has slots, the representation defines a default prototype: each slot of the prototype contains a new object from the class specified for that slot in the representation. Alternatively, an explicit prototype may be provided as the `prototype=` argument to `setClass`. You will need to supply a prototype when the default object in a slot needs to be something other than the default object of the corresponding class, or when the slot is specified as a virtual class. The default object from any class is the value of `new` given only the class name:

```
> new("numeric")
numeric(0)
```

A prototype can also be specified explicitly after the class has been defined:

```
setClassPrototype(Class, object)
```

stores `object` as the prototype for the given class. The `object` must be interpretable as a valid object from the class; prototypes themselves never have `Class` as their class, but `setClassPrototype` will unclass the object if you want to provide a previously generated object from the class as the new prototype.

Classes may be defined without explicit slots by supplying the prototype but not the representation:

```
setClass("sequence", prototype = numeric(3))
```

As with classes having slots, a new object of class `sequence` can be created as a copy of the prototype. Methods using the class will interpret the object in special ways, but not by having named slots. The `sequence` example consists of three numbers to be interpreted as defining a sequence of equally-spaced numeric values. We do *not* want a `sequence` object to be interpreted as a numeric vector, because the three numeric values have special meaning (the start, end and step value of the sequence). Doing arbitrary arithmetic on the object, for example, would be disastrous.

The definition of sequence above is therefore what we want, not:

```
setClass("sequence", "numeric") # NO!
```

The use of prototypes without representations allows the class designer to limit the legal computations on objects made up of numeric data (or anything else).

A new object from any class can be generated by calling the function `new`. The first argument to `new` is the name of the class; other arguments can be supplied to specify the data you want in the new object. With only the class supplied, `new` returns a copy of the prototype for that class; for example, `new("matrix")` returns the prototype of class `"matrix"`, which need not be the same as a call to `matrix` with no arguments.

```
> new("matrix")
NULL matrix: 0 rows, 0 columns.
> matrix()
     [,1]
[1,]   NA
```

If additional, named arguments are supplied to `new`, they are interpreted as values for the slots with the corresponding names. The class `track` defined on page 281 had numeric slots `x` and `y`, so we can generate an object with a call such as:

```
new("track", x = 1:length(newData), y = newData)
```

If the arguments to `new` are *not* named, they are assumed to provide some data to be coerced to the specified class. Assuming that class `sequence` is represented by three integers (start, end, and step values), the expression below will generate an object representing the sequence $1, 11, \cdots, 51$:

```
new("sequence", c(1, 51, 10))
```

If the new class extends another class, an object from the extended class can be supplied, along with optional named arguments for any additional slots.

When writing a generating function, you can call `new` and then assign the individual slots with the appropriate information, for classes defined by slots. Alternatively, you can construct an object that has the right information in it, and then assign the object's class or coerce the object to the desired class:

```
object = as(object, "myClass") # or ...
class(object) = "myClass"
```

For this to work, `object` must either be suitable as a prototype for `myClass` or belong to a class that can be coerced to `myClass`.

The opposite operation is also useful: going from an object in a particular class to something that looks like the prototype for that class. The function `unclass` is defined to return its argument, but with class set to the class of the prototype. If `seq1` is an object of class `sequence`:

```
unclass(seq1)
```

is the corresponding numeric object. Unclassing an object is the way to compute with objects that don't have slots. You can replace an object's class by that of the prototype by setting the class to NULL. The following turns seq1 into a numeric object with the same contents:

```
class(seq1) = NULL
```

This idiom is largely retained for back-compatibility with earlier versions of S, but as long as you realize that no object actually has NULL as its class, it's a useful shorthand.

7.1.4 Computations with Slots

The representation of a class defines how data can be extracted or replaced for objects belonging to the class, by using the slot names or by extracting the objects of other classes included in the representation. The operator "@" followed by a constant name or string extracts the corresponding slot:

```
track1 @ y
```

extracts the slot named "y" in the representation. An error results if "y" is not a slot in the representation of track1. Slots can be replaced by putting the same expression on the left of an assignment. Use the function slot if the name of the slot is not a constant.

```
for(what in slotNames(x)) {
    ## print out all the slots of 'x'
    cat("\n", what, "\n"); dput(slot(x, what))
}
```

In the call to slot, the second argument can be anything that evaluates to a slot name for the corresponding class. Both "@" and slot can also appear on the left of an assignment to replace the slot in the object.

```
track1 @ y = log(track1 @ y)
```

The replacement value has to be coercible to the class of the slot.

For users of earlier versions of S, the slot operations may look familiar. Slots tend to resemble named components of a list and the "@" operator resembles in appearance the "$" operator:

```
fuel.frame$Type
```

The resemblance is not coincidental, but the two ideas are very different. Named components of lists were used as an informal approach to classes of objects (in the 1988 book, [1], that is often the meaning of "class"). With the availability of actual class definition, however, named lists should not be used as a substitute. The class `named` is still very useful, but not for defining classes. It provides objects, rather like databases, in which elements are associated with arbitrary character string names.

Component names can only be found by looking in each individual object and no information is available about what the component should contain. Slot computations work from the definition of the class, and slot names are not inserted in the object. Slot names must match those in the definition exactly, whereas named components are extracted with name completion. Slot computations can be more efficient, particularly if the class definition is read into C code. But the most important difference is *validity*: class and slot computations use and maintain valid objects. Much is done automatically that with a more informal approach would be done by hand (or, more likely, not at all). The `"$"` operator should be thought of as a convenience to save typing when you, as a user, know the names of some elements in a named object, and not as an alternative approach to classes.

To see all the slots of a class, use `getSlots`, which returns the classes of the slots, named by the slot names. For the slot names alone, use `slotNames`.

```
> getSlots("file") # names, classes of slots
 description  pars        mode
 "character" "raw" "character"
> getSlots("character")
character(0)
> slotNames("file") # names of slots
[1] "description" "pars"        "mode"
```

When given a single string, as here, these functions interpret their argument as the name of a class. Otherwise, the slots of the object passed in as an argument are used. So the first example above returns the same result as `getSlots(file())`.

The function `hasSlot` can be used to query the existence of a slot of a given name; however, in most cases being uncertain about an object's slot names suggests confusion in the program design.

For representations that include other classes, the slot names are the union of the included slots and slots specified directly. As noted, inclusion also implies methods to coerce an object from the new class to any of the included classes. For example, if `tx` is an object of a class that extends `matrix`,

```
as(tx, "matrix")
```

extracts the portion of its representation forming the included `matrix` object. The same operation on the left of an assignment replaces those slots.

```
as(tx, "matrix") = xNew
```

7.1.5 Virtual Classes

The concept of a *virtual* class provides a flexible way to associate several actual classes that share some properties but may not share a representation. Usually, the virtual class itself may or may not have a specified representation; in either case we don't expect to encounter any objects from the class. Actual classes will be defined that extend the virtual class.

Surprisingly, this vague-sounding concept is powerful and useful, with at least three important applications.

1. Methods can be defined that use the virtual class in their signature. These methods then apply to any of the actual classes that extend the virtual class.

2. The name of a virtual class can be supplied as a slot in the definition of another class. Any of the extending actual classes is a candidate for this slot, when an object is created from the new class.

3. A virtual class can form part of the representation of other classes that extend it. Methods for the virtual class use the shared representation but take and return objects from the actual classes.

These and other applications share the common flavor that makes virtual classes useful: they group together some actual classes to express behavior these classes have in common, but without tying down the representation of the actual classes in any way.

S comes with a number of virtual classes and you are encouraged to consider defining your own when appropriate. Many large-scale programming projects benefit from defining virtual classes to capture some essential aspects of the data they use. Of the existing classes, the most commonly encountered are `vector`, the class of all objects that behave like S vectors, and `structure`, the class of objects that add structure to vectors. Class `vector` has no representation; it is defined essentially by the methods defined for it. Class structure has a representation, with one slot to hold the vector that the structure object contains; structure methods use and modify that slot. Section 5.1 discusses both these classes as they are provided with S. Section 7.1.7 discusses defining new structures and structure-like classes.

Vector classes share the ability to be treated as S vectors; that is, they all have a concept that they contain length elements and that these elements can be extracted or replaced by indexing their positions with positive integers. Asserting that an actual class extends vector is to assert that these operations are available for the actual class.

Once that assertion applies, methods can be written for class vector using functions that only depend on these common features. On page 34, we defined a method for vectors to implement our one-line summary function:

```
setMethod("whatis", "vector",
 function(object) paste(class(object), "vector of length",
   length(object)))
```

This only assumes that length is meaningful for this object. Notice also that we included the object's actual class in the summary; the actual class will *not* be "vector".

A virtual class can be included in the representation of any of the actual classes that extend it. The class string is a vector class and would be defined as follows.

```
setClass("string",
representation("vector", offsets = "integer", text = "raw",
  stringTable = "indexLookup", levelsTable = "levelsLookup"
  )
)
```

The inclusion of "vector" as an extended class has no effect on the representation of the string class, because vector has no slots. But it is a crucial part of the definition of class string, because it means that vector methods will apply to string objects.

Extending a virtual class by an actual one is slightly special for two reasons. First, the coercion, say from string to vector, does nothing at all, not even changing the class of the object. The method selection computations, for example, make sure not to change the class of a string object before calling a vector method.

Second, the extends relation is two-way. An implication of an extends relation from a non-virtual class to a virtual class is that a conditional extends relation applies in the opposite direction. For example, when an object is asserted to have an is relation to class vector, it will actually belong to one of the non-virtual classes such as numeric or string. Whether this is the case, obviously, can only be determined by testing the actual class of the object. This relationship is handled automatically in method selection and elsewhere.

As an example of the use of virtual classes for slots, look at the representation of the class array.

```
> getClass("array", complete = F)
An object of class "classRepresentation"

Slots:
     .Data       .Dim       .Dimnames
  "vector" "integer" "list or NULL"

Extends:
Class "structure" by direct inclusion
Class "matrix" by method for test
```

(The complete=F restricts the extended classes to those explicitly defined in the representation of array).

Notice the class for the .Dimnames slot: "list or NULL". You might guess (correctly) that this is a virtual class that is extended by both list and NULL. The goal in this example was to accommodate a large body of existing code that sometimes used a list and sometimes a NULL object for this information. Specifying the slot as this virtual class allows either, but prevents some other meaningless class to be used for the slot.

Creating virtual classes is easy. Any class that has no representation is a virtual class automatically. The "list or NULL" class can be specified in three steps:

```
setClass("list or NULL")
setIs("list", "list or NULL")
setIs("NULL", "list or NULL")
```

The first expression defines the new virtual class, and the other two enumerate the classes that extend it. We'll need to say a little more about setIs used with virtual classes, but for nearly all applications the obvious way of saying things works as you would expect.

Virtual classes often have no representation, but they can if they want to. Classes can be specified to be virtual by including in their representation the special class VIRTUAL. Class structure, for example is defined as follows:

```
setClass("structure",
    representation(.Data = "vector", VIRTUAL))
```

Other classes that extend structure by direct inclusion will then all have a `.Data` slot, plus whatever other slots are wanted.

Class `structure` and other virtual classes with a representation illustrate a key property of virtual classes. A method defined for class `structure` does not strip away the additional slots of the actual object it receives as an argument. Computations use the `.Data` slot and often return an object with a modified `.Data` slot, but the class of the object usually stays whatever it was (`matrix`, `named`, and so on). In contrast, a method for a *non-virtual* class must coerce the object it gets to that class. Except by some special programming, it cannot use or return the un-coerced object. Virtual classes, therefore, are a key tool when we want to manipulate part of the representation of objects but leave the rest alone. By using virtual classes in this way, we as designers *assert* that this sort of computation makes sense.

7.1.6 Validity-Checking Methods

Objects are checked for validity when permanently assigned, or explicitly by the function `validObject`. For a class with slots, this at least requires that the object's slots agree with the representation. The class representation, though, may only be part of the implicit requirements for a valid object from a class.

An array, for example, must satisfy the requirements that each element of the `.Dim` slot be positive and that the product of the elements equal the length of the data. Those elements of the `.Dimnames` slot that are not empty have to be of length equal to the corresponding `.Dim` element.

The call to `setClass` to define the class can include a `validity` argument; that is, a function that tests for validity. A validity function can also be specified later by a call to `setValidity`. The function will be called with a single argument, the candidate object from the class.

Validity functions return either a logical `TRUE` or a character string message describing the problem found. S uses the functions to prevent permanent assignment of an invalid object. Specifying a function to check validity of an `array` object, for example, can be done by:

```
setValidity("array", validArray)
```

We would have to own the definition of class `array` to be allowed to set this, but let's assume we do. One definition of `validArray` might be:

```
validArray =
function(object)
{
    d = object@.Dim
    if(!all(d >=0))
        return("Negative elements in dim")
    else if(prod(d)!=length(object@.Data))
        return("Length of data != product of dim")
    dn = object@.Dimnames
    if(length(dn) >0 && length(dn)!=length(d))
        return("Bad dimnames length")
    else {
        dnl = sapply(dn, length)
        if(any(dnl>0 & dnl!=d))
          return("Elements of dimnames of wrong length")
    }
    return(TRUE)
}
```

Notice that the validity method is defined at a low level, referring to the slot `.Dim` rather than using the function `dim`. When we write code for general computations, we usually are better off with the functional form, since our computations then often adapt to other classes or circumstances. But in writing validity functions we specifically do *not* want the computations to adapt; the validity function wants to be explicit about what it expects.

With this validity method in place, suppose we generate an invalid array object:

```
> x = new("array", .Dim = c(2, 2), .Data = 1:5)
Warning messages:
  Invalid object of class "array": Length of data != product
        of dim (will replace with unclass(object))
```

The S evaluator called the validity method, just before committing the permanent assignment. Getting back a character string message, it reported the problem, but only as a warning. In order not to lose all the data (the assignment might have been the last step in a long computation), S goes ahead with the assignment but removes the class of the object to avoid problems when the object is used later on.

For classes including other classes, validity checking is handled from the inside out. For example, given that the representation of `matrix` includes `array`, the validity check,

```
validMatrix = function(object)
{
    if(length(dim(object)) != 2)
        return("Must be two-dimensional")
    TRUE
}
```

can be written assuming that the contained object classes are already validated.

Explicit validity methods are only needed for nontrivial tests. Checking that the number and class of slots in a particular object conform to the specified representation of the class is carried out automatically once validity checking is invoked. Such checks will have been performed before any explicit validity methods are called.

Validity checks are applied to virtual classes when these classes are contained in non-virtual classes. For example, the class of structures asserts that the data slot is a vector. This assertion will be part of a validity check for any class that contains `structure`.

The question of *when* to check the validity of an object always arises in any language that supports classes. Because validity checks can be relatively expensive, validity checking will not be done automatically during evaluation; normally, any object is accepted to be a valid member of its class. Validity checking is provided at certain special times, mainly when making a permanent assignment. The notion is that allowing an invalid object to persist is bad enough to spend some effort on checking. For example, the following little function tries to create a square matrix from a vector, but forgets to adjust the length of the data slot and so returns an invalid array when given a vector whose length is not the square of an integer.

```
squareMatrix =
function(x)
{
    n = floor(sqrt(length(x)))
    new("array",
        .Data = x,
        .Dim = c(n, n))
}
```

If we assign the result of this function validity checking will be called, but if the result is passed on to another function, the invalid object will not be detected.

```
> diag(squareMatrix(1:4))
[1] 1 4
> diag(squareMatrix(1:5))
[1] 1 4
```

Since the object returned by `squareMatrix` is only assigned in the frame of the call to `diag`, no validity check is made. The result then just depends on what subsequent computations do (in this case `diag` just ignores the actual length of the `.Data` slot).

Temporary assignments of invalid objects (in the frames of function calls) can certainly cause problems, but there are so many of them in a typical S task that checking them all would be very time-consuming. To produce such general validity checking, turn on the `check` option:

```
options(check = TRUE)
```

The `check` option causes automatic validity checking at various points during evaluation, and a great deal of other extra computation as well. It can slow evaluation down by an order of magnitude or two. However, it does pick up subtle problems.

```
> options(check=T)
> diag(squareMatrix(1:5))
Problem in new("array", .Data = x, .Dim = ..: bad object
    returned from .Call to "new_object": Length of data
    != product of dim (check option turned off)
```

The option can be essential for really tough debugging problems, but is overkill for most situations.

Validity checking can also be invoked explicitly. To force a validity check, call the function `validObject`:

```
validObject(object)
```

Called with one argument, `validObject` will generate an error if there is some internal problem in `object`, that is, if `object` fails either an explicit or implicit validity check for its class. Including the argument `test=T` causes `validObject` to return either `TRUE` or a character vector describing the problem (see page 149 for discussion of `identical` and `all.equal`, functions that may be helpful in using `validObject`).

7.1.7 Structures and Structure-like Classes

Structure objects provide a good example of the use of class definitions, even though structures in S have been around much longer than the current approach to classes. This section discusses the definition and use of structure classes, as well as some non-structure classes that use a similar technique to extend vector classes.

The `structure` class forms the basis for all the classes such as arrays, matrices or time-series, implementing the concept of structured data in S. The concept is that the data values in the object are arranged somehow, in a *structure*, and that this structure is unrelated to the data values themselves. The structure may lay the values out on some spatial or temporal dimension, as in time-series; or it may imply some kind of indexing, as in multiway arrays. The class is defined by a mapping from the vector-style indexing, `1` to `length(x)`, onto this other structure. Since it is the indexing that is being mapped, the nature of the structure remains the same whether the data values are numbers, character strings, or other S objects. But any computation that rearranges the data, such as taking a subset or sorting, breaks the mapping to the structure. The result of any such computation should be the data part, with all structure dropped.

The fundamental structure concept then consists of three points:

1. the data can be *any* vector object;
2. element-wise transformations of the data do not alter the information in the other slots;
3. any subsetting or rearrangement of the data throws away the structure information.

The `structure` class is implemented as a virtual class with a representation:

```
setClass("structure",
    representation(.Data = "vector", VIRTUAL))
```

Classes that extend `structure` will then contain a data slot, which `structure` methods manipulate independently of the other slots. Non-virtual classes, such as `array`, are defined to include `structure` and contain some other slots that define the structural properties of the object. These other slots implement the concept of structure attributes in earlier versions of S; see, for example, Chapter 5 of reference [1]. Specifying `structure` to be a virtual class discourages creating an object belonging to class `structure`, rather than to one of these specific classes. For back compatibility, the function `structure` will still generate such objects, but you should not need them.

The class `array` is defined as follows.

```
setClass("array",
    representation("structure",
        .Dim = "integer", .Dimnames = "list or NULL"))
```

This defines `array` objects as structures with two slots of the name and class shown, plus the `.Data` slot.

```
> getSlots("array")
    .Data       .Dim       .Dimnames
 "vector" "integer" "list or NULL"
```

In defining new classes, you may consider making the class an extension of `structure`. If the fundamental concept applies to the new class, structures are a powerful way to inherit a number of methods. However, the concept does not always apply.

Avoid using `structure` for classes that share the idea of adding some information to a vector, but not the fundamental structure concept. For an example, suppose we want to add to some numeric data a measure of the precision of the data; that is, of the observational uncertainty in recording the data. Defining and using such uncertainty is very challenging, but it is also an important part of computing with data, and distinctly neglected in practice. At any rate, passing over these details, suppose we are prepared to assign a numeric value, `fuzz`, to the uncertainty, and use this in later computations to guide numerical summaries, visualization, or models. Let's define a class `measured` to represent such data.

Measured data clearly extends class `numeric`, but we would be wrong to rush in with a `structure` class. Indeed, `measured` objects violate all three of the points in the structure concept. The data need to be numeric, rather than an arbitrary vector; transforming the data either transforms the `fuzz` or throws it away; and rearranging the data probably leaves the `fuzz` unchanged.

We can still get much of the convenience of the `structure` approach, however, by defining class `measured` in the natural way, as an extension of `numeric`.

```
setClass("measured",
  representation("numeric", fuzz = "numeric"))
```

The numeric data being measured enter, not as a slot, but as an extension of class `numeric`. All the numeric computations can be applied to a `measured` object, by transforming it to class `numeric`. This fits the meaning of the class: if we can't do anything more sensible, we can throw away the `fuzz` information and deal with just the numbers.

Let's look at what the class definition did.

```
> getClass("measured", complete=F)

Slots:
      .Data      fuzz
  "numeric" "numeric"

Extends:
Class "numeric" by explicit test and/or coercion
```

The `setClass` call constructed the new class much like a structure—with a `.Data` slot and with methods to convert a `measured` object to `numeric` by extracting or replacing that slot. We can then automatically apply any method for `numeric` data to a `measured` object: the method will be applied to the `.Data` slot. However, this slot is not class `vector` and `measured` does not extend `structure`. If it did extend `structure`, then the `fuzz` slot of a measured object, `xm`, would remain the same in `log(xm)`, which we certainly don't want. With the actual definition, `log` will use the vector method, not the structure method. The value of `log(xm)` will be of class `numeric`, a more faithful reflection of the concept behind the `measured` class. We could go on to define `measured` methods for those computations which we decide should retain the `fuzz` slot, perhaps suitably recomputed.

7.1.8 Classes with Fixed Definitions

Although most S classes can be edited in much the same way as functions or methods, the classes on library `main` are considered part of the language's definition. S will not allow the classes on the main S library to be redefined except by someone who is editing that library. An inconsistent definition of one of the classes S depends on for its own operations is one of the few ways to kill the language completely. For the names of those classes, look at `getClasses("main")`. There are an alarming number of them, but in fact nearly all are old-style classes or the basic vector classes.

For classes other than those on the main library, S applies its usual rules of searching for objects. The first chapter in the search list containing a definition of the class is the one chosen. The S evaluator does *not* try to merge class definitions.

Since the explicit extensions of a class are stored in the class's definition, this must be done in the database that holds that definition. If you use `setIs` to redefine the *extensions* of a class on a library, you won't be allowed to, because the library is attached read-only.

```
> setIs("list", "tree",
+   coerce = listToTree)
Problem in setIs("list", "tree"): can't set the 'is'
    relation:no write permission on database main
```

For a class, such as `list`, defined on library `"main"`, S prohibits redefining the class. If you really do want your own version of class `list`, you can get very close, simply by defining a new class with the same representation:

```
setClass("myList", "list")
```

Admittedly, you still can't convert a `list` object to `myList` without an explicit call to `as`, but that is the intention of protecting the classes on library `main`.

For classes other than those in the main library, you can develop your own version *in place of* the current version in some database or library. You just need to define the class on your own chapter, and attach that chapter ahead of the library holding the standard version. The simplest approach to create the new definition is to call `dumpClass` and then insert a `where=` argument into the `setClass` call and any `setIs` calls in the dump file.

This restriction does not prevent extensions of virtual classes, such as

```
setIs("quaternion", "vector")
```

This relation can be stored in the definition of either class; the evaluator will infer the relation in both directions. You only need to own the definition of either the actual or the virtual class to create the relation.

7.2 Relations Between Classes

Much controversy about object-oriented programming systems concerns interclass relationships. What are the natural concepts relating classes and what are the implications for defining methods? Statements such as $\mathcal{B}$ *inherits from* $\mathcal{A}$ or $\mathcal{B}$ *extends* $\mathcal{A}$ are often made when a class $\mathcal{B}$ in some sense is a derivation of another class $\mathcal{A}$. Class $\mathcal{B}$ typically contains all the information in $\mathcal{A}$, and adds some additional information as well. Class relations are important, but sometimes ambiguous or confusing. Partly, the confusion reflects difficult design decisions; choices often balance convenience against unintended consequences. In some languages, the relations are only implicit, defined by the source code but not accessible.

S provides an explicit and controllable mechanism for defining inheritance, the *extension* relation between two classes, or more precisely, the *is* relation between an object and a class. We say a class $\mathcal{B}$ extends a class $\mathcal{A}$ if any computation applicable to class $\mathcal{A}$—in particular any method for class $\mathcal{A}$—can also be applied with the object of class $\mathcal{B}$ substituted where the class $\mathcal{A}$ object was required. S provides both automatic and explicit ways to create such relations. The relations are generated automatically if class $\mathcal{B}$ is defined to include $\mathcal{A}$. In addition the programmer can explicitly generate (or prohibit) relations; this explicit control is the main topic of this section.

S includes the notion that some but not all objects in one class can be substituted for another class. To make this explicit, we talk about an `is` relation, this time between an object and a class. Is relations can represent conditional extensions. The S function `is` tests exactly this relationship. For example,

```
is(x, "matrix")
```

returns `TRUE` if object `x` can be substituted for an object of class `"matrix"` anywhere, and `FALSE` otherwise, based on the information S has about the object and class. Often, the relation applies for all objects from class $\mathcal{B}$, and the distinction between `extends` and `is` can be ignored.

For an example in which the distinction matters, if `x` is of class `array`, the relation is true if `x` is a two-way array, but not otherwise. The `extends` function, which takes two classes as arguments, cannot resolve conditional relations. The third argument to `extends` is `maybe`, the value you want S to return when the relation is conditionally true.

```
extends("array", "matrix", NA)
[1] NA
```

By default, `extends` returns `TRUE` for conditional relations.

7.2.1 Specifying `is` Relations

Substitution relations can be specified or prohibited directly, using an `is` specification, a call to the function `setIs`. As an example, consider matrices and general multiway arrays. Intuitively, any matrix `is` an array. Whatever array computations we do can be applied to the matrix. This information could be specified by an `is` specification:

```
setIs("matrix", "array")
```

This says that any object of class `matrix` may be substituted when an object of class `array` is wanted. Actually, by specifying the representation of `matrix` as we did on page 284, we already implied this relation:

```
setClass("matrix", "array")
```

says that the representation of a matrix contains that of an array—in this case, the two classes have the same representation. The class representation (section 7.1.2) implies extension if class $\mathcal{B}$ is contained in the representation of class $\mathcal{A}$. That is, a valid object of class $\mathcal{A}$ can be uniquely defined by extracting the appropriate slots from any object of class $\mathcal{B}$. Such relations are generated automatically when the class representation is specified by a call to `setClass`.

Extension in one direction may or may not mean extension in the other direction. In this example, an array is *sometimes* a matrix, but the relation must be tested. We can express this by an `is` specification as well, supplying a test function to `setIs`. In this example:

```
setIs("array", "matrix",
    test = function(object)length(dim(object))==2)
```

The `test=` argument should be a function of one argument (the object) that returns `TRUE` or `FALSE`. The `is` relation in this direction is not implied by the representation and therefore is not generated automatically, even in the case that the two representations are identical.

Here is a hint for writing test functions. These functions must *always* return a single logical value, so they must use only conditional expressions guaranteed to do so. For example, the test above is reliable because the `length` function always returns a single number for any object. Therefore, the comparison with 2 cannot, for example, evaluate to `NA`. Generally, testing with the `"=="` operator is not a good idea. A better test is the function `identical`. The test function

```
function(object) identical(length(dim(object)), 2)
```

would reliably return `TRUE` or `FALSE` regardless of the behavior of the `length` function. Casting test functions in terms of `identical` is good strategy.

Test functions are very like control expressions appearing in `if` or `while` expressions; see section 4.2.4 for a discussion of these.

Extension is transitive. The representation of `array` was defined to include `structure`, which automatically generates the relation:

```
extends("array", "structure")
```

Transitivity then implies that a `matrix` object can be substituted where an object from class `structure` was specified. The consequences of transitivity are worked out automatically by the analysis of class representations and `setIs` specifications. Direct extensions of a class are stored with the class definition itself. When the class definition is first needed by the S evaluator, the implications of direct extensions are worked out.

The indirect extensions of a class depend on what other classes the evaluator has encountered, and so on what libraries are currently attached. A call to `getClass` will show the currently known extensions. If you only want the direct information included with the definition of this class, supply the argument `complete=F` to `getClass`.

```
> getClass("matrix")

Slots:
     .Data       .Dim       .Dimnames
  "vector" "integer" "list or NULL"

Nonstandard prototype for slot .Dim

Extends:
Class "array" by direct inclusion
Class "structure" indirectly through class "array"
Class "vector" indirectly through class "structure"
Class "complex" indirectly through class "vector"
    and many more ...

> getClass("matrix", complete=F)

Slots:
     .Data       .Dim       .Dimnames
  "vector" "integer" "list or NULL"

Nonstandard prototype for slot .Dim

Extends:
```

```
Class "array" by direct inclusion
```

To get only the names of the extensions, use `extends("matrix")`. The order in which the extended classes appear is relevant: the direct inclusions come first, then relations needing to be tested or coerced, then indirect relations (see section 7.6 for how the information is computed and stored and section 4.9.6 for its use in choosing methods).

Transitivity plus conditional extension can lead to some unwanted extensions. If a class has mutually exclusive specializations, incorrect conditional extensions may appear to exist. Such extensions will always fail when tested, but waste time in identifying methods. For example, if someone decided to make three-way arrays a class, `threeWay`, this class would, like `matrix`, extend `array`. The specification of the class might look as follows:

```
setClass("threeWay", array)
setIs("array", "threeWay",
   test = function(object)length(dim(object))==3)
```

Because of the conditional extension of `array` for `matrix`, a conditional path appears to exist from `matrix` via `array` to `threeWay`. We know that a `threeWay` is never a `matrix`, but analysis of the specifications for the classes cannot automatically determine this. The `is` specification may be used to prohibit such extensions, by supplying the test argument as `FALSE`:

```
setIs("matrix", "threeWay", FALSE)
```

Whenever several mutually exclusive classes specialize the same class, and conditional extensions are allowed from the general class to one of the specialized classes, such explicit prohibitions may be needed.

You might worry that a similar problem arises in the very common situation that several actual classes extend a virtual class; for example, `numeric`, `string`, and so on extending `vector`. The situation seems formally the same: several classes specializing `vector`. Fortunately, because `vector` is a virtual class directly included in each of the actual classes, the S evaluator detects this case and does not generate superfluous tests when trying to find methods. For example:

```
> extends("string", "numeric")
[1] F
> extends("numeric", "string")
[1] F
```

No special action was needed to avoid spurious relations between `numeric` and `string` via `vector`.

Extension relations also may include a `coerce` function whose argument is the object and whose value is the corresponding object in the target class. Usually, the extension implied by an `is` relation is trivial to carry out. For example, if the target class representation is contained in the current class representation, the evaluator just notes which slots to copy (see page 317). A conversion function can be supplied in the specification for cases that need explicit coercion. Recall the class `sequence` defined by

```
setClass("sequence", prototype = numeric(3))
```

This class does not have an automatic `is` relation to class `numeric`; in fact, it is important in such cases that the prototype *not* be treated as an ordinary object of class `numeric`. Instead, an explicit `is` relation could be specified:

```
setIs("sequence", "numeric", coerce = expd.sequence)
```

says to use the function `expd.sequence` as a method to convert a `sequence` object where a `numeric` is wanted. The notion of class `sequence` is that of representing, say, the result of the expression `1:1000` by the parameters of the sequence, not the entire set of numbers. Then an object of class `sequence` can be substituted without ambiguity wherever a `numeric` is wanted, but to deliver the numeric we need to call an explicit conversion method.

7.2.2 Coercing: `as` Relations

Extensions convert objects between classes automatically, whenever needed for applying a method. An alternative is to provide a method for coercing an object to another class, which programmers can invoke explicitly but which does *not* happen automatically. The function `as` converts the object given as its first argument to the class given as its second.

```
labels = as(x, "character")
```

Explicit coercion is in a sense safer than automatic extension, in that it has fewer unexpected side effects. The choice between power and safety, as anywhere, must be made according to the needs of the application.

To define an `as` relation, use the function `setAs`. For example, there is no automatic extension *from* class `vector` *to* class `array`. To define an explicit way to coerce a vector to be a (one-dimensional) array:

```
setAs("vector", "array",
    function(object) array(object, length(object))
)
```

An arbitrary vector can *not* be substituted naturally wherever an array is wanted, but the coerce method allows an unambiguous transformation to an array when the programmer wants one.

The function supplied to `setAs` works like a `coerce=` argument to `setIs`: it should be a function of one argument, expecting an object of the `from` class and returning one from the `to` class.

Although you should rarely need to be aware of the mechanism, `as` relations are stored by mapping them into the standard S method selection mechanism. The `as` relation depends on a pair of classes, but the actual method is one for the function `coerce`, taking `from` and `to` objects as arguments.

7.3 Generating Objects from a Class

Once a new class has been defined, users need to generate objects from that class. This usually means a generating function with the same name as the class, to make it easy for users to remember. The generating function often takes arguments that specify the structure of the object; for example, the length of a vector, the dimensions of an array, etc. This function is completely up to the designer of the class: there can be any number of them, they can have any arguments that make sense, and the essential criterion is that they help users create objects from the class easily, and without errors. In this section we discuss tools for writing such functions as well as alternative ways to generate objects from a class.

For simple classes defined in terms of slots, the generating function often just assembles the data in the slots. The `track` class defined on page 281 has two slots, `x` representing the positions at which some measurements are made, and `y` for the measurements themselves. The natural arguments to the generating function are just `x` and `y`. So a trivial generator is:

```
"track"= function(x, y)
{
  new("track", x = x, y = y)
}
```

Notice that we did nothing to ensure that the arguments were actually numeric. The computation that sets the slot in the object will check that `x` is a numeric, in the S sense that it has class `numeric` or a class that extends numeric. For example,

```
tEqual = track(1:10, yVals)
```

works fine, since `1:10` has a class that extends `numeric` and will be coerced automatically. However:

```
> t1 = track(xVals, yVals)
Problem in value@x = x: Not a valid object of class
    "numeric"

Debug ? (y|n): y
Browsing in frame of track(xVals, yVals)
Local Variables: value, x

R> x
[1] "0.098" "0.368" "0.430" "0.720" "0.825"
R> class(x)
[1] "character"
R> as(x, "numeric")
[1] 0.098 0.368 0.430 0.720 0.825
```

The user forgot that `xVals` was a character vector containing the `x` positions.

This raises a general question: How hard should a generator function try to interpret the arguments appropriately? Should the function try to coerce the arguments to the necessary class? The change to the function is simple:

```
"track"= function(x, y)
{
  new("track",
    x = as(x, "numeric"),
    y = as(y, "numeric"))
}
```

Now `track(xVals, yVals)` will work fine. Which version is better? The answer really depends on the circumstances, such as whether serious errors by users will be caught or perpetuate themselves in apparently valid objects that contain wrong information. My general suggestion: define fairly permissive functions to create the objects and, if needed, define also a validity method to check the objects after they are created.

Although putting such checks *into* the generating function might seem attractive, it is usually not a good idea. The checks should be applied to any object from the class, and a validity method does that (see section 7.1.6

for the details). Since `track` objects represent measurements of values `y` at positions `x`, the two slots should have the same length. We will also require the `x` values to be monotone and prohibit `NA` values in `x`.

```
setValidity("track",
    function(object) {
        if(length(object@x) != length(object@y))
         return("bad lengths")
        if(any(is.na(object@x)))
         return("missing values in positions")
        d = diff(object@x)
        if(any(d < 0.) && !all(d <= 0.))
         return("positions not monotone")
        return(TRUE)
})
```

As in all validity and test methods, the function must be careful that all the tests will compute without errors. In general, `all` and `any` can cause problems, because they may return `NA` if they encounter `NA`'s. We anticipated the problem by checking for missing values first.

The example raises another general question: does the object have to contain some data, or is an empty object meaningful? For the `track` case, empty vectors in both slots are perfectly reasonable and quite likely to be useful if, for example, we want to build up the object as observations are recorded. It is *not* obvious, but in fact the validity method we wrote works fine when both slots have length zero. We could work this out from the behavior of the `any`, `is.na`, and `diff` functions, but in this and in most cases, it's easier to test the computation by doing it. In addition, we should build up a set of tests for our new class (section 6.2.3 discusses how). Several generator examples, including the zero-length case, should be included.

A possible extension to our `track` function might be a default for the tracking positions. If we want to treat ordinary numeric vectors as being "tracked" at `1`, `2`, etc., we could extend the generator by something like:

```
if(missing(y)) {
  y = x
  x = seq(length = length(x)
}
```

However, let's consider what we really want to say here. When the generator is called with only one argument, we intend to coerce that argument to be a `track` object. Rather than just doing that for the case that the one argument is numeric, a better technique is to state what we mean directly:

```
"track"= function(x, y)
{
  if(missing(y))
    return(as(x, "track"))
  value = new("track")
  value@x = as(x, "numeric")
  value@y = as(y, "numeric")
  value
}
```

Now any methods defined by a call to `setAs` will be picked up by the generating function. In particular, specifying a method for coercing a `numeric` object to class `track` will give the result we wanted.

```
setAs("numeric", "track",
  function(object)
     track(seq(length=length(object)), object)
)
```

7.4 Updating Classes; Version Management

Nothing is perfect. Just as you will need to revise functions or methods, the initial definition of a new class of objects is unlikely to be the last. There is no problem in *doing* this; just call `dumpClass`, edit the file this produces and source the result back to create the new definition.

However, problems can occur if S handles an object based on the wrong definition of that object's class. If objects still exist created from a previous definition when a class is redefined, they may not produce a valid object when accessed. S faces this problem more seriously than other object-oriented systems because it provides persistent storage for objects and also keeps the definition of a class itself as a persistent object. To avoid disaster, S tries to handle multiple versions of classes carefully.

Objects are verified against the class definition when they are read from or written to a permanent database. If they fail to verify, the object is replaced by an "unclassed" version, containing the same data but not having the asserted class. S provides tools for managing successive versions of a class. Using these, you can update objects from older versions to the current version of a class. The version/updating tools are the main topic of this section.

There is no practical way to find all the objects with a given class at the time its definition is updated. For one thing, the objects created from the old definition may be on any database, not just the database that owns the definition of the class. These databases need not be attached when the class definition is updated.

The version management tools maintain a version object for the class, stored in the metadata as is the class definition itself. The version object contains successive representations of the class and, most important, methods for converting from one version to the next. Version management is optional. You should seriously consider it for any class that is being used extensively. The conversion methods required by version management can often be generated automatically, depending on the way in which the updated version of the class differs. Not every update can be converted automatically; if not, the programmer needs to define the conversion method explicitly.

To commit the current version of a class to the version management object, call `setClassVersion`. As an example, suppose we decide to update the `track` class. This was originally defined on page 282 to contain two numeric slots, to track some observable data, `y`, at specified positions `x`. To save the current version before updating the class:

```
setClassVersion("track")
```

If no version management currently exists for class `track`, this will initialize the version management object with the current definition. Suppose the new version imagines that `y` can consist of several variables. If there are three variables to track, `y` should be three times as long as `x`. An additional slot, `variables`, is provided to store the character string names of the variables.

```
> setClass("track",
+ representation(x="numeric", y="numeric", variables = "string))
```

When we're ready to commit this new version, we repeat

```
setClassVersion("track")
```

After the second call, the version management for `track` contains three things: the previous class representation, the new representation, and a method for converting between them.

To see the versions of a class, use `getClassVersions`:

```
> getClassVersions("track")
----     2 versions  ----
```

```
Class "track", version 2 saved: Mon Apr 28 11:38:44 EDT 1997:
  (added slots: variables)

Slots:
         x         y variables
 "numeric" "numeric" "string"

----
Class "track", version 1 saved: Mon Apr 28 11:38:44 EDT 1997:
  (Initial Version)

Slots:
         x         y
 "numeric" "numeric"

----
```

To update the objects from a previous version, call `updateObjects`.

```
updateObjects("track")
```

By default, this looks for all the objects of class `track` in the database with a definition for the class. If these match the current definition (or no definition!) they are left alone. If they match a previous version, they will be updated by successive calls to the methods that are stored in the version object, and then re-assigned. (In particular, caution suggests dumping the objects before running the update, in case you don't like the way the conversion turns out.) There are arguments to `updateObjects` that control which objects and which database the update works on.

In generating the method to convert between versions of the class, S applies three heuristics to classes with slots:

1. a slot that was in the old representation but not in the new will be dropped;

2. a slot that is in the new but was not in the old will be initialized to its default value (the value in the new prototype);

3. if a slot that is in both versions has changed its class, a call to `as` will be generated to convert the slot in the old object.

If the class representation has no slots (for example, if the prototype is just numeric or character string data), S constructs a method that coerces between the old and new version. The `setClassVersion` call will store the constructed method, along with the new representation and some comments on the date and what changed. You can provide an additional description of the change in the call to `setClassVersion`.

The class version object then contains, for each revision, the new representation and a method for converting objects to that version from the preceding version. The heuristics for constructing the method will fail, of course, if there is something more substantive going on. In this case, the programmer needs to provide a method if updating is to work. The method provided is a function of one argument. Expect the argument to be an object from the preceding version of the class, but with the class unset. Specifically, the argument will be the equivalent of

```
unclass(object)
```

if `object` is a member of the class, previous version. For a class with slots, this object is a named list, with components named according to the slot names. The method should convert `object` to the new representation in the same format, but *not* set the class (this will be done by the updating code in S).

In our `track` example, suppose we insist that the `variables` slot have length equal to the number of variables. Then the automatically generated method will produce an invalid object. We need instead to provide a method that generates a single character string for this slot. This method can be provided as an argument when defining the update, or can be specified explicitly later:

```
> setClassVersion("track", method =
+     function(object){ object$variables = "Y"; object})
```

By default, the method applies to converting to the latest version of class `track`, either the new one being committed now, or the version last committed if the class hasn't changed its representation since then. The optional argument `which` controls which version the method applies to; be sure to supply that argument if you have a new definition of `track` but do *not* want to commit it.

7.5 New Vector Classes

Because vector classes (section 5.1.1) are so fundamental to many S computations, it's worth discussing how to define new vector classes. Other than the fundamental importance of vector classes, there is nothing special going on, so you could also regard this as a case study in some of the techniques discussed in this chapter.

All the vector classes are glued together by the virtual class of the same name, "vector". As usual with virtual classes, this class provides a conceptual grouping reflecting properties shared by several actual classes, in this case essentially the ability to extract and replace numerically indexed subsets.

If the class "vector", which has an empty representation, is included in the representation of another class, then the new class is a vector class. For example, the class string is a vector class with explicit slots.

```
setClass("string",
   representation( "vector",
     offsets = "integer", text = "raw",
     stringTable = "indexLookup", levelsTable = "levelsLookup"
  )
)
```

The inclusion of "vector" has no effect on the slots in the representation, but serves to generate an is relation to class "vector".

If the new class does not have a representation with slots, it must be explicitly specified to be a vector class. Suppose we define a class "quaternion" to have the same representation as numeric vectors:

```
setClass("quaternion", prototype=numeric())
```

(The idea here is that quaternions need more than one number per element; in particular, the length of the quaternion vector will not be the length of the numeric prototype.) Specifying the prototype of a quaternion to be a numeric vector does not automatically cause it to have an is relation to numeric vectors, but we could have done so:

```
> setClass("quaternion", "vector", prototype=numeric())
> getClass("quaternion")

No slots; prototype of class "numeric"

Extends:
Class "vector" by direct inclusion
```

Notice that this does not specify that quaternions are numeric. It does imply that we will provide enough methods, such as those for extracting and replacing elements and subsets, and for extracting and setting the length, so that our new class can be sensibly used as a vector.

As a slightly different example, consider class "sequence". This also uses a numeric prototype, in this case always of length 3 (containing the start, end, and step for the sequence). Sequences can substitute for numeric, but we want the substitution to be by computing the actual numeric vector corresponding to the sequence, *not* as the three numbers in the prototype. This is accomplished again by an explicit is specification:

```
setIs("sequence", "numeric", coerce = expd.sequence)
```

Transitivity of the is relation then means sequence is a vector class, but always via the numeric expansion.

Defining a new vector class asserts that existing methods having vector in their signature will work with the new class. Vector methods can assume the use of length, el, and the operators "[" and "[[", along with the corresponding replacement functions for each of these. Methods for all of these should exist for the new class, unless it can use existing methods by extension (that's not very likely in most examples).

Methods will very likely be needed for some other functions that, while not absolutely implied by the definition of vector, are extremely useful. The operators "==" and "!=", in particular, are nearly always important. You can define these directly, or via a definition of the Ops group generic.

Look at the methods for class string to see some examples of what gets defined. Many of the methods for this class are written in C, since the class is central to S, but viewing *which* methods are defined will help plan for other vector classes.

7.6 Metadata for Classes

Classes take much of their meaning from the context in which they are used. The approach to classes and methods in S emphasizes the *chapter*—relations among classes and between classes and functions are recorded at the level of the chapter database. The intended programming style is to create a library of classes and methods in a chapter database. The database then represents the class/method definitions, and attaching that database makes the software available during an S session. These are all stored as metadata in the chapter specified as the where argument to functions setClass and setIs; by default, where=1, the working database.

The essential side effects of the specifications discussed here are to create and modify various objects in the *meta* database of the corresponding chapter. The meta databases are accessed by supplying an argument `meta="classes"` in calls to `get`, `assign`, and other database functions. Assignments to them have the semantics of ordinary S assignments, but the names used never conflict with ordinary assignments of the same name. User programs should generally avoid assigning directly to the metadata, since they risk interfering with the conventions used by S for class and method assignments. On the other hand, user programs are free to get any of the metadata and use it. In most cases, this can be done indirectly without knowledge of the actual objects stored there.

For example, `getClass` returns the definition of a class as stored on the metadata of an S chapter. By default, `getClass` returns the full information about the class, but supplying either the `where` argument to specify the chapter or just the argument `complete=F` will restrict the information to exactly what was in the metadata.

Each class specification generates an object of class `classRepresentation` in the metadata for the database to which the `setClass` applies (the argument `where` to `setClass`, by default the working data). The slots of the class representation include the following.

`slots:` the slots defined for this class. A named vector of class names; that is, if `z` is a class representation for class $\mathcal{A}$, then `names(z@slots)` is the vector of slot names, and `as.character(z@slots)` is the vector of the class names for the slots.

`contains:` the classes contained in this class. This is also a named vector. The names are the names of the contained classes. The elements of the vector define how the class extends the corresponding class. See below for an example.

`virtual:` a flag, `TRUE` if this is a virtual class.

`prototype:` the object that functions as a prototype for the class. This is defined whether or not a prototype was used explicitly in specifying the class. The expression `new("thisClass")` always returns a copy of the prototype for `"thisClass"`.

`validity:` the validity test method, if any.

The information in these slots is determined by the setClass call that defined the class, and perhaps by later calls that specify particular information about the class, such as setIs or setValidity. You can go backwards from the class to S expressions that will define it by calling dumpClass to write out the expressions. A call to dumpClass is the preferred way to revise a class definition, but we can also use it just to display the S definition.

Consider the class array. To see its definition, we can call dumpClass:

```
> dumpClass("array", stdout())
setClass("array",
representation=representation("structure", .Dim = "integer",
   .Dimnames = "list or NULL")
)

setIs("array", "matrix", test = function(object)
length(object@.Dim) == 2)
```

The class extends class structure and has in addition two slots, .Dim and .Dimnames. The separate setIs specification is used to test whether an array can be used in the case that a matrix was wanted.

The class representation object in the metadata corresponding to any class can always be obtained by calling the function getClass:

```
> x = getClass("array", F)
> x

Slots:
     .Data       .Dim      .Dimnames
  "vector" "integer" "list or NULL"

Extends:
Class "structure" by direct inclusion
Class "matrix" by explicit test and/or coercion
```

We supplied the second argument to getClass to tell it we did not want the complete information about "array", only the information explicitly in the metadata object. This excludes classes extended by "array" indirectly through "structure". After evaluating this expression, we can examine individual slots to dig out detailed information about the class. For example, the contains slot is a named list; the named elements define how the extension happens:

```
> names(x@contains)
```

```
[1] "structure" "matrix"
> x @ contains $ structure
[1] 1
```

When an element of the `contains` list is an integer vector, it is the starting position of the slots for the contained class. An integer 0 means that the contained class was virtual and had no slots.

If the extension involved an explicit coercion or test, the corresponding element of the `contains` slot is itself a named list.

```
> x @ contains $ matrix
$test:
function(object)
length(object@.Dim) == 2

$coerce:
NULL

$replace:
NULL

$forward:
[1] F
```

The `test`, `coerce`, and `replace` elements are the corresponding functions to test, coerce, or replace the extended object, or `NULL` if no such method was defined. The `forward` element says whether S will look for further extensions through this relation. It's normally `TRUE` but must be `FALSE` in this case because not only does `array` extend `matrix` conditionally, the reverse relation also holds, unconditionally. Loops in the `is` relations would result if both relations could be used indirectly. (Specifying such two-way relations is potentially confusing. Better not to do so unless there is a good reason; the `matrix` and `array` relations were allowed for the sake of back compatibility.)

When a class definition is needed during evaluation, the representation information is expanded to include `contains` information about indirect extensions; for example, class `array` extends class `vector` indirectly through `structure`. This additional information is returned by `getClass` unless the argument `complete=F` is supplied, as we did above.

The order of the contains information is significant. Elements of the `contains` slot are stored in order of closeness to this class, using a pseudo-distance measure. Distances are computed using a penalty that reflects in a heuristic way the notion that direct relations are closer than indirect ones and that relations requiring testing or coercion should be penalized. If c_1, c_2 are any two classes, the distance between them can be interpreted as the penalty for substituting an object of class c_1 where one from class c_2 is wanted. The penalty is very large if c_1 does not extend c_2. All classes extend class `ANY`, but are farther from it than from any nontrivial relationship. If class c_1 extends class c_2 directly, the distance is 4 plus a penalty of 1 for coercing and 2 for testing. If class c_1 extends c_2 through c_3, the distance is the sum of the distance from c_1 to c_3 and the distance from c_3 to c_2. The definition ensures that direct relations are closer than indirect ones. The particular heuristic used has no unique argument in its favor, but will order the extensions of a class in terms of the number of steps of indirection needed, and within that to favor unconditional and direct inclusions over those requiring testing or coercion.

The main application of the ordering is in matching calls to the method with the classes most like the actual arguments. If the evaluator cannot find a method that matches the exact class of the arguments involved in method signatures, it cycles through the classes extended by each of the arguments involved until it finds a combination of classes for which a method does exist.

Suppose, for example, that an argument has class `matrix` and that methods are available with corresponding classes `structure`, `numeric`, or `integer` in their signature. All these are possible candidates because `matrix` extends them all. Type `extends("matrix")` or `getClass("matrix")` to see the classes extended by `matrix`, briefly or in detail. The method involving `structure` will be preferred, because `matrix` extends `structure` more directly than `numeric` or `integer`; in fact, `matrix` only extends the other two classes indirectly through `structure`: you first turn the matrix into a structure, and then extract the `.Data` part as a vector; this vector then may or may not have an `is` relation to `numeric` or `integer`.

If the `structure` method cannot be used, say because there is another argument involved in the signature and that argument does not match, the evaluator then needs to test whether the actual object is numeric or integer, and if so, which is closest. If the data in the matrix is of class `integer`, both `is` relations are true, but the `integer` method is closer, because the `is` relation from the integer object to class `numeric` requires coercing the data.

Section 4.9.6 discusses other aspects of method selection.

Chapter 8

Creating Methods

Methods customize the evaluation of S function calls, according to the classes of the arguments. You can supply methods for existing S functions or for your own; likewise, the classes can be existing S classes or your own. The method definitions are themselves functions, and you can use the standard S techniques in creating them. This chapter describes tools for editing, documenting, and debugging methods. Section 8.2 discusses methods for some important functions, likely to be relevant for implementers of new classes. We also discuss the creation of generic functions, which occasionally need to behave differently from ordinary functions.

8.1 Basic Techniques

A method is a function definition to be used to evaluate a call to an S function. But, unlike ordinary functions, it is only to be used *under certain circumstances*—specifically, when one or more of the actual arguments in the call match particular S classes. Specifying an ordinary function only requires two things: the function definition, and the name by which users will call the function. The expression

```
sqrt = function(x) x ^ .5
```

assigns the function definition object (the expression on the right of the assignment arrow) with the name `"sqrt"`.

If the S evaluator encounters a function call with this name and finds the assigned definition, it evaluates the call by matching arguments to the definition and then evaluating the body of the function.

S methods just refine this process by associating different methods for the function with particular classes for the arguments. The arguments to the function and the users' concept of *what* the function achieves remain the same, but the programmer can refine the definition of *how* the function works through methods responding to different arguments, without merging all those methods into one bulky function definition.

8.1.1 Method Specification

A method is specified by calling `setMethod`, which will save the necessary information in the metadata of an S chapter. The arguments to `setMethod` are the name of the function, the signature for which this method should be used, and the definition of the method. A signature is, in general, a named object with names taken from the function's argument names and corresponding elements specifying the classes these arguments should match.

Suppose we want to redefine the function `sqrt` for matrix objects, to call the function `chol` instead of the standard computation.

```
setMethod("sqrt", "matrix",  function(x) chol(x))
```

The signature here is given as `"matrix"`. When signatures are supplied to `setMethod`, the programmer can use several shortcuts; `setMethod` will match the signature using the same rules the evaluator uses to match function arguments. Since `sqrt` only has one argument, there's not much chance for ambiguity in this example; see page 324 for the general case.

After the method has been stored, future calls to `sqrt` will evaluate the body of the new method definition if the argument matches class `matrix`. Just what "match" means, precisely, is discussed in section 4.9.6, but intuitively it means that the actual argument either has class `"matrix"` or its class extends `matrix` and there is no other method that matches the class more closely.

Functions for which methods have been specified are called *generic* functions. A generic function is a function whose purpose and formal arguments are fixed, but for which the *method* used to produce the desired result may depend on the classes of the actual arguments. Any ordinary function can be converted into a generic by simply setting a method for it: the interpretation is that the arguments stay the same and the initial, unique method (the body of the function) becomes the *default* method, to be used when no specific method applies.

When you convert a non-generic function to a generic, S will print a message telling you what happened.

```
> setMethod("sqrt", "matrix", function(x)chol(x))
redefining function "sqrt" to be a generic function on
database 1
```

Some of the messages are warnings; for example, in the `sqrt` example we would actually be warned that there were now two different versions of `sqrt`, a generic version on the local chapter and a non-generic version on library `main`.

Occasionally, you may need to have more control over the generic function definition, such as defining a generic for which there is no default method. The function `setGeneric` provides such control: see section 8.3 for more discussion of generic functions.

The third argument to `setMethod` is the method itself, a function object. The body of that function becomes the method. Strictly speaking, the method supplied must have the exact same formal argument list as the generic, because only the body of the method is substituted by the evaluator after the method has been selected. However, if you do supply a function with different arguments, `setMethod` will construct a valid method that calls the function you supplied, passing down all the arguments of the generic. For example, suppose we decide to use `print.matrix` as the method for `show(object)` when `object` has class `"matrix"`:

```
setMethod("show", "matrix", print.matrix)
```

Since `print.matrix` has several arguments, its body cannot be the body for the `show` method. Instead, S constructs a new function that has a local copy of `print.matrix` and ends by calling that function. A warning message is also printed, since there is no guarantee that the call `print.matrix(object)` was what the programmer wanted. In this case, the constructed method is fine. One subtle point should be mentioned: `setMethod` makes a local version of `print.matrix` inside the newly constructed method. If `print.matrix` changes in the future, the method will still call the previous version.

If the programmer wanted the method to always call the *current* definition of `print.matrix`, the method must be specified as the call, with no local version:

```
setMethod("show", "matrix",
  function(object) print.matrix(object))
```

Either could make sense; just keep in mind the difference. For safety, S has to assume you want a local copy, since there is no guarantee the called function will still exist when the method is invoked.

So far, we specified the method corresponding to the first argument to the function, by giving a single character string as the second argument to `setMethod`. A general signature, however, is a named list matching the classes to the formal arguments of the function. In fact, the same rules are used as in matching positional and named arguments in function calls. Any arguments, and any number of arguments, can appear in the signature. Arguments that are missing from the signature correspond to `"ANY"`, that is, they will match any class. For example, if we wanted to specify a method for `plot` that matched argument `y` to `"track"`:

```
setMethod("plot", signature(y = "track"), ....
```

To see the classes implied, give the function definition and the signature to the function `matchSignature`. The arguments of `plot` are `x` and `y`, so the actual signature corresponding to this specification is:

```
> matchSignature("plot", signature(y="track"))
[1] "ANY"   "track"
```

Not supplying the `x` argument in the signature is quite different from saying that `x` must be missing in the *call*. That's also a valid, and useful, signature, but is done explicitly by using `"missing"`:

```
> matchSignature("plot", signature(x="missing", y="track"))
[1] "missing" "track"
```

A number of examples appear in section 8.2.3.

The `setMethod` call has several side effects, assigning an object in the chapter and telling the evaluator to revise its definition of this function. By default, these side effects do not take place immediately, but rather at the end of this task (specifically, on exit from the top-level frame). The distinction usually makes no difference, but occasionally the consistency matters. If the current task involves calls to the function being modified, you usually do not want some of the calls to use one interpretation and some another, particularly in S, where lazy evaluation can muddy the sequence in which evaluations occur. Delayed commitment of the changes ensures that the function gets its new definition only for subsequent tasks.

If you do need to use the revised method right away, set the `immediate` option before redefining the method:

```
options(immediate = T)
```

This would be required, for example, if you used `source` on a file that both defined and then used a new method. But if the source file was processed as separate tasks, by a `setReader` or `sink` call, delayed commitment may be preferred.

8.1.2 Editing Methods

To see what methods are defined for a particular generic function, use `showMethods`:

```
> showMethods("whatis")
       Database    object
[1,] "./course" "Default"
[2,] "./course" "array"
> showMethods("Ops")
      Database      e1    e2
[1,] "main"      "ANY"    "ANY"
[2,] "main"      "raw"    "raw"
  ...
```

For each function, the call to `showMethods` computes a matrix whose first column gives the name of the database on which the method is stored, and the remaining columns show the signatures for each method.

The `showMethods` function can be used in a very general way: you can ask for any selection of functions, classes, and databases, corresponding to the three arguments to the function. The result will show all methods for any of the selected functions, when any of the selected classes appears in the signature of the method, and when the methods are stored on any of the selected databases. In the examples above, we selected a single function each time, and left the classes and databases unspecified. If we wanted, say, to see all the methods including class `"probeTest"` in the signature that were stored on the working database:

```
> showMethods(where = 1, classes = "probeTest")
$"[":
     Database            x
[1,] "."       "probeTest"
```

```
$scan:
     Database  file        what
[1,] "."       "ANY" "probeTest"

$waferLayout:
     Database        code
[1,] "."       "probeTest"
```

When more than one function is requested, the value returned is a named list whose elements describe the methods for the individual functions.

To edit a particular method for a particular generic, call `dumpMethod` to dump the specification of the method to a text file, suitable to be edited.

```
> dumpMethod("show", "connection")
Method specification written to file "show.connection.S"
```

The call to `dumpMethod` takes the same first two arguments as `setMethod`: the name of the function and the signature of the method. The effect is analogous to using `dump` to dump the definition of an ordinary function; in both cases, S will write to a file some S code that will recreate the function or method when sourced back into S. The actual S expression generated is different, being here a call to `setMethod` rather than an ordinary assignment. The default file name chosen by `dumpMethod` concatenates the name of the generic and all the classes in the signature, so the files for different methods will very likely not overwrite each other.

An important point of strategy is that you don't need to have *any* method defined for this signature in order to use `dumpMethod`; in fact, a good way to create the initial version of a method is to call `dumpMethod` as if a method already existed, and edit the result. The method dumped to the file will be whatever the S evaluator would have used, maybe the default method.

The file can be edited in any way you want. Once you are satisfied with the editing, you can load the dumped code in any of the usual ways, most often using the `source` function.

```
source("show.connection.S")
```

For alternatives to `source`, see page 258 in section 6.2.1.

To dump *all* the methods for a particular function from a database, call

```
> dumpMethods("show")
Methods written to file "show.S"
```

By default, methods are dumped from the first database in the search list having methods for this generic. Supply an argument `where` to select the database.

As with `dump`, you can supply a `connection` argument to `dumpMethod` or `dumpMethods`. If you open the connection before the dump call, and supply the opened file as the connection argument, multiple dumps will go to the same file. This allows dumping multiple selected methods, or the methods from several libraries, or other customized selections of methods.

```
> f = file("polySplineMethods.S","w")
> dumpMethod("polySpline", "bSpline", f)
> dumpMethod("polySpline", "nbSpline", f)
> close(f)
```

By opening the file beforehand, we have successive dumps concatenated on the file; by opening it with mode `"w"` (write only), we truncate anything that might have been on the file before.

Giving several names as the second argument to `dumpMethod` does *not* mean to dump multiple methods; instead, it defines a single method signature:

```
dumpMethod("Arith", c("track", "track"))
```

This dumps the `Arith` method for both arguments being of class `track`.

8.1.3 Examining Methods

You can enquire about single methods for a generic, about all the methods for the generic, or about all the methods on a particular database. A particular method will correspond to a signature; e.g.,

```
signature(e1 = "numeric", e2 = "integer")
```

where `e1` and `e2` must be formal argument names of the generic you're interested in, with `"numeric"` and `"integer"` the classes for these arguments. The functions `getMethod`, `existsMethod`, and `findMethod` are analogous to `get`, `exists`, and `find` for ordinary objects (see page 221 in section 5.2.4). They each take as arguments `f`, the name of the generic, and `sig`, the signature. The first two also take, optionally, the argument `where`, the database.

```
> existsMethod("+", signature(e1 = "numeric", e2 = "integer"))
F
```

If sig is missing, S looks for the default method. If where is missing, S looks on all databases, otherwise only on database where.

These functions look only for methods that match the signature exactly. S will also select methods indirectly, using both extends relations for the classes and group relations for the generic. You can ask about methods selected with the full current information by using selectMethod instead of getMethod and hasMethod instead of existsMethod. These use the method selection mechanism described in section 4.9.6; a method for classes extended by the signature, for either this generic or its group generic, qualifies as a selected method.

Similar functions can be used to look for and retrieve the complete methods definition for a generic:

```
findMethods(f); getMethods(f, where)
```

Related, but slightly different is findGeneric(f): this asks only which databases assert that f is a generic. None of the databases need actually have any methods for f.

8.1.4 Removing Methods

To remove methods from a library, use one of removeMethod, removeMethods, or removeGeneric. The removeMethod call removes the metadata for a single method:

```
removeMethod("show", "track", where = 1)
```

This removes the method for generic show and signature track from the working database. If there are no more methods for this generic on this database, but there *are* methods on other databases, removeMethod will clean up by removing the generic definition from the metadata:

```
> removeMethod("show","track")
No more methods for "show" on database "."; removing
the generic from this database
```

If there are no other databases with this generic, however, the definition is left in the metadata.

To remove *all* the methods for a generic, use either removeMethods or removeGeneric. Their arguments are the name of the generic function and optionally the chapter from which to remove the methods.

```
removeMethods("whatis")
```

By default, the chapter is the first writable database on the search list with methods for this generic.

The two functions differ only in that `removeMethods` will try to restore the non-generic definition of the function, if that makes sense, while `removeGeneric` removes all trace of the function, generic or not, from this chapter. In either case, all the methods for this generic will be gone from the chapter.

Let's take an example to see what is actually happening. Suppose we defined the one-line summary function `whatis` initially as

```
whatis = function(object) class(object)
```

Later on, we defined methods for `whatis` for class `"vector"` and `"matrix"`. The situation is now:

```
> showMethods("whatis")
     Database          x
[1,] "."      "ANY"
[2,] "."      "matrix"
[3,] "."      "vector"
```

When the first method was defined for `whatis`, the initial function definition became the default method:

```
> showMethod("whatis")
function(object)
class(object)
```

The current function object for `whatis`, on the other hand, is now the standard call to the generic dispatcher. It needs to be, because this object might be supplied as an argument to one of the `apply` functions, and we want to ensure that the generic function will be invoked in this case.

The strategy of `removeMethods` is to restore the situation as it was before the first method was defined. So it replaces `whatis` with a function that corresponds to the default method.

```
> removeMethods("whatis", ".")
> get("whatis", where = ".")
function(object)
class(object)
```

In other cases, there may be methods for the generic on several chapters and the generic function may not be on this chapter. If not, or if the function is not the standard generic dispatcher, `removeMethods` will not assign a new function object in the chapter. A call to `removeGeneric` is interpreted as wanting to remove any trace of generic function `f` from this chapter. It will not assign an ordinary function object, and will remove one if there is one already. Otherwise, both `removeMethods` and `removeGeneric` behave the same; that is, they remove the object in the metadata for the chapter that defines all the methods for `f`. If there are no other attached chapters with methods for this generic, it is no longer a generic function to the S evaluator:

```
> isGeneric("whatis")
[1] F
```

8.1.5 Tracing Methods

S provides tools within the language to help understand and debug the evaluation of function calls. These apply equally to generic functions; for example, you can trace all calls to the generic function `whatis` by

```
trace(whatis, browser)
```

which will invoke the interactive browser as soon as a call to `whatis` starts evaluation. Similarly, a trace call with `exit=browser` will let you interact after the call completes.

When debugging methods, however, you usually want more specific control, tied to the particular method you are currently studying. The function `traceMethod` lets you establish some debugging for a particular method. Like `setMethod` and `dumpMethod` its first two arguments are the name of the generic function and the signature of the method. These take the place of the name of the function in the call to `trace`. The remaining arguments to `traceMethod` are exactly the same as those to trace, and specify where in the method to put the trace code and what function or expression to insert there. To call the interactive browser on exit from the method for function `whatis` and objects of class `matrix`:

```
traceMethod("whatis", "matrix", exit=browser)
```

This works exactly as the `trace` function does. It's all done in S and the easiest way to understand how to use the mechanism is probably to know how it works.

A call to `traceMethod` is effectively a call to `setMethod`. The difference is that `traceMethod` finds the current method, applies the ordinary `trace` to this method, and then sets the traced function to override the current method, but *only* in the session database, not on the working data. So as long as the trace remains in effect, the method search will find and use the tracing version. So `traceMethod` has all the arguments to `trace`, such as `exit` and `at`, because it actually uses `trace`.

8.2 Designing Methods for Some Important Functions

When you define a new class of objects, you should consider providing methods for some essential functions: to create objects in various ways, to print and plot the objects, to modify the objects and extract useful pieces of them, perhaps to do arithmetic or other numerical computations. Users will work more easily with your new class of objects if they can continue to use functions that perform similar tasks for older, more familiar classes of objects. You can make this happen simply by defining methods for these functions when one or more of the arguments are from the new class. The set of relevant functions may vary depending on what the class of objects means to users, but the functions discussed in this section are among the most likely candidates for most classes.

Many of the examples illustrate a common style: the method for a particular function for a new class often invokes the same function for some portion of the object or for some derived object. A method for `show` to print an object may call `show` on some slots, for example. Using this style helps to hide details and also to let the method adapt automatically to changes.

8.2.1 Scanning a File

You will likely have provided a generator function for the new class, and the function `new` can also be used to create objects from the class out of the underlying data. A related question is how to set up human-editable versions of the objects: how can users modify an existing object or enter the data to define a new one? If we can, we should provide a way to read the data for an object of the new class from ordinary text input, the kind that humans could type or that some program that knew nothing about S could generate. We should also make it possible to dump an object from the class in a form that is reasonable for humans to edit.

The two functions that correspond, and for which methods are useful, are `scan` and `dput`. The `scan` function reads a file of data that was produced, outside of S, and interprets it as a particular class of object. The `dput` function writes a current object to a file, in a format designed to be editable by a human and then sourced back into S to redefine the object.

Methods for `dput` are discussed in section 8.2.4; methods for `scan` in this section. The `scan` function takes two main arguments: the file or other connection from which to read, and an argument `what` that is intended as a prototype for the object to be returned. Thus

```
scan(what=string())
```

tells S to read some data and interpret it as an object of class `string`. (The `file` argument was omitted, and will be `stdin()` by default.)

You can provide a `scan` method for any class of objects. For example, the following sets up the function `read.table` as a method for scanning in `data.frame` objects:

```
setMethod("scan", signature(what="data.frame"),
    function(file, what, ...) read.table(file, ...))
```

Methods for `scan` are a little odd in that they ignore the object `what`, even though it appeared in the signature; that argument served its purpose just by arranging that the method will be called.

For such scan methods to be useful, there needs to be a way to generate an object from the class. Typically, the generator function for the class can be called with no arguments to create an "empty" object from the class. In the `data.frame` example, this is true, so the method defined for `scan` for data frames could be used by

```
scan("myfile", data.frame())
```

Users may take a while to get used to the style of supplying an *object* to indicate the class, rather than the character string name of the class, but this style is equally simple to type.

8.2.2 Printing

Methods for automatic display of objects are specified as methods for the function `show`. This function takes only one argument, the object to be shown. The standard S evaluator calls this function to display the result of a task. In general, `show` can do anything it wants: print the object or plot it if that would be more informative. The functions `print` and `plot` are more specialized and take extra arguments related to the details of printing or plotting.

In designing display methods for a class, you can define methods for `show`, `print`, or `plot`. Here is a method for `show`, but little would change if we were defining the method for `print`.

These methods are intended for general-purpose output for humans. They should give enough information to tell the reader all the useful information about the object, but there is no point in printing information that humans can't easily understand. For example, if your class has some `raw` data used internally, there isn't much point in printing that data.

Consider our class `track`: the printed data should show the values `y` close to the corresponding positions `x`. Here's a method that creates a 1-row matrix from `y` and labels the columns with `x`.

```
setMethod("show", "track", function(object)
{
    vals = matrix(object@y, nrow = 1)
    dimnames(vals) = list("y:", object@x)
    cat("An object of class \"", class(object), "\"\n",
        sep = "", file = stdout())
    show(vals)
})
```

It announces the class of the object and then prints the data:

```
> tr1
An object of class "track"
   156 182 211 212 218 220 246 247 251 252 254 258 261 263
y: 348 325 333 345 325 334 334 332 347 335 340 337 323 327
```

A small but useful detail is that the printed class is computed from the object, not assumed to be `"track"`. If a new class is defined to extend `track`, it should get its own `show` method if it contains additional data, but until then it will be printed using the inherited method for `track`: at least this will show the correct class.

8.2.3 Plotting

The generic function `plot` takes one or two objects, arguments `x` and `y`. When called with one argument, the user expects `plot` to display that object graphically; it is this situation that most `plot` methods handle. Methods can also be defined when an object from the new class is plotted "versus" another object. For some classes, it makes sense to go farther and define how the objects should be interpreted in a whole range of plotting functions, including those in basic S graphics to add points, lines, and other graphical output to an existing plot.

Let's begin with the simpler case, specifying a method for function `plot` only. To see the arguments for `plot`:

```
> functionArgNames("plot")
[1] "x"   "y"   "..."
```

The `"..."` arguments are used to provide auxiliary information for the plot, such as titles, labels for the axes, and control over the detailed appearance of the plot. In the case to consider first, `y` is missing. The task in this case is just to display object `x` in some suitable way. For simple `x` (e.g., numeric vectors), the plot looks much the same, but now the object provides the y co-ordinates, and the x co-ordinates are, somewhat arbitrarily, the sequence `1:length(x)`.

When we define analogous methods for other classes of objects we can choose anything we like: the plot need not be a scatter plot as it is for numeric vectors, and in fact need not be a single plot at all. We can choose the display that makes the most sense for visualizing data from this class. In section 1.7.2, a plotting method for test results for manufacturing generated an image rather like the physical appearance of the wafer itself.

To specify the method, remember that we are only defining it when `y` is missing, so the specification is of the form:

```
setMethod("plot", signature(x = "myClass", y = "missing"),
  function(x, y, ...) { .....}
)
```

The method has exactly the same arguments as the generic, even though we know that `y` is missing.

Simple plotting methods usually amount to extracting some pieces from the object and then using plotting methods suitable for those pieces. Suppose the objects in `myClass` are list-like objects and we decide just to plot each of the components in a separate plot.

```
setMethod("plot", signature(x = "myClass", y = "missing"),
  function(x, y, ...) {
    for(el in as(x, "list")) plot(el, ...)
  }
)
```

This is a little too simple (it doesn't even label the separate plots uniquely), but let's leave it as is for now and go on to a more general situation, which will also provide a good way to write better `plot` methods in some cases.

In the general situation, we have a class of objects that can be interpreted as points in the plane, that is, as pairs of (x, y) co-ordinates. Examples include a vector of complex numbers and a two-column matrix. The class `track` that we have been using as a running example, starting on page 38, also falls in this category. Track objects have slots `x` for the position of the measurements, and `y` for the measurements themselves. We even chose the slot names to suggest the interpretation as points in the plane: a track is conceptually a set of points on the curve representing the measurements.

In such cases, we would like to make all functions that naturally relate to points in the plane work with our class of objects, not just `plot`. To do this in one step, S provides a generic function, `xyCall`, whose purpose is to call some other function (provided as an argument to `xyCall`), with the appropriate (x, y) co-ordinates as the first two arguments. The S functions `points`, `lines`, and `text`, as well as many others, including some non-graphical functions, all use `xyCall`, passing down to it the appropriate function to handle the pairs of numbers.

To adapt all these functions to your new class, you write methods for one generic, `xyCall`:

```
> functionArgNames("xyCall")
[1] "x"     "y"     "f"     "..."   "xexpr" "yexpr"
```

The argument `f` is the function to be called; that function will have arguments `x, y, ...`. Let's ignore arguments `xexpr` and `yexpr` for the moment, and show how typical methods can be written for `xyCall`.

First, a method for `y` missing. This is directly analogous to the `plot` method, but now with the task of defining the appropriate points in the plane and supplying them as arguments in a call to `f`. For `track` objects this is easy, we just use the corresponding slots. A method for `xyCall` corresponding to `x` from class `track` and `y` missing could be specified as follows.

```
setMethod("xyCall", signature("track", "missing"),
  function(x, y, f, ..., xexpr, yexpr)
    f(x@x, x@y, ...)
)
```

For some objects, this is all that makes sense. For others, including the `track` class, we can go farther. Suppose one of the `x` or `y` arguments is a track object and the other argument is anything at all (but *not* missing). Then we would need to interpret the track object as a numeric vector. No problem here: we decided when designing this class that we would just drop the positions and use the measurements (the y slot) to represent the object when numeric values were needed.

If there is an is relation from the class at hand to class numeric, nothing more needs to be done, so far as getting the co-ordinates right. If we do want to be explicit about the methods, we need to define two more methods, matching the x and the y arguments to this class. For track objects, the specification would look like this:

```
setMethod("xyCall", signature(x = "track"),
  function(x, y, f, ..., xexpr, yexpr)
    f(x@y, y, ...)
)

setMethod("xyCall", signature(y = "track"),
  function(x, y, f, ..., xexpr, yexpr)
    f(x, y@y, ...)
)
```

We are making a decision here: if we did *not* think it was sensible to represent the object by its y slot when we wanted one plotting co-ordinate, we would not define these two methods, and calls corresponding to them would generate an error. As so often, the important questions relate to what the objects should *mean* to users.

If you want to keep your methods for plotting simple, you can stop reading here; if you are willing to devote a little more effort to making the methods more user-friendly, keep going. We have not used arguments xexpr or yexpr yet. These arguments are supplied by the function plot when it calls xyCall. They are the S objects representing the x and y arguments to plot, unevaluated. Why? Because plot would like xyCall to construct arguments to f for labelling the axes, if the user did not supply such arguments in the call to plot. All the rest of this subsection is just a discussion of how methods for xyCall can carry out this service. If you don't bother, then plot calls using your method may have uninformative axis labels, but all the data will be shown as you want.

An xyCall method that provides x-axis labels has to do two things: First, check whether xexpr was missing. It will be, for all calls except those that want xyCall to construct the labels. (This has nothing to do with x being missing.) Second, the method checks whether xlab was among the "..."

arguments, and if not uses xexpr to construct something useful. The same steps are taken for the y-axis labels, but the check for xexpr missing can serve for both axes: either both xexpr and yexpr are supplied, or neither. For track objects in the case that y is missing, here is a method that constructs the labels.

```
setMethod("xyCall", signature("track", "missing"),
  function(x, y, f, ..., xexpr, yexpr) {
    if(missing(xexpr)) f(x@x, x@y, ...)
    else {
        xexpr = deparseText(xexpr)
        dots = list(...)
        if(hasArg(xlab)) xlab = dots$xlab
        else xlab = paste(xexpr, "(Positions)")
        if(hasArg(ylab)) ylab = dots$ylab
        else ylab = paste(xexpr, "(Measurements)")
        f(x@x, x@y, xlab = xlab, ylab = ylab, ...)
    })
```

The if branch is just the same as before. In the else branch, we construct labels from the argument expression for x, followed by "(Positions)" and "(Measurements)". Two functions used in this example deserve mention, as they are frequently useful. The function deparseText converts an object representing an S expression into a single character string, for use as text in printing and plotting. The string will be truncated if it is too long; see ?deparseText for details.

The function hasArg returns TRUE if the name supplied to it corresponds to an argument in the call, but unlike the function missing, it looks at arguments that match "...", so it can be used in this example to look for xlab among the "..." arguments.

8.2.4 Dumping Data for Editing

We can customize how the dump function treats objects from any class. It may be very helpful to do so, because users may then be able to edit objects from the class much more easily. The trick to customizing dump is to note that although it takes a vector of object *names* as its argument, it then in effect calls the function dput for each object to write that object to the output file or connection. By writing methods for dput, the output from dump will be customized as well.

The default dput method for a class with slots is to generate a call to the extended version of the new function, with arguments specifying the values of the individual slots (these will be printed by calling dput recursively for each slot).

```
> class(tr1)
[1] "track"
> dput(tr1)
new("track",
x = c(1., 1.25, 1.5, 1.75, 2., 2.25, 2.5, 2.75, 3.)
, y = c(-0.22, -0.02, -0.88, 1.32, 0.05, -2.37, 1.48, -0.49, 0.67)
)
```

This is not too bad, but knowing that a generator function exists, we can specify a method that uses the generator:

```
setMethod("dput", "track",
# output a call to the 'track' generator function
function(x, file) {
    cat("track(x =\n", file=file)
    dput(x@x)
    cat(", y =\n", file=file)
    dput(x@y)
    cat(")\n", file=file)
})
```

This hides the actual slot names, admittedly not a big deal with this class.

```
> dput(tr1)
track(x =
c(1., 1.25, 1.5, 1.75, 2., 2.25, 2.5, 2.75, 3.)
, y =
c(-0.22, -0.02, -0.88, 1.32, 0.05, -2.37, 1.48, -0.49, 0.67)
)
```

Let's push the design of the method one step further, to illustrate a couple of useful S programming tools for printing things, and also a way of thinking through the design question.

Users would probably find it easier to edit the data if the corresponding elements of the x and y slots of the object lined up. One reason they don't is that the standard dput method for dumping a numeric vector formats each number separately. But there is an S function, format, whose purpose is to find a single format for converting anything to character strings, so that each string is the same width. By calling format on the combined data from the two slots, we'll have a fixed-width format for all the data.

The next step is then to recreate the dumped version of the individual slots. Notice that `dput` dumps numeric vectors (actually, all atomic vectors) as a call to the `c` function. That seems reasonable, so we need to convert the formatted data values into a comma-separated list of arguments, which brings up the second useful tool. The `paste` function specializes in pasting things together, as its name suggests. In particular, it has an argument, `collapse=`, designed for just our current need. If we supply a string as this argument, `paste` will paste together all the data values into a single string, using the `collapse` argument as the separator. (It's no accident that `paste` has this feature: it was built into it precisely to create printed lists of the sort we need here.)

With these two ideas incorporated, here is the next version of our method:

```
setMethod("dput", "track",
function(x, file = stdout())
{
  xy = format(c(x@x, x@y))
  n = length(xy)/2
  cat("track(x = c(\n",
      paste(xy[1:n], collapse = ", "),
      "),\n y = c(\n",
      paste(xy[-(1:n)], collapse = ", "),
      ")\n)\n",
      fill=T, sep="", file=file)
})
```

The `fill=T` argument to `cat` causes it to break lines when they get too long. Here's the same `tr1` object, printed with the new method:

```
> dput(tr1)
track(x = c(
 1.00,  1.25,  1.50,  1.75,  2.00,  2.25,  2.50,  2.75,  3.00),
y = c(
-0.22, -0.02, -0.88,  1.32,  0.05, -2.37,  1.48, -0.49,  0.67)
)
```

The alignment of the numbers should make this version easier to edit. Try reading through the arguments to `cat` in the method and relating them to the pieces of the output, to see the design in detail.

8.2.5 Extracting and Replacing Subsets

S users are accustomed to using the `"["` operator to extract or replace subsets—portions of an object specified by a range of indices, by a logical expression or by some other more specialized argument. You may want to make such expressions meaningful for your new class of objects as well. To do so, define methods for function "[". The first argument to this function is the object from which a subset is to be extracted. There usually follows one index argument, saying which elements to extract. The function is written for arbitrarily many index arguments, using "..." as the second argument, to accommodate matrices, multiway arrays, data frames, and other objects for which multiple index arguments make sense. Subsets of an object can be replaced by using "[" on the left of an assignment.

Extraction methods are natural for many classes. The result will either be an object from the same class or else a simpler object if the special structure of the class is naturally lost when subsetting. For our example of the `track` class, which has slots `x` and `y` for the positions and values, the natural way to interpret subset extraction is that it extracts the same subset of each slot.

```
setMethod("[", "track",
  function(x, ...) track(x@x[...], x@y[...]))
```

This allows us to use all the standard kinds of S subsetting operations without special programming.

```
tr1[1:3]
tr1[yy > 100]
```

A slightly more user-friendly implementation of this method might check legality of the "..." arguments, only because an invalid expression would produce a confusing error message.

The same basic approach to defining extraction methods applies to replacement methods, for similar expressions using the "[" operator on the left side of an assignment. As with all replacement methods, remember that the replacement function corresponding to "[" has a final formal argument, `value`, which will correspond to the expression on the right of the assignment.

Methods to replace subsets of specific classes are defined by a call to `setReplaceMethod`. For example, we might want a method to replace subsets of the `y` slot of an object of track class, when the user supplied an expression such as

```
tr1[yy > 100] = 100
```

The following call to `setReplaceMethod` will allow such expressions.

```
setReplaceMethod("[", "track",
  function(x, ..., value)
    x@y[...] = as(value, "numeric")
    x
  )
```

A replacement method can depend on the class of the right-hand side, as well as the class of the argument `x`. For the `track` class, suppose we only want to allow replacement of subsets of the `y` slot by `numeric` data. The method above was designed to be forgiving: it tries to turn anything into numbers.

```
> tr1[3] = "abc"
Warning messages:
  1 missing values generated coercing from character to
numeric in: as(value, "numeric")
```

If we had omitted the call to `as`, the computation would only have succeeded in replacements when the `value` object had class `numeric` or some class extending `numeric`.

We might also want to apply a special method when the right side of the assignment is itself a `track` object. Just how general such a method should be would in practice require quite a bit of thought. Here's one that is simple. The `track` object supplying the replacement values is required to include the positions that need to be replaced.

```
setReplaceMethod("[", signature(x = "track", value = "track"),
  function(x, ..., value) {
    xold = x@x[...]
    repl = match(xold, value@x)
    if(any(is.na(repl)))
      stop("some of the positions to be replaced ",
           "are not in the replacement: ",
           paste(xold[is.na(repl)], collapse=", "))
    x@y[...] = value@y[repl]
    x
  })
```

If we use both this method and the previous method, which left `value` unspecified, S will use the second method when value has class `track` or any class that extends `track`. Why? Because the rule is that the evaluator selects the method for which the classes of the actual arguments best match the signature of the method. An unspecified argument is equivalent to specifying class `ANY` in the signature. Class `ANY` is defined to extend any other class, but very weakly. In particular, class `ANY` is farther from class `track` than is any other class that extends `track`. Section 4.9.6 goes through the rules for method selection.

8.2.6 Mathematical and Summary Functions

S comes with many functions that take numeric data and return new objects with the same "structure" as the argument, but with the data values altered in some way; in other words, they perform *element-wise* or *one-to-one* operations. For example, `abs` takes the absolute values of the data values, `exp` computes the exponential function. All the usual trigonometric functions are included as well as functions for rounding and other elementwise numerical operations.

These functions all have the same group generic, `Math`, which is a slightly misleading name since some of the transformations are not very mathematical, but it's a less clumsy name than, say, `Elementwise`. Two closely related functions, `round` and `signif`, operate similarly except that they take a second argument, `digits`, to control how much numeric information to retain. All functions in a particular group have to have the same set of formal arguments, since S evaluates the body of a method once it selects that particular method—there is no further matching of arguments between the call and the method. Therefore, `round` and `signif` have group `Math2`, behaving much like `Math` except for the second argument.

When a new class is defined, the designer needs to consider whether such functions *should* have methods for all, some, or none of the `Math` and `Math2` functions. If objects from the class contain some (numeric) values and if transforming those values makes sense, then a method can be defined for these group generics. Typical methods will either return an object from the same class with the data values transformed, or will throw away everything but the data values and return the transform as a vector. If transforming the data values produces an object that can reasonably be interpreted as coming from the new class, that's what we return. If transforming the values makes sense, but the object no longer behaves as it should to belong to the class, we return the values only. If transforming the values makes no sense, we should prohibit application of these functions (usually they won't apply unless we write a method).

Consider our `track` class: these objects represent tracks, or curves, taken at some positions in space or time. Taking an element-by-element transform of these makes sense; we would expect the result to also to be a `track` object with the values transformed, but with positions the same as the original. We could write a method for the `abs` function, looking something like this:

```
function(x) {
    x@y = abs(x@y)
    x
}
```

Simple enough, but since all the `Math` functions would have essentially the same method, we can write one method for the group generic, which will then be dispatched for each of the functions belonging to the group:

```
setMethod("Math", "track",
    function(x) {
        x@y = callGeneric(x@y)
        x
    })
```

When this method is invoked, `callGeneric` will determine the name of the current generic function (e.g., `"abs"`) dynamically, and will call that function with the arguments given to `callGeneric`.

The `Summary` group of functions all take some object, again usually containing numeric values, and return one, or occasionally two, numbers giving some summary property of those numbers. For example, `sum`, to sum numerical values, belongs to this group, as do `min`, `range`, and other similar functions. The functions are all formally defined to take an arbitrary number of objects, with the definition that the summary is applied over all the objects; for example, `max(x1, x2)` should be the maximum value in either `x1` or `x2`. The functions also take an argument `na.rm`, an optional flag to tell the function that missing values should be removed (i.e., ignored) in computing the summary.

The value of a call to a `Summary` function is generally numeric, regardless of the class of the object, although that is not obligatory. For class `track`, summary computations are simple: we return the corresponding function of the values in slot `y`. A method can be defined as follows.

```
setMethod("Summary", "track",
  function(x, ..., na.rm=F) callGeneric(x@y), na.rm=na.rm)
```

Whether summary methods make sense for a particular class of objects usually amounts to whether the class contains some implicit numeric or at least ordered data. If the data are only ordered, but not numeric, summary functions `max`, `min`, and `range` might make sense but `sum` and `prod` would not.

We have ignored the possibility of multiple objects in the `Summary` method, and handling arbitrary numbers of objects would be somewhat challenging. It's not enough to assume that all the arguments belong to the same class; instead, we have to be prepared for arbitrary combinations of objects, such as `min(track1, 100)`. Fortunately, the methods do not have to deal with multiple objects at all. The generic function detects multiple objects and handles them by a recursive technique, only invoking methods on single objects. The ability to have generic functions do more than just invoke methods is important here; see page 351 for the details.

8.2.7 Arithmetic and Other Operators

S follows the traditional style of languages, other than those in the Lisp family, by writing arithmetic expressions, comparisons and logical operations as "infix" expressions:

```
x * y; y > x; tested | (x > 0)
```

The operators used in this way are not, however, really different from other S functions. Each operator invokes a function of the same name, `"*"`, etc. The names contain special characters, but this is irrelevant except for parsing. Once we do look at the function objects, they turn out to be ordinary S functions:

```
> getFunction("*")
function(e1, e2)
.Call("S_c_use_method", "*")
```

The operators all have arguments `e1` and `e2` and belong to groups `"Arith"`, `"Compare"`, and `"Logic"`. In turn, these three group functions belong to group `"Ops"`.

```
> getGroupMembers("Arith")
[1] "-"    "+"    "*"    "^"    "%%"    "%/%" "/"
> getGroupMembers("Compare")
[1] "!="        "<"         "<="        "=="        ">"
[6] ">="        "compare"
> getGroupMembers("Logic")
[1] "!" "&" "|"
> getGroupMembers("Ops")
[1] "Arith"   "Compare" "Logic"
```

To implement arithmetic and other operator methods for a new class, you can usually write methods for some or all of the first three group functions, or perhaps just methods for group "Ops", rather than for all the individual operators. The existence of three specialized groups within "Ops" is often useful, since many classes may only have valid methods for some of the groups, or may interpret the different kinds of operators very differently. So the first questions to decide in designing operator methods for a new class are: which operators should be defined, and should they produce objects from this class or something else (often just the vector of values).

A second class of questions has to do with what *combinations* of classes we want to support. That is, since operators mostly take two arguments, the signatures defining the methods can involve both arguments. There are many different strategies that may make sense for different classes. Sometimes, the operators are being interpreted in very special ways, quite unlike their interpretation for numeric data or other vector classes, in which case methods may only make sense if *both* arguments match the new class. A very inconsistent interpretation of the purpose of operators should be approached cautiously, because users may become confused. For example, it might be tempting to interpret "+" for strings as meaning string concatenation, as in Java. On balance, this seems a bad idea; while it saves some typing over the paste function, it muddies the concept of the operator.

The more common case is that the new class retains essentially the usual interpretation of those operators that make sense, but takes care to retain some extra information or to check certain conditions. In this case, we may need additional methods, to handle mixed expressions where one argument is from the new class and the other is more general. We may also need methods for *unary* operator expressions, such as -x; these correspond to argument e2 having class "missing".

Let's examine these questions for our track class. This was defined to only handle numeric values at the positions. If that really was the intention, then we can't have a track object with logical data as values, so while Arith operators can return track objects, Compare operators will return vectors, and Logic operators don't apply. Since track objects are interpreting the operators in very much the same way as for numeric data, we want to handle computations such as

```
tr1 * 2; tr1 > 0
```

as well as computations with two track objects. Here's an Arith method for two track objects:

```
setMethod("Arith", signature(e1 = "track", e2 = "track"),
  function(e1, e2) {
    if(!identical(e1@x, e2@x))
        stop("only identical positions allowed")
    e1@y = callGeneric(e1@y, e2@y)
    e1
})
```

We're being very picky (or lazy) and don't try to interpolate or match non-identical positions. Having verified that these are identical, we just replace the y slot by applying the same operator to the y data from the two arguments, and return the resulting track object. As in many applications, identical is the way to test: do *not* use the "==" operator, which will be confused by missing values.

To handle mixed expressions, we can define methods for one track object and one numeric object.

```
setMethod("Arith", signature(e1 = "track", e2 = "numeric"),
  function(e1, e2)
    {e1@y = callGeneric(e1@y, e2); e1})
```

We would need a similar method for the case that e2 is a track object and e1 is numeric:

```
setMethod("Arith", signature(e1 = "numeric", e2 = "track"),
  function(e1, e2)
   {e2@y = callGeneric(e1, e2@y); e2})
```

There is no checking in these methods for the length of the vector. If it's too short, the vector method for the operation will extend it; if it's too long, however, the resulting track object will be invalid, because it will have different lengths for the two slots. In situations like this, you will be doing your users a favor by adding checking, and an appropriate action or error message when something seems confused. But, even without that, a validity method for the class will protect the user from serious damage in most cases.

By now, the method for unary arithmetic operators should be fairly obvious:

```
setMethod("Arith", signature(e1="track", e2="missing"),
  function(e1, e2){ e1@y = callGeneric(e1@y); e1})
```

The methods for comparison operators are also straightforward; they work essentially the same as the arithmetic methods, but they just return the logical vector, instead of inserting it into a track object.

```
setMethod("Compare", signature(e1 = "track", e2 = "numeric"),
  function(e1, e2) callGeneric(e1@y, e2))
```

The other signatures are left as an exercise.

Leaving some methods for the class unimplemented means that the computations will fall through either to some class that the new class extends or to the default method, which will look for vectors. If the operators do not apply, an error will result. The error is probably reasonable, but you may want to ask whether the error message is confusing. If it is, you can implement a method that produces a more helpful comment. For the group `"Logic"` and class `track`, the messages will usually complain that the `track` object is not a vector. Not too confusing, but we might expend the effort to define a method such as:

```
> setMethod("Logic", "track",
+   function(e1, e2) stop("logical operators are meaningless ",
+      "for track objects"))
> tr1 & tr1 > 0
Problem in tr1 & tr1 > 0: logical operators are meaningless
for track objects
```

The method only has to be defined for each argument matching the new class, leaving the other argument to be `ANY`; in addition to the above, we need the same definition for `signature(e2="track")`.

8.3 Generic Functions

A function is generic whenever there are distinct methods for it, so that the function itself can be thought of as the *generic* definition of what the function should do, with the methods giving the specific definition of how the function does it. Occasionally the function needs to be considered as a whole, meaning its generic definition and all the methods.

8.3.1 Generic Functions as Objects

The usual way to create a generic function is to convert an existing function, turning that function into the default method for the new generic. No explicit action is required in this case: just calling `setMethod` with the name of the function as first argument implicitly defines the function as a generic. S then creates in your chapter a *methods definition* for that function, assuming one was not there before. You can ask whether a particular function name corresponds to a generic (meaning, usually, whether there are any methods defined for it on any currently attached library), by calling `isGeneric`. Suppose we start off with an ordinary function, `polySpline`:

```
> isGeneric("polySpline")
[1] F
> setMethod("polySpline", "nbSpline", polySpline.nbSpline)
> isGeneric("polySpline")
[1] T
```

S now knows that methods exist for `polySpline` and any calls to it pass through a "method dispatch" computation to find and evaluate the appropriate method.

The creation of methods for a function on a particular database via `setMethod` assigns an object in the metadata of that database, containing the function and the local methods. When the S evaluator needs to call a generic function, it merges the information about methods for the function from all the currently attached databases. The object that stores the definition of the methods and the generic function has class `"methodsGeneric"`. You rarely need to deal with these objects directly, and should *very* rarely consider modifying one. The `setMethod`, `setGeneric`, `removeMethod`, and related functions manage the `methodsGeneric`.

A call to `getMethods` will return the `methodsGeneric` object corresponding to a particular function's name. An optional argument `where` selects the methods on a particular database. The show method for this object will display the generic function's definition, the default method (if there) and all the known methods with their corresponding signature. For common generic functions, this is a lot of printout. However, if you are designing a new generic function, it will occasionally be worth looking at the specifics. The function `dumpMethods` is an alternative; it writes out S expressions that would recreate all the methods. Showing `getMethods` is a bit more explicit. We'll show an example in the next section, on page 350.

8.3.2 Specifying the Generic Function

So far, we have stuck to the standard notion that a generic function is created automatically by specifying a method for any existing function. This is by far the more common situation, but you can also make this process explicit. Two reasons for doing so are to *avoid* having any default method or to do some extra computation in the generic function other than selecting a method. Whatever the reason, we can create a generic function explicitly by a call of the form:

```
setGeneric(name, def)
```

The call to `setGeneric` needs the name of the generic, but also a function definition, `def`. Usually the definition is only important in that it establishes the formal arguments, which previously S took from the existing function. The body of the function will normally be

```
standardGeneric(name)
```

meaning that the generic function just dispatches a method for generic `name` to do the actual computing. For example,

```
setGeneric("polySpline",
   function(object)standardGeneric("polySpline"))
```

In this case, the body of the generic has the same standard form that S had previously created itself; all it means is that S will look for a method for `polySpline`.

Generic functions without default methods can arise when we want to define a new function that *only* makes sense for some very specific classes of arguments. We could just write an ordinary function that checked the arguments, but it's often cleaner logically to define a new generic function plus one or a few methods. That way, if the concept of the function grows new methods can be added conveniently. We may even eventually add a default method.

An example arises in dealing with the `stringTable` slot of an object of class `string`. This is a very special table that speeds up a search for a particular character string among the elements of a string object. The slot may be empty or it may contain a table, computed by some internal C code and containing integer values. Having a string table is important for efficiency in large string objects used for lookup, so we want to "assign" a string table in the sense of making sure an object has one.

The catch is that computing these depends on the exact string and is done only by some rather obscure C routines. So essentially we want to assign string tables *only* in the sense of

```
stringTable(obj) = T # add a string table, or ...

stringTable(obj) = F # delete one
```

The tables take a substantial amount of space, so we may want to make sure there isn't one for a large object that doesn't need one.

To implement this idea, the generic replacement function for `stringTable` is defined, with no default method and with a method only for the case that the two arguments match classes `string` and `logical`.

```
setGeneric("stringTable<-",
   function(object, value) standardGeneric("stringTable"))
```

There was no such function before creating the generic. After defining the desired method for object from class `string` and `value` from class `logical`, the complete methods definition for `"stringTable<-"` looks as follows:

```
> getMethods("stringTable<-")
Methods and Generic Function for "stringTable<-"

Generic Function:
function(object, value)
standardGeneric("stringTable<-")

Default method:
function(object, value)
stop("No default method for stringTable<-")

Method for signature "string", "logical":
function(object, value)
{
  if(value && length(object@stringTable) == 0)
    .Call("set_string_table", object)
  else {
    if(!value)
      object@stringTable = new("string")@stringTable
    object
  }
}
```

Although we spoke of the function as not having a default method, actually the `setGeneric` call *created* a default method that generates an error if called.

So far, the generic function itself is still just the form that would have been created automatically, containing only a call to `standardGeneric`. However, other computations may be done in the generic function; it will still call `standardGeneric`, to invoke the appropriate method explicitly at some point in the body of the generic function. The value of the call is the value of the method. The call is just an ordinary S expression, and you can do anything else you like before or after.

The `Summary` group of functions provides an example, illustrating also an approach to methods for functions with arbitrarily many arguments. Summary functions, such as `max`, `sum`, and `range`, are designed to work on any number of objects. Their formal arguments are `x` for the first object, `"..."` to absorb any additional objects, and a special flag `na.rm` for dealing with missing values. The problem with methods for such functions is that the method has to be selected for each object and may be different, but the function needs to combine the results over all the arguments.

It would be inconvenient (and hard to express) to leave all this up to the designer of a specific method. Instead, we take advantage of what we know about the *result* of these functions to recast the generic computation recursively, with methods being applied only to single objects.

We can formulate a rule for combining the results of each summary function over more than one object. For example, `max(x1, x2)`, the maximum of the data in both objects `x1` and `x2`, is identical to

```
max(max(x1), max(x2))
```

It is this recursive reformulation that we can use in the *generic* function, not in the individual methods. The definition of the generic function for `max` illustrates the technique.

```
setGeneric("max",
function(x, ..., na.rm = F)
{
  if(nDotArgs(...) > 0)
    max(c(max(x, na.rm=na.rm) , max(..., na.rm=na.rm)))
  else standardGeneric("max")
}
)
```

The utility function `nDotArgs` just returns the number of actual arguments matching `"..."` in the call. The standard method selection is invoked only if there are no additional arguments; otherwise, the function invokes itself recursively, both on the first object and on all the remaining objects. The style here is sometimes called *tail recursion*.

Writers of methods for the `Summary` functions can now assume that only one object will be supplied. This is the more commonly encountered case anyway, so the extra computation involved in recursion when more than one object is supplied should not be a serious burden. As a technique for handling methods with arbitrarily many arguments, tail recursion helps fairly often. It does require that the generic function know how to put together the results of applying the function recursively to individual objects, or perhaps to pairs of objects.

8.3.3 Group Generic Functions

Generic functions may be organized into *groups*. Each generic function can belong to a group; more precisely, the specification of a generic can name another generic function to represent the group to which the first generic belongs. The group generic is intended to be a template for all the member generics in the group. The usefulness of the concept is that methods written for a group generic will be used for any member of the group, if there is no more directly applicable method defined for the member. This section discusses how group generic functions work, in case you want to define a new one or (perhaps more likely) would like to understand how existing group generic functions work.

The group generic `"Math"` is supplied with S, and the one-to-one mathematical functions, such as `exp`, `abs`, etc., described in section 8.2.6, belong to this group. You can find out the functions currently sharing a particular group by calling `getGroupMembers`:

```
> getGroupMembers("Math")
 [1] "abs"     "acos"    "acosh"   "asin"    "asinh"
 [6] "atan"    "atanh"   "ceiling" "cos"     "cosh"
[11] "cumsum"  "exp"     "floor"   "gamma"   "lgamma"
[16] "log"     "sin"     "sinh"    "tan"     "tanh"
[21] "trunc"
```

The extension of methods in the group to a new class usually works the same, or almost the same, for all members of the group. In this case, the method can be written once for generic `"Math"`, typically using `callGeneric` to re-invoke the actual function on some modified data.

A group generic is specified much like a regular generic, but by calling `setGroupGeneric`. Since group generics do not correspond to ordinary functions, the definition of the generic has to be supplied. Only the arguments in the function definition matter, however, since group generics are never called directly. The body of the defining function is ignored, and is always inserted as an error message, telling the user that group generic functions are not meant to be called directly. The "Math" group generic is specified as follows:

```
setGroupGeneric("Math", function(x)NULL)
```

The arguments to the group generic *must* agree with those of the members of the group, for the same reason that methods must agree with the generic function in their formal arguments. The technique for dispatching a method in S is to evaluate the body of the method in the evaluation frame corresponding to the call to the generic function. Therefore, methods for a group generic have to have the same formal argument names as the members, so this evaluation will make sense. For this reason, a second group, Math2, contains the functions round and signif, which resemble the Math functions but take a second argument, digits.

Once a group generic is defined, other generics can be specified as belonging to this group. The group is an optional argument to setGeneric:

```
setGeneric("sin", group = "Math")
```

works in the usual way, assuming there was an ordinary function definition for sin. Most of the time, we don't need explicit calls to setGeneric, so you can set the group of a function directly, after some methods have been defined:

```
setGroup("sin", "Math")
```

Once group membership is established, the method selection mechanism will include methods for function "Math" as well as for the generic function itself. If no matching method can be found for "sin", methods for "Math" will be tried. Group membership is transitive:

```
setGroupGeneric("Arith", function(e1, e2) standardGeneric("Arith"),
     group = "Ops")
setGeneric("+", group = "Arith")
```

implies that "+" belongs to group "Arith" and indirectly to group "Ops".

Any generic function can be specified as belonging to any other as its group. That is, groups are essentially ordinary generic functions, except that they are never invoked directly. Methods for them are unusual in that they are written to apply to any of the member generic functions. The general mechanism that makes this possible is the presence of the local variable .Generic in the evaluation frame of a method. The value of .Generic can be used to distinguish which function is really being called. More conveniently, the built-in function `callGeneric` can be used in most cases to call the actual generic with some new arguments. As a typical example, objects of class `"track"` have a slot `"y"` containing the numeric values associated with the object. We want to implement math functions for this class by just applying the generic to the `"y"` slot, and storing the result back into the object. (Presumably, the other information about the object is unchanged by the one-to-one function.) A method to do this is as follows:

```
setMethod("Math", "track",
  function(x) {
    x @ y = callGeneric(x @ y)
    x
    })
```

Sometimes a group generic method for a particular class makes sense for some members of the group but not all. There is a choice in this case: either to define separate methods for the individual functions, or to weed out the inapplicable functions in the group method, using `.Generic`. For example, if equality comparisons for a particular class made sense, but the object didn't understand ordering, a method for group generic `Compare` could select only those that made sense.

The class `raw` is an example. This class stores raw bytes of data as an atomic vector object. But S does not want to make assumptions about the contents of the bytes (that's why the class is called `raw`). Therefore, the only comparison operators that make sense are equality and inequality, interpreted as comparing the two objects overall. A method for the `Compare` group generic can implement both the methods and the restriction.

```
setMethod("Compare", signature("raw", "raw"),
# comparisons of equality, inequality only are allowed
# for raw bytes, no ordering is implied.
function(e1, e2) {
  switch(.Generic,
  "==" = return(identical(e1, e2)),
  "!=" = return(!identical(e1, e2)))
  stop("comparison (\"", .Generic,
       "\") not meaningful for raw data")
})
```

Chapter 9

Documentation and Documentation Objects

> This chapter describes how to use and develop online documentation for S functions, classes, and methods. Utilities for using online documentation include the "?" operator and other more specialized tools. These read or generate *documentation objects*, initially through self-documentation. Programmers can edit documentation through a dump, edit, and source cycle similar to that for other S objects. Documentation involves both S objects and the information that describes other S objects; for the information itself, we use the SGML language, the basis for most web documents. Programming with S documentation is a combination of programming in S with the S objects, plus editing and programming with the dumped SGML files. This chapter deals mainly with the S side; the book's web site describes additional tools, particularly for the SGML side.

S provides online documentation to the user as part of the language. Information about functions and other topics is stored in *documentation objects*, which in turn are kept in documentation meta databases, corresponding to `meta="help"` in calls to database tools. Users see online documentation through utilities, such as the `"?"` operator, that find documentation objects and display information contained in them. Section 9.1 discusses how online documentation can be viewed by the user. Functions, classes, and methods are *self-documenting*: initial documentation is generated from the corresponding function, class, or method (see section 9.2). Programmers can dump, edit, and source back documentation objects, starting from existing objects or from the self-documentation, and using tools similar to those for editing other S objects. Documentation objects are dumped to text files, with the structure in the documentation converted to commands (tags) in the standard markup language SGML. Section 9.3 discusses editing documentation; section 9.4 considers the few changes for documenting classes rather than functions or methods; section 9.5 describes the documentation objects and tools to manipulate them directly.

The structure of documentation objects in S maps directly into analogous structure in SGML. As a result, the S programmer has access to a rich and growing range of possibilities for developing interactive documentation facilities. Facilities such as browsers based on HTML or XML can have access to the inherent structure of the S documentation and, indeed, to much of the structure of S itself. At the same time, the power of S programming can be applied to examine and restructure the content of the documentation objects. A simple example is the function on page 371. As this book is written, we are just beginning to exploit the synergy between the two views, S and SGML, of the documentation. For the ongoing developments, consult the web site for the book,

```
http://cm.bell-labs.com/stat/Sbook
```

The rest of the chapter discusses viewing documentation from S itself, but understand that this is only one aspect of a richer documentation programming environment.

9.1 Viewing Online Documentation

S has an operator, "?", for viewing documentation. Typing the operator followed by the name of a topic causes S to print the documentation for that topic:

```
> ?sum
Title:
        Sums and Products
Usage:
        sum(..., na.rm=F)
```

```
        prod(..., na.rm=F)
Arguments:
  ...:  numeric or complex objects.  Missing values (‘NA’s)
        are allowed, but will cause the value returned to be
        ‘NA’ unless ‘na.rm’ is ‘TRUE’.
  na.rm:  a logical value, default ‘FALSE’. If ‘TRUE’,
        missing values are ignored.

Value:
        the sum (product) of all the elements of all the
        arguments. If the total length of all the arguments
        is 0, ‘sum’ returns ‘0’ and ‘prod’ returns ‘1’.

Examples:
        meanX = sum(x)/length(x)
Key Words:
         math
```

Typing the operator as an expression by itself, without arguments, prints the detailed documentation on "?".

The topic needs to be quoted if it is not a name. To see the online documentation for the operator "+", for example:

```
?"+"
```

The subscript and element operators are represented by the opening single or double bracket:

```
?"["
?"[["
```

All S functions have some online documentation. If nobody has written any special documentation for the function, documentation is created automatically. This may only give the arguments and their default values, but is better than nothing. Functions can also be self-documenting more explicitly, if the author of the function has included some comments at the beginning of the function definition (see page 359). If a function has methods for different classes of arguments, these methods can have separate online documentation, which can also be self-generating from the definition of the method. S classes also have online documentation, with automatic documentation inferred from the definition of the class.

Documentation topics do not need to correspond to any specific function, method, or class:

```
?Syntax
```

prints the online description of the syntax of S. You can ask whether *explicit* documentation for a topic exists by calling `existsDoc(topic)`, but self-documentation, as discussed in section 9.2, will not show up from this call.

The `"?"` operator can be used in other, specialized forms. If an argument precedes the operator, it requests only the corresponding type of documentation:

```
class ? file
```

requests only documentation on the *class* `file`, as opposed to the function of the same name. The built-in documentation types are `"function"` (the standard type), `"class"`, and `"method"`.

Documentation for classes is stored specially, so a function and a class can have the same name, but distinct online help. For example, while `?matrix` documents the `matrix` function,

```
class ? matrix
```

produces documentation on the `matrix` class of objects.

Similarly, to get the methods documentation for a particular function:

```
method ? whatis
```

This shows the user what methods are defined for function `whatis` and then starts up a dialog in which the user is prompted for the particular method(s) for which documentation is wanted. A more extended expression for displaying method documentation specifies both the function name and the objects that will be supplied as arguments, or their classes. To see the documentation for the method that `whatis` would use for an object of class `matrix`:

```
method ? whatis("matrix")
```

You can also supply an object instead of the class name:

```
method ? whatis(xm)
```

would produce the method documentation for whatever method `whatis` selects for argument `xm`. If `xm` has class `"matrix"`, the result will be the same as for the previous example. If the argument evaluates to a single character string, though, `"?"` interprets that as a class name.

There doesn't need to be a method explicitly defined for the classes supplied. S applies the rules that are used in selecting the method for computation. For example, if there is an `array` method for `whatis` but not a `matrix` method, the documentation for the `array` method will be used, since class `matrix` extends `array`.

9.2 Self-Documentation

It's very desirable to document new functions, classes, and methods, but also somewhat tedious. To make the work easier and to fill in until the author gets around to it, all these are self-documenting.

If you ask for documentation on a function and none has been explicitly created, S generates a documentation object from the function object itself, containing four things:

1. the calling sequence;
2. the individual arguments with any default values;
3. optionally, a `Description` paragraph, made up of any comment lines that appeared just before the function definition on the source file;
4. stubs for the other important sections of documentation.

Most of the stubs are omitted when a brief form of the documentation is displayed online. The third item means that initial comments can (and definitely should) be used to provide a rough description of the function, pending more serious documentation.

As an example, let's look at a function, `reScale`, whose purpose is to rescale its first argument to the range of its second argument. A version of the function was described in section 6.1.1 on page 247. Here is a slightly more refined version, revised to ignore missing values in argument `xrange` and to use two convenient S functions, `range` and `diff`.

```
"reScale" =
# the data in x, rescaled to the range of xrange.
#
# Missing values in xrange are ignored
function(x, xrange)
{
  xr = range(x, na.rm=T)
  yr = range(xrange, na.rm=T)
  x = (x - xr[1])/diff(xr)
  yr[1] + x * diff(yr)
}
```

The best place for the comments is between the assignment operator on the top line and the function definition. Other positions at the beginning may work, but this position is guaranteed, and the comments will not get moved around. Assuming there is no explicit documentation for reScale, then a request to get online documentation for it would create a self-documentation object and print the useful parts of it:

```
> ?reScale
Title:
        Function reScale
Usage:
        reScale(x, xrange)
Arguments:
  x:    argument, no default.
  xrange: argument, no default.
Description:
        the data in x, rescaled to the range of xrange.

        Missing values in xrange are ignored
```

Notice that the initial comment lines have been extracted and formatted as the Description section of the documentation. The printing will fill in the text in the description section, but empty comment lines will be retained to separate paragraphs within the description.

One convention, used to save typing, is that text that refers to S objects or expressions will be highlighted if it is enclosed in single back- and forward-quotes, such as ‘x’. This is an easy way to show off such text, and adds to the readability of the documentation. When you come to edit documentation, S will automatically convert such text to an SGML tag for displaying it in "computer output" style. When documentation is plain-printed (what "?" produces by default), the text is again enclosed in quotes. The text will be highlighted properly in other forms, for instance in offline printing or by browsers. We can revise the comments in the example to follow this style:

```
"reScale" =
# the data in ‘x’, rescaled to the range of ‘xrange’.
#
# Missing values in ‘xrange’ are ignored
  ....
```

If you are willing to do a little more work, there are alternative tags that identify the text as a reference to an S argument, another S function, or some other type of reference. Using such tags makes the documentation better for browser use; see the book's web site for details.

Function comments are usually formed from editing the function, but a call to functionComments will extract or replace them explicitly.

```
> functionComments(reScale)
[1] "the data in 'x',"
[2] "rescaled to the range of 'xrange'."
[3] ""
[4] "Missing values in 'xrange' are ignored"
```

As a replacement function on the left side of an assignment, `functionComments` replaces the comments with the right side of the assignment, one comment line per character string.

When you are ready to create an actual documentation object for the function, a call to `dumpDoc` will generate a file containing the self-documenting information. This file can then be edited to produce more informative documentation, and the function `sourceDoc` will read the file and create the documentation object. See section 9.3 for details on editing the documentation.

Self-documentation for methods works basically the same way. In this case, function comments in the function definition for the particular method produce the description. Suppose we dump the method for function `show` and class `track`, described on page 333, by a call to `dumpMethod`. We can then add some lines of comments to the function definition:

```
setMethod("show", "track",
# this method turns the object into a 1-row matrix of
# the values, with columns labelled by the positions.
function(object)
{
    vals = matrix(object@y, nrow = 1)
    dimnames(vals) = list("y:", object@x)
    cat("An object of class \"", class(object), "\"\n",
        sep = "", file = stdout())
    show(vals)
})
```

As with functions, the initial comments are turned into a `Description` section, and the signature is shown as a separate section.

```
> methods?show("track")
Title:
        Method show
Usage:
        show(object)
Arguments:
  object: argument, no default.
```

```
Signature:
  object:  "connection"
Description:
       this method turns the object into a 1-row matrix of
       the values, with columns labelled by the positions.
```

9.3 Editing Documentation

The current documentation on any number of topics can be dumped to a text file by `dumpDoc(topics, file)`, where `topics` is a vector of the topics and `file` is any file or writable S connection (see chapter 10). The topics do not need to have been documented previously. If there is self-documentation as described in section 9.2, this will provide the initial contents of the dumped documentation. When the file has been edited to your satisfaction,

```
sourceDoc(file)
```

will read the file and turn its contents into a documentation object.

The programming cycle is similar to the dump-edit-source style that applies to programming with functions, classes, or methods. The main difference is that the file written is not by itself made up of S expressions; instead, it is written in the SGML language. As a result, the source command has to be specialized, to translate the file in the markup language back to S documentation objects.

Suppose we begin with the function `reScale`, on page 359, and dump its self-documentation:

```
> dumpDoc("reScale")
[1] "reScale.Sgml"
```

By default, the file is given the name of the function with a suffix of `".Sgml"`, but you can supply a filename, or any S connection object, as the second argument to `dumpDoc`.

The file contains all the text of your documentation, supplemented by special text in the SGML markup language. The special text is in the syntax of *tags* in SGML: the tag is enclosed in angle brackets, possibly with *attributes* (similar to named arguments in S) preceding the closing `">"`. Arbitrary text follows, possibly including additional tags recursively, and the section is usually closed by the same tag, this time preceded by a slash. So, for example, the tag `s-section` might open with

```
<s-section name="Note">
```

Text follows, and the section is closed by

```
</s-section>
```

There are a vast number of books and online documents on SGML and related topics; the "official" web site for its definition is:

```
http://www.w3.org
```

It's not necessary to know much about SGML to edit the S documentation; just following the example of the file generated from the self-documentation should be enough to get going.

Some of the sections generated by `dumpDoc` should *not* be edited by hand, unless you're feeling bold. The usage section, for example, is generated from an S expression by some special code that creates specialized tags for nearly every part of the expression. These tags are useful in developing interactive documentation tools, but they make the dumped version of the section hard to read and harder to edit without making mistakes. The usage section for `reScale` prints as just:

```
Usage:
       rescale(x, xrange)
```

but the corresponding SGML has the form:

```
<s-usage>
<s-call>
 <s-function-name name="reScale"> reScale </s-function-name>
 <s-argument name="x"> </s-argument>
 <s-argument name="xrange"> </s-argument>
</s-call>
</s-usage>
```

The structure of the S call has been mapped into corresponding SGML structure, useful but not for humans to edit unaided. A better alternative when such documentation sections must be edited is to edit the documentation object in S. For example, suppose we wanted to merge documentation for another function, `reScale2`, into the same documentation object. One way to do this is as follows:

```
setDocSection("reScale", "s-usage",
    expression(reScale(x, xrange), reScale2(x, y, xrange)))
```

The `setDocSection` function will convert the `expression` argument into a suitable SGML form and insert it in the metadata object for this documentation.

Human editing of the dumped documentation file can then be kept largely to the plain-text portions, such as the descriptions of arguments, of the value returned, etc. After editing the file, you can restore the revised documentation object by, in this case, `sourceDoc("reScale.Sgml")`. The tags will be reinterpreted in terms of S documentation objects and the resulting object will be assigned in the `meta="help"` part of your chapter. As with classes and methods, you should get and set documentation objects through the tools, not by working directly with the metadata objects.

The documentation file is organized into top-level sections, corresponding to the sections shown when viewing the documentation (`Title`, `Usage`, etc.). Each of these is marked off by a corresponding tag, `s-title`, `s-usage`, etc. The sections are bounded by the corresponding tags in SGML-style; for example, the title section starts with `"<s-title>"` and ends with `"</s-title>"`.

The dump file is produced from self-documentation if there is no current documentation object for the given name. Rather than omitting some sections, as is done in ordinary printing, empty sections are left as stubs to be replaced by the author. Here's a portion of the file `"reScale.Sgml"`:

```
<!doctype s-function-doc system  "s-function-doc.dtd">
<s-function-doc>
<s-topics>
<s-topic>
reScale
</s-topic>
</s-topics>
<s-title>
Function reScale
</s-title>
<s-description>
the data in x, rescaled to the range of xrange.
<p>
 Missing values in xrange are ignored
</p>
</s-description>
<s-usage>
  ......
</s-usage>
<s-args>
<s-arg name="x">
```

```
argument, no default.
</s-arg>
 ....
</s-args>
<s-value>
<!--Put description of return value in this section -->
</s-value>
 ....
<s-docclass>
function
</s-docclass>
</s-function-doc>
```

Comments in SGML—the tags beginning `"<!"`—indicate sections to be filled in during editing. At least to start with, only a few of the sections in this SGML file should, or need to be, edited by hand, mainly the following:

`s-title`: the overall title for the documentation;

`s-arg`: the individual sections describing each argument;

`s-value`: the section describing the value returned by the function;

`s-example`: a section containing examples of the use of the function.

As mentioned before, some sections of the file (e.g., `s-usage`) will be a jungle of nested tags, usually an indication that you don't need to edit this section, but instead can use an S tool, such as the `setDocSection` function, to extract and replace the information in S.

Within the `s-args` section, individual arguments each have their own subsection, enclosed in the `s-arg` tag, with a `name=` attribute (in SGML terminology) giving the argument name. As always, in SGML the attribute string supplied with the `name=` argument appears *inside* the angle brackets around the tag's name. This string is used by tools in S that extract documentation for a particular argument to an S function. It's best just to think of it as the way of representing, in the SGML syntax, the name of an element in the documentation object, the element itself being the text between the opening and closing appearance of this particular `s-arg` tag. If we wanted the text describing argument `x` to be `"numeric vector"`, the corresponding piece of SGML syntax would be:

```
<s-arg name="x">  numeric vector
</s-arg>
```

The s-arg tag takes its text in free format; that is, we expect the methods for displaying the text in this section to move words around and fill lines as needed. If the argument names are those produced by the self documentation, the SGML syntax will have been generated automatically. Only the plain text needs to be edited.

Inside the s-examples section, examples of S expressions will be enclosed in one or more s-example tags. This tag arranges for display of the S expressions, without text filling. The s-example tag displays S expressions, set off from the rest of the document. With a suitable interactive interface and browser, these sections could also be delivered to an S process to be tried out.

In editing the file, we will replace the stubs with real information, delete any irrelevant sections, move pieces of the description section to other appropriate sections, and delete the description section. The special tags starting and ending the file are generated automatically and are crucial to processing it correctly. They should not be edited.

The edited version of the file contains the following.

```
<!doctype s-function-doc system  "s-function-doc.dtd">
<s-function-doc>
 ....
<s-title>
Rescale A Vector
</s-title>
 ....
<s-args>
  <s-argument name="x">  numeric vector
  </s-argument>
  <s-argument name="xrange">
     desired range, or arbitrary other numeric vector.
     Missing values are ignored in computing the range.
  </s-argument>
</s-args>
<s-value>
the data in <code>x</code>, rescaled to the range of
<code>xrange</code>.
</s-value>
<s-examples>
<s-example>
plot(x0, y0)
gridLines = range(stdGridLines, y0)
abline(h = gridines)
```

```
</s-example>
</s-examples>
<s-see>
<s-function-ref>pretty</s-function-ref>
for choosing grid-line positions.
</see>
 ....
<s-docClass> function </s-docClass>
</s-function-doc>
```

In filling in the `s-see` section (the `"See Also"` section), we used a special tag, `s-function-ref`, which allows any browser or other documentation tool to treat this material specially, as a reference to an S function. See the detailed documentation on the web site for a fuller description of such special tags.

Inside the `s-topics` section of the documentation, each `s-topic` tag means that the text becomes an identifying name for this documentation. Suppose we add a section of the form:

```
<s-topic>Grid Rescaling</s-topic>
```

Then users can access the same documentation by referring to the (complete) text of this topic as they could previously for topic `"reScale"`:

```
help("Grid Rescaling")
```

will produce the same documentation as `?reScale`. The common use of multiple topics comes when we want to document closely related functions together, such as `exp`, `log`, and `logb`. The corresponding documentation has three topics and is accessible by any of them.

Other tags in the documentation file are generated automatically in processing the documentation object. The `s-docClass` tag, for example, defines what kind of documentation the whole document contains. There are special documentation classes for `method` and `class` documentation. General-purpose documentation has `docClass` equal to `"function"`. The S functions that manage documentation objects may use the documentation class to treat some documentation specially; for example, `method` documentation includes an `s-signature` section for information about the signature of the method.

When we have edited the file to our satisfaction we can source the documentation into S by a call to `sourceDoc`. The suffix `".Sgml"` generated for the file can be omitted; the `sourceDoc` function works with multiple files having different suffixes, and will construct the file names as needed. In our example, to create the modified documentation object:

```
> sourceDoc("reScale")
[1] "reScale"          "Grid Rescaling"
```

The value returned by sourceDoc is the vector of all documentation topics associated with this file. The essential side effect of sourceDoc is to store the documentation objects in the meta database associated with documentation in an S chapter, in the working data by default. You would normally want the documentation in the same chapter as the definition of the function or class being documented: the argument where= to sourceDoc can be used to specify the chapter.

The online documentation for reScale now reflects the changes.

```
> ?reScale
Title:
        Rescale A Vector
Usage:
        reScale(x, xrange)
Arguments:
  x:    numeric vector
  xrange: desired range, or arbitrary other numeric vector.
        Missing values in 'xrange' are ignored.

Value:
        the data in 'x', rescaled to the range of
        'xrange'.
Examples:
        plot(x0, y0)
        gridLines = range(stdGridLines, y0)
        abline(h = gridines)
See Also:
        'pretty' for choosing grid-line positions.
```

Method documentation is dumped and edited by an extension of the same dumpDoc function. The arguments can include special="method" to indicate that methods documentation is wanted. In this case the argument sig= supplies the signature for the method.

To dump the documentation for function whatis and signature "vector":

```
> dumpDoc("whatis", special="methods", sig="vector")
[1] "whatis.vector.Sgml"
```

This file is edited and restored as before. Because it contains "method" for its documentation class, and an s-signature tag, sourceDoc will restore it suitably as method documentation.

9.4 Documenting Classes

Aside from a few details, class documentation is handled much like function documentation. Classes have self-documentation, produced from the stored representation of the class. There is currently no convenient shortcut for adding comments to the class documentation.

To dump the class documentation, include `special="class"` as an argument to the `dumpDoc` function. The class documentation has different tags (`s-slots` but not `s-args`, for example), and the `s-docClass` is `"class"`.

The edited documentation is installed by `sourceDoc`. This does not need any `special` argument; the `s-docClass` determines what documentation object to create. To see the structure of class documentation, dump documentation for one of the standard classes, for example `sgmlTag`.

9.5 Documentation Objects

To edit S documentation you should dump the documentation to a file, edit the file, and source it back to create or recreate the corresponding documentation object. This object will then be stored in the `"help"` metadata of the chosen database. The documentation object is used by all the tools to print or dump online documentation. For ordinary editing, you don't want to assign these objects directly; the documentation utilities maintain consistency among topics, for example, and direct assignment risks breaking that.

Documentation objects correspond to topics; most of the time, there will only be one topic per object, but several topics can be documented together, as illustrated in our example. The documentation utilities will manage this, by "aliasing" the corresponding documentation objects. Aliases should be removed or introduced by dumping and editing the corresponding documentation object, adding a topic. The documentation utilities maintain information about the documentation aliases; that information will be updated when `sourceDoc` is called. The one exception is that when *removing* a topic with aliases, you have the opportunity to only remove some of the topics, as a shortcut, by giving `removeDoc` the argument `all=F`.

You can see all the topics aliased together by calling `getDocTopics` with one of the topics as argument:

```
> getDocTopics("sin")
[1] "cos" "sin" "tan"
```

You can call the `setDocSection` function to modify pieces of documentation objects, as stored on the database. For constructing other information (e.g., tables of cross-references) or for advanced manipulation, the objects can be accessed directly. The utilities in the rest of this section help you examine them.

Documentation objects can be manipulated directly by functions `getDoc`, `existsDoc`, and `findDoc`, which work much like the functions `get`, `exists` and `find` for ordinary objects. Along with `dumpDoc` and `sourceDoc`, they support a programming cycle and style analogous to that for functions. Function `removeDoc` removes documentation objects. As with class and method metadata objects, it's strongly recommended to deal with documentation objects only through the standard tools. They work much as the analogous functions for other objects do, with a few differences.

A documentation object has class `"sgml"`. The `body` of the object contains all the actual information. This information is organized by analogy with SGML tags; specifically, as a tree of `sgmlTag` objects. Each `sgmlTag` object has a tag, the character string specifying the corresponding tag in SGML; such as `"s-title"`. Each `sgmlTag` object has its own body, which is always a list. The elements of the list are either ordinary text or other `sgmlTag` objects. Functions `docTag` and `docBody` extract the tag name and the body.

Some tags, such as `"s-arg"`, include attributes in the SGML sense, which are naturally represented in S as an object of class `named`. The standard documentation objects only make limited use of SGML attributes, mostly to provide argument and slot names, but the mechanism in the `sgmlTag` class is quite general, for possible future extensions. The function `docAttributes` extracts the attributes, which will be of length zero if this tag does not use SGML attributes.

The `sgmlTag` class is a complete tree-structure representation of the corresponding documentation in SGML. As a result, S computations have the whole of the documentation structure available. A fundamental design goal for the documentation objects in S was for the programmer to have full information to allow any computations to be done either in S, using the objects, or outside, using the corresponding SGML file.

Computations that manipulate documentation objects in S may or may not need to work with all the `sgmlTag` objects in the tree. For working with the top-level sections of an S documentation, only the first level of the tree will be needed. For this purpose, the `docSection` function can be used to extract or replace sections, using the standard tag names described in section 9.3. More general computations will need to work down the tree; for example, to find any reference to a particular term or topic.

The function `rapply` may be useful for such computations: it applies a computation recursively to all the nodes of a tree-like structure, providing a filter to restrict computations to certain classes and doing the recursive tree walk in C to increase speed. As an example, here is a little function to extract all the unique strings enclosed in `"<code>"` tags.

```
"allCode" =
## all the material contained in '"<code>"' tags
## in the documentation for 'topic'.
function(topic)
{
  doc = getDoc(topic)
  codeText = function(obj) {
    if(identical(docTag(obj), "code"))
      unlist(docBody(obj))
    else NULL
  }
  text = rapply(docBody(doc), codeText, "sgmlTag")
  sort(unique(text))
}
```

Applying it to the documentation for `reScale`:

```
> allCode("reScale")
[1] "x"      "xrange"
```

The same logic could be used in a more general function to manipulate various cross-reference tags, such as `s-function-ref`, matching text against known argument or documentation topic names, for example. You can also use `rapply` to replace the nodes of the tree structure, allowing more sophisticated editing of documentation objects in S.

The `allCode` function takes a topic, rather than a documentation object, as its argument. Where it makes sense, dealing with topics helps to insulate the user's computations from details of how documentation objects are organized. The philosophy is similar to that for programming with classes and methods: the explicit objects from the metadata can usually be kept out of user-level computations.

Chapter 10

Connections: Reading and Writing Data

The `connection` class of S objects connects an S process to other entities in the computing environment, for the purpose of reading or writing data. S manages these connections internally to preserve consistency. The input/output operations in S that involve byte streams, such as parsing, printing, and reading data, are defined in terms of connections. This chapter provides details on the tools available to handle connections, as well as the connection class itself. In addition, section 10.6 discusses some facilities for handling events; these mainly, but not entirely, involve reading from a connection when input is waiting.

10.1 Reading and Writing with Connections

Connection objects unify the *connections* between S and the rest of the computing environment, so far as input and output is concerned. A connection object represents something to which streams of bytes will be written or from which streams of bytes will be read.

Connections provide the facilities of *stream* input/output as used in C programming, but with simplicity, security, and generality through the use of S programming concepts. The most common example of connections is a file in the file system, given as a character string.

```
xx = readLines("/tmp/namesFile")
```

The character string is interpreted as a reference to a file, the S evaluation manager opens this as an S `file` connection, and `readLines` reads from this connection.

You have very likely used routines or commands in C, Java, awk or other systems to do this kind of computation. Connections provide similar features but try to be smarter, doing more for the programmer. S manages connections internally, keeping information about the connection and its current state to preserve consistency from one operation to another. Functions can access the evaluation manager's internal information about currently open connections. The S virtual class `connection` allows methods to be written for connections, using utilities defined for all connection objects.

Files are a particular class of connection; in the terminology of S classes, the class `file` extends the virtual class `connection`. Other connection classes connect S to special kinds of files, to other processes via pipes, or to S character vectors directly. Objects of other connection classes are supplied as arguments to functions operating on connections the same way files are, but are created using a function that specifies the class of connection object.

For example, `pipe(description)` says to create a pipe connection to the shell command given by `description`:

```
writeLines(text, pipe("tr '\012' '\010' |mailW"))
```

This writes `text` not directly to a file but instead to a shell command. The command gets the output written by the call to `writeLines` as *its* standard input.

The special type of files known as `fifo`'s or named pipes can also be used: `fifo(description)` takes a file name as argument but creates it, if necessary, as a fifo. S text vectors can themselves be used as input connections, by supplying them as the argument to the function `textConnection`. Section 10.2 gives details of the properties of the different classes of connections.

Functions that read data from a connection include

`parse`	*Read (parse)* `n` *S expressions.*
`parseSome`	*Read (parse)* `n` *lines or 1 expression.*
`scan`	*Read* `n` *data items.*
`readLines`	*Read* `n` *lines, one string per line.*
`read.table`	*Read a two-way table of data.*
`readRaw`	*Read binary data items.*
`dataGet`	*Read the S symbolic dump format.*

The read functions all take the connection as the first argument. They may have other arguments that control the interpretation of input.

Most read functions have an optional argument `n` that says how many "things" to read. The things will be the appropriate units for the particular function: lines for `readLines`, S expressions for `parse`, and fields (individual items) for `scan`. A convention says that supplying a value of `-1` for `n` means "read all the data available on the connection". The functions don't mind if there is no data available, they just return an empty (zero-length) object of the suitable class. In particular, the read functions do not generate an error on encountering an end-of-file (in contrast to typical C routines, for example).

Some examples of function calls to read data are:

```
scan("myData", what = numeric())
readLines(n = 1)
parse("newFuns.S", n = -1)
```

The first call reads fields from file `"mydata"` and tries to interpret each field as a number. The second call reads one line from standard input, returning a character vector of length 1. The third example expects to find S expressions on file `"newFuns.S"`; it will read the entire file, parsing all the expressions and returning a vector of class `expression`, with length equal to the number of expressions found on the file. In the first and last case, the file will be opened by the functions `scan` and `parse`; when the function returns, the file will be closed. Another call with the same argument will read the file again, from the beginning. Standard input is already open, and the call to `readLines` will leave it open. This behavior is part of a consistent general pattern that you can use to have essentially complete control over the connection (see section 10.3), but the default behavior is usually right for simple examples.

There are analogous functions for writing to a connection, including:

`dput`	*Write (deparse) S expressions.*
`cat`	*Write data items.*
`readLines`	*Write lines, one string per line.*
`writeRaw`	*Write binary data items.*
`dataPut`	*Write the S symbolic dump format.*

These functions typically take one or more arguments defining what S data to write out plus an argument to specify the connection. The connection often defaults to standard output. The function `cat` prints arguments in a direct style, with the call controlling details of what appears; `writeLines` corresponds to readLines, and writes its argument out as lines of text. The functions dput and dataPut write objects in a deparsed form (usually for human editing) and in a symbolic dump format (usually to dump S objects for archive or moving around). The function writeRaw writes output in internal binary form. See the online documentation of all these functions for more detail.

Some functions (print and message, for example) do not have a connection argument. Instead, they write to the standard connections associated with the S session. There are three of these, corresponding to standard input, output, and messages (or error). Output or input for these files is redirected by the sink function (see section 10.4).

Other functions exist to prepare connections for reading or writing. The functions open and close open and close connections explicitly. A connection opened explicitly is not automatically closed, allowing programs to use the connection in much more general ways (see section 10.3). The function pushBack allows character strings to be "pushed back" onto a connection for later reading; while this is a subtle function, it is also a powerful one (see section 10.5.1 for its use). The function seek allows the user to position a file for reading or writing.

Currently open connections are kept internally by the evaluation manager. As a result, they have a life outside the actual object representing them. The function getConnection returns a connection corresponding to its argument, or NULL if the argument does not correspond to an open connection. The function getAllConnections, called with no arguments, returns a list of all explicitly opened connections; showConnections prints a table of all the connections. These functions can be useful both in programming and for the interactive user, especially if we've lost track of just what connections *are* currently open.

Connections also play a key role in event-handling: setReader sets up a reader to look for input on a connection and invoke an action function to read a task from the connection. See section 10.6 for setReader and also for setMonitor to set up tasks performed after a specified timeout.

If the connections you deal with are ordinary files and you just want to read all the information from the file or to write a file in one step, you can ignore the remainder of this chapter. Just assume that when you identify the file (for example, as a character string), S will take care of the details. The S functions called will open the file and read the S expressions. When the computations are finished, the file will be closed. Similarly, the writing functions will write at the end of the file, and close it when done.

You may want to learn more about connections to do any of the following:

- Read from or write to a connection in stages, perhaps using more than one S function, or in other ways have more control over the input or output. In this case you will want to open the connection explicitly (see section 10.3).
- Write new functions to encapsulate such specialized treatment of connections. In this case you need to follow a paradigm to deal cleanly with connections provided as an argument to your functions (see page 384 in section 10.3). Once you do so, you can use your own functions as simply as the built-in functions.
- Use connections to objects other than files, such as to special file-like objects (pipes or fifo's) or to S character vectors. You may then need to define the connection more explicitly (see section 10.2).
- Make sure your output or input remains attached to the standard connections. About all you need to know here is that calls to `stdin`, `stdout`, and `stderr` will always return the standard connections for input, output, and messages: use them instead of explicit file names (see section 10.4 for the details).
- Push back data to cause a connection to reread data, or to treat other data as if it had been on the connection (see section 10.5.1).
- Deal with input in separate tasks, perhaps as input becomes available; or, schedule tasks at some time in the future (see section 10.6).

10.2 Connection Classes

Most connections are to files, but there are several other connection classes predefined. All connection objects are treated identically, so far as makes sense, and all can be supplied to the functions expecting a connection argument. The connection object can represent:

- a `file`, defined by a path in the file system;
- a `pipe`, defined by a shell command;
- a `fifo`, defined by a path, but acting like a pipe and sometimes referred to as a named pipe;

- a `textConnection`, using S character strings as lines of text input;

- a `terminal`, connected to the user's terminal or display.

The first three classes correspond to facilities supplied by the operating system, and so their properties are a combination of what S provides and the underlying operating system definition.

Files and fifos correspond to paths; that is, to a character string made up of directories, subdirectories, and a final file name. The names are separated by the character `"/"`: if the first character is not a `"/"` the path is assumed to start in the local directory, where the S session was invoked. These are similar to the rules that the command shell and other interfaces to the operating system apply. The path must be a legal file path in the same sense as it would for use by the shell. By default, S will create the file if it does not exist. If the file does exist, S will not truncate the file but will begin reading at the start of the file and writing at the end (see section 10.5.2 for other modes when opening the file).

File connections do not have to correspond to *any* path. If the connection's only purpose is to provide a place to write data out and then read it back in, perhaps in a different form, call `file` with no arguments:

```
ff = file()
```

The connection `ff` is unlinked from the file system, but can be used for reading and writing throughout the session, or until it is explicitly closed. The following little function reads from its `con` argument and copies lines to a temporary file until it gets a special marker; then it parses the result.

```
parseToMark =
## read 'con' until a line equal to '"<<MARK>>"',
## or to 'end-of-file'.  Parse this much of the input.
function(con) {
    ff = file()
    on.exit(close(ff))
    repeat {
        line = readLines(con, 1)
        if(length(line) == 0 ||
           identical(line, "<<MARK>>")) break
        writeLines(line, ff)
    }
    parse(ff, n = -1)
}
```

The test `length(line)==0` catches an end-of-file. The call to `on.exit` arranges to close the temporary file on exit, even if an error occurs. Readers familiar with similar programming in languages such as C may believe that `parseToMark` has a fundamental bug: it doesn't rewind the file before starting to parse. The S evaluation manager, however, maintains separate read and write positions for connections, and as a result writing does not move the initial read position, at the beginning of the initially empty temporary file. Page 393 in section 10.5.2 discusses file position generally.

The call `file()` creates and opens a temporary file. Don't confuse it with an older S function having the unfortunate name `tempfile`. This does not actually produce a file connection at all (it dates from before connection classes); instead, it returns a character string that can be used as the argument to a call to `file`, and which is guaranteed writable and not previously used in the S session. For most purposes, `file()` is the right choice, but if you will need to hand the path of a file to a shell command or other program outside S, the string returned by `tempfile()` is suitable. Once finished with the file, you need to both close it, if it is open, and also to unlink it so as to recover the file space it occupies. The following fragment suggests the style:

```
tempName = tempfile()
 .... ## do something with a shell command
tempFile = open(tempName)
 .... ## do something with tempFile
close(tempFile); unlink(tempName)
```

In contrast, `file()` has already unlinked the file path it used to create the connection. At most, the programmer needs to close the file.

Fifo's appear to be files and their description is a path just like a file. However, they only hold data written to the file until that data is read. A read on a fifo gets the first data that has been written but not yet read, at which point that data effectively "disappears". This gives them their name *f*irst-*i*n-*f*irst-*o*ut. No matter how much data flows through a fifo, it only appears as large as the data currently written but not read. The `fifo` class comes into its own in dealing with asynchronous events; for example, suppose some data is written every so often by another program (outside of the S session) and we want to capture this data and append it to an object, say `testResults`, in the current working database.

Since the data can arrive anytime, we don't want the user to have to look for it explicitly. Instead, we can arrange for the other program to write its output to a `fifo`, say `"./test.out"`. Then the `setReader` function will set up a reader that responds to any input on the fifo by calling an action function. Readers are discussed in detail in section 10.6, but the following example shows a simple version of the idea:

```
> assign("testResults", character(), where=1)
> setReader(fifo("./test.out"),
  function(con) {
    data = get("testResults", where = 1)
    assign("testResults", where = 1,
        c(data, readLines(con, -1)))
})
```

The action function picks up the `testResults` object, initialized to an empty vector, and appends to it all the lines of text currently available on the `fifo`. The `fifo` class is natural for such arrangements, since the data disappears from the file once written and then read.

Pipes, in the S view, are shell commands whose standard input can be written from S and whose standard output can be read by S. The description string in this case is the command to give the shell. S opens the pipe connection by starting up the command. Data written to the pipe becomes its standard input. Output from the pipe does not come back until some function in S reads from the pipe. If you want the standard output of the command to be printed (that is, to go to the standard output of the S session), you can open the pipe for write only (see section 10.5.2). Opening the pipe for *read* only implies that input to the command comes from the input to the S session: this is usually *not* a good idea in a standard S session, since confusion arises whether typed input is meant for the command or S itself.

A pipe connection differs from the `shell` function in providing more flexibility in dealing with output. With `shell`, all the command's output comes back at once, one string per line; with `pipe`, any S read function can read from the pipe. Similarly, input for the pipe's command can be sent in stages, by any of the writing functions.

A recommended style of programming with both shell commands and pipes is to get the command output back into S as simply as possible, and then manipulate it into the form needed. Those familiar with shell tools or related languages tend to massage the command output, perhaps through several tools, to make it "easy" for S to read. This is not wrong, but often unnecessary, and can both make the computation more complicated to understand and reduce its portability if the tools are not universally available. The following example involves matching file serial numbers ("i-nodes") to a vector of file or directory names. File serial numbers are used in S to identify databases, so this is a step in summarizing internal information about databases. The relevant shell command works like this:

```
> cmd = paste("ls -1 -id", paste(searchPaths(), collapse=" "))
> shell(cmd, output=F)
35915 .
27824 ../S
27841 ../book
25772 /usr/s/library/models
27916 /usr/s/library/main
```

The command prints lines with serial numbers followed by the name of the file or directory. Let's resist the urge to apply tools that split the output and/or rearrange it into simpler form. Instead, we can just read the n lines into a character vector of length $2n$, every other element being a serial number. If we use `cmd` to define a pipe, and give the pipe connection to `scan`, the character vector is generated by:

```
txt = scan(pipe(cmd), what = "")
```

Now we can put all this into a function, `matchInodes`, that takes some numbers and a vector of file or directory path names and matches the numbers to the i-nodes of the names.

```
"matchInodes"=
function(numbers, files)
{
    cmd = paste("ls -id", paste(files, collapse = " "))
    txt = scan(pipe(cmd), what = "")
    seq1 = seq(1, length(txt), 2)
        ## seq1 matches the numbers, seq1+1 matches the names
    inodes = txt[seq1]
    ## return the names as they match (or "" where they don't)
    txt[seq1+1][match(numbers, inodes)]
}
```

Here's an example of how it might be used, to check reliably for a chapter:

```
> !ls -id ../book
 135210 ../book
> matchInodes(135210, searchPaths())
[1] "/home/jmc/book"
```

As a programming detail, notice that the string returned is the expanded directory name from the shell, not the original element of `searchPaths()`. If some databases are not chapters, the corresponding element of `searchPaths()` will be empty, but `matchInodes` will still work.

There may be operating system limitations on how much data can "sit" in a pipe before it is read. If the command is going to generate more than, say a few thousand bytes of output before any of it is read back, the safer strategy is to write to a fifo.

Text connections are a convenience to make it easy to use an object containing character strings in a computation that expects to read from a connection. The call

```
tcon = textConnection(x)
```

coerces `x` to a character vector and creates a connection that appears to contain the resulting data. Each element of the vector acts as a line of input, when `tcon` is given to any read function. Text connections are a simple way to use existing data as input, but temporary files or fifo's are more general, if less convenient. You cannot write to text connections, by definition, but you can push back data onto a text connection (see section 10.5.1), which is just as good for moderate amounts of data, and has the small advantage that it does not involve any physical output. For greater flexibility in writing, then reading, use `file()` to create a temporary file.

Terminal connections represent connections to the user's display or from the user's keyboard. They cannot be created, but testing for them allows a function to behave differently when interacting with the user (for example, by prompting for input). If `con` is some connection object:

```
is(con, "terminal")
```

provides the needed test. The main use is with the standard connections; see page 390.

10.3 Opening and Closing Connections

The functions in S that read and write data need a connection that is *open*; that is, ready to read or write as appropriate. Connections can be opened in various *modes*; section 10.5.2 discusses some possibilities. More often, though, the important question is who should open and close a connection. For example, suppose a file is passed to the `scan` function, and no connection has been opened to that file:

```
scan(n = 3, file = file("newData"))
```

The call to `file` had no argument telling the evaluator to open the connection, so `scan` will open it and arrange that the connection is closed again when it exits.

All S functions that want to read or write with connections are expected to behave similarly. If you plan to write such functions, they should obey the following rule.

> If a function needs to read or write from a connection, and the connection is not open in the appropriate mode when the function is called, the function opens it and ensures that the connection is closed when the function returns. If the connection is already open, the function leaves it open.

Think of this as a good-citizen rule for connections, analogous to the good-citizen rule for hikers and visitors to parks: "If you brought it in, take it out. If you found it here, leave it here." Functions that prepare a connection for *later* reading or writing, on the other hand, will not follow this rule—`sink` and `setReader` are examples.

For the programmer calling the functions, the implication is that you should open the connection explicitly if you want several functions to read from data on the connection, without starting over again at the beginning.

An example will make this clearer. Suppose the file `"model.data"` contains an S expression for a model, followed by a table of data, to which we want to apply that model:

```
Fuel ~ .
                     Price Country Reliability Mileage    Type
    Acura.Integra.4 11950   Japan            5      NA   Small
         Audi.100.5 26900 Germany           NA      NA  Medium
         BMW.325i.6 24650 Germany            4      NA Compact
 Chevrolet.Lumina.4 12140     USA           NA      NA  Medium
     Ford.Festiva.4  6319   Korea            4      37   Small
       Mazda.929.V6 23300   Japan            5      21  Medium
   Mazda.MX.5.Miata 13800   Japan           NA      NA  Sporty
    Nissan.300ZX.V6 27900   Japan           NA      NA  Sporty
Oldsmobile.Calais.4  9995     USA            2      23 Compact
  Toyota.Cressida.6 21498   Japan            3      23  Medium
```

We want to use the function `parse` to interpret the model, since it is an S expression. Then `read.table` should read the data. However, we can't just pass `"model.data"` to each function, as in:

```
model = parse("model.data")
data = read.table("model.data")
```

The call to `read.table` will reopen the file and try to interpret the first line as part of the table.

To continue reading without the file being closed, we rely on the good-citizen rule, and open the file before either function reads from it. This is done most easily by calling the function `open`. The argument to `open` is a file or other connection: `open` opens the connection and returns the opened connection as its value. A character string is, as usual, interpreted as a file.

```
myfile = open("model.data"")
model = parse(myfile, n=1)
data = read.table(myfile)
close(myfile)
```

The call to `parse` reads lines from the connection until it has parsed one S expression (one line, in the example). The connection is then positioned at the beginning of the second line. The call to `read.table` reads from there to the end of the file.

The same rule applies to writing to a connection. The file `"model.data"` in the previous example could be created by:

```
outfile = open("model.data", "w")
dput(theFormula, outfile)
write.table(theData, outfile)
close(outfile)
```

The call to `dput` writes a formula and the call to `write.table` writes a data frame. Opening the connection explicitly before the first output ensures that the second function call will write data following that written by the first. Opening `outfile` with mode `"w"` truncates the file if it exists.

Section 10.5.2 discusses in detail the options for opening connections. For many applications, the details are not important. One key concept *does* need emphasis. Once a connection has been opened, the evaluation manager keeps it open until it is explicitly closed, or the session ends, even if no corresponding S connection object exists. Forgotten connections can hang around. They don't usually cause major problems unless a very large number accumulate, but organizing your computations to be tidy avoids confusion and saves system resources.

Good functional programming style will reward you here. In the examples above, we used the *object* returned by `open` in subsequent operations on the connection. Our general programming style says that the value of the function call is key and that, like everything, an open connection is an object. Assign the object, use it and, when you're done, hand it to the `close` function to close the connection. Those familiar with more procedural languages, such as C, might be tempted to think that the side effect of the `open` call was all that mattered, and to ignore the value. This can cause more problems than you might expect. Consider the two similar approaches below.

```
textFile = file("data.inout"); textCon = open(textFile) # Okay

textFile = file("data.inout"); open(textFile) # Trouble!
```

In the first version, we will use `textCon` as the argument for reading, writing, and other operations. In the second case, we probably intend to use `textFile`, but we're in trouble. The object `textFile` is an *unopened* file object, so the rules are that functions will open the file and close it on exiting. As a result, for example:

```
> cat("Now is the time\n", file=textFile)
> scan(textFile, "", n=2)
[1] "Now" "is"
> scan(textFile, "", n=2)
[1] "Now" "is"
```

We probably meant the second call to `scan` to get the third and fourth words, but `scan` followed the general rule for unopened connections.

Fortunately, we can get connections back after we have forgotten about them. The function `showConnections` creates a table summarizing all the currently open connections. Suppose we had been doing a variety of computations and wanted to see what's going on.

```
> showConnections()
              Class Mode  State   Description
5 "textConnection" "r"   "Read" ""
7 "file"           "rw"  "Read" "<temporary>"
8 "fifo"           ""    "Read" "./test.out"
9 "file"           "*"   "Read" "model.data"
```

We had created a text connection, opened a temporary file, started up a reader on a `fifo`, and, apparently, forgot the `close(myfile)` call after reading in the model data.

The numbers labelling the rows of this table are unique identifying tags created by the evaluation manager when a connection is opened. They can be used in a call to `getConnection` to get back the corresponding connection, sometimes a crucial convenience since temporary files, text connections, and pipes have no other global identifier. Files and fifos can be identified by the path in the file system.

We could clean up in our example by closing the text connection, if it's no longer needed, and the `"model.data"` file. The text connection needs to be retrieved from its tag.

```
> close(getConnection(5)); close("model.data")
[1] T
[1] T
> showConnections()
   Class Mode  State   Description
7 "file" "rw" "Read" "<temporary>"
8 "fifo" ""   "Read" "./test.out"
```

Let's package up the example of reading a formula and data into a function, say `readWithFormula`, to illustrate the style of opening and closing connections. The function will take a connection as an argument, will read formula and data, and will package the two up as a model frame (see section 3.3 of reference [3]).

```
readWithFormula = function(con) {
  if(!isOpen(con)) {
    con = open(con)
    on.exit(close(con))
  }
  model = parse(con)
  data = read.table(con)
  model.frame(model, data)
}
```

The discipline for opening connections is simple: we check whether the connection passed as an argument is open. If it is not, we open it, and set an action to close it on exit from our function. The call to `on.exit` works better than a direct call to `close`, because it guarantees to close the connection even if an error occurs before `readWithFormula` returns.

10.4 Standard Input and Output Connections

S follows a tradition of organizing the default interaction with the user into three streams, one for input and two for output, roughly for standard printing and for messages to the user. These streams are connections, referred to as the standard input, standard output, and standard message (or error) connections.

The three function calls `stdin()`, `stdout()`, and `stderr()` return the current connections associated with input, output, and message. These functions are the way to ensure that reading or writing refers to the corresponding standard connection. To parse a line from standard input:

```
nextParseLine = parseSome(stdin(), n = 1)
```

and to write `myObject` in dump form on the message connection:

```
dataPut(myObject, stderr())
```

When S is running in a standard interactive session, the three connections are all initially associated with the user's terminal; that is, `stdin()` reads from the keyboard and both `stdout()` and `stderr()` write to the display.

The function call `auditConnection()` returns a fourth predefined connection: the connection to which auditing information about the session is being written, or `NULL` if there is no audit output (see section 4.1.3 for the use of the audit file).

Programs can associate (sink) the input, output, or message connection with another connection (usually a file). Such associations remain in effect over subsequent tasks, until another call explicitly alters them. Sinks can be used to capture output and/or messages; sinked input is an alternative to `source` or to creating a reader (see page 389).

A standard connection is associated with a file by calling the function `sink`; for example,

```
sink("/tmp/junk.out")
```

associates standard output with the file `"/tmp/junk.out"`. The `sink` function exists for its side effect: after the call is evaluated, the implied association continues until another call to `sink` alters the association. In general, the function takes two arguments: a connection and a specification of what standard connection to sink, with `"output"` the default. For example, the following sinks both output and message to the same file, having turned on echoing of input expressions. (The effect is to record the image of the following interaction on the file, although the function `record` does a better job if you really want this.)

```
> options(echo = T)
> Tf = file("recording", "a")
Tf=file("recording","a")
> sink(Tf)
sink(Tf)
> sink(Tf, "message")
date()
    ... various commands
```

The first call to `sink` associates standard output with the file; the second associates message output with the same file. Notice that after the second call, the echoed commands and also the prompt string, `"> "`, no longer appear. These went to the message connection, now diverted.

As with other functions, `sink` interprets a character string as a file name. The `sink` function treats its connection argument slightly differently from functions like `cat` or `scan`: to behave consistently with underlying C usage of the standard streams, `sink` reopens the connection as the standard input, output, or message file regardless of whether it was originally open or not. Giving `sink` an output connection that is already open for writing is allowed, but you should then be careful with any explicit writing to the connection until the `sink` is over, because in effect both connections will now refer to the same file. Interleaving of the two sources of output may not work as expected. It is safer to only use unopened output connections for `sink`.

The evaluator remembers the successive calls to `sink`: omitting the connection argument causes `sink` to revert to the previous setting. Each of the three standard connections is treated independently. To restore both message and output to the default in the previous example:

```
sink(, "message"); sink()
```

(Watch for the easy error of giving `"message"` as the first argument to `sink`, rather than the second. This means to use file `"message"` as the sink for standard output.)

Sinking standard *input* is obviously rather different than sinking the other two connections, but it too can be useful. Given a text file containing S expressions, say `"recipes.S"`, sinking standard input to that file causes the main S reader to read these expressions as tasks, as if they had been typed on the terminal. Suppose `"recipes.S"` contains the following:

```
date()
rnorm(10)
cat("That's all Folks!\n")
```

The effect of sinking standard input to this file would be somewhat as follows:

```
> sink("recipes.S", "input")
[1] "Mon Dec 22 14:28:55 EST 1997"
 [1] -1.30966984  2.54603382 -0.24942895  1.30105832 -0.33010325
 [6] -2.14317216 -0.09785605  0.57139833  0.40783789 -0.46872209
That's all Folks!
Reader connection "recipes.S" closed
```

The output and message have not been redirected, but the parser does not write out prompt strings because the input is no longer associated with the terminal. The reader connection opened to the file is closed on the end of the sink.

There are three alternative function calls that will cause S to parse and evaluate the contents of a file:

1. `source("recipes.S")`;

2. `sink("recipes.S", "input")`;

3. `setReader("recipes.S")`

The first choice parses the entire file and evaluates the resulting expression, all within the current task. Both the other functions use the standard S reader to parse, evaluate, and (maybe) print each expression in the file as a separate task. The differences between 2 and 3 are more subtle: with `sink` the evaluator diverts the current reader on standard input to the file `"recipes.S"`, whereas `setReader` starts up a second reader. As a result, `sink` causes the evaluation to be done synchronously—no further tasks will be taken from the previous standard input until the sink completes—but with `setReader` the tasks are done asynchronously, possibly interleaved with more tasks from standard input.

The `setReader` version is more flexible, in that it can use other readers than the standard one. See section 10.6 for details on `setReader`.

Sometimes a function wants to know whether one of the standard connections is currently connected to the user's terminal; for example, to decide whether to engage in a dialogue. For this purpose, the special connection class `terminal` exists. The value of `stdin()`, or any of the standard connection functions, will be an object of this class if S believes that the connection is a terminal. The function `dialogue` writes to `stderr()` and reads from `stdin()`, so we might precede using it with the following test:

```
if(is(stdin(), "terminal") && is(stderr(), "terminal"))
    answer = dialogue( .... )
```

Testing individual connections is slightly more precise than the general call to `interactive` in deciding whether to interact with the user.

10.5 Manipulating Connections

The techniques described in this section provide additional control over connections. The connections can be opened with various *modes* to specify how and whether reading and writing are allowed and with *blocking* set on or off to control what happens when no input is available. Data may be pushed back onto a connection for later reading. Files (only) may be positioned for read or write. Data may be written and read in raw (binary) form as well as the character-oriented forms discussed so far.

10.5.1 Pushing Data Back onto a Connection

Data can be pushed back onto any connection open for reading. If a function reads the connection immediately afterwards, the character strings pushed back appear to be the first input available from the connection. Consider the following example:

```
header = readLines(con)
pushBack(header, con)
data = scan(con, ....)
```

The first line of input from connection object `con` is assigned, and then pushed back onto `con`. The call to `scan` then goes on as if the first line had never been read, but the programmer can use the first line for other purposes.

Push-backs are somewhat subtle, but very powerful. The power comes largely from their generality. No matter what input functions are called later on, the pushed back text appears just as it would if read directly from the connection. Typical uses include a function that reads input of a certain pattern until it encounters a line that does not match the pattern. This line is then pushed back onto the connection, after which the function can return, confident that whatever further input is read will not have been disturbed.

Although the term "push back" and the example suggest that the text pushed back is something just read, there is no constraint to this effect. A function can push back *any* character vector, to insert new, artificial input. This technique can be used to insert new expressions to be parsed, by pushing back the corresponding text onto the standard input connection:

```
pushBack("options(warn = 2)", stdin())
```

If no other text is pushed onto standard input during this task, the string pushed back here will appear to be the input for the *next* task. Push-backs provide a way for a task to start up another, following task.

A second push-back to `stdin()` during the current task will appear to come *before* the previous one. Push-backs are a stack, in computing terminology. To get something into the push-back stack of a connection anywhere other than the top, we need to read as many lines of input as needed from the connection, then push back both that data and the new data. The function `pushBackLength` returns the number of lines of pushed-back text on a particular connection. Here's a simple function to insert some text `n` lines after the top of the pushed-back text.

```
insertInput =
# insert 'text' as pushed back text on connection 'con'.
# Leave 'n' lines of previously pushed back text
# ahead of the new text.  By default, insert after
# all currently pushed-back text.
function(text, con, n = pushBackLength(con))
{
  previous = readLines(con, min(n, pushBackLength(con)))
  pushBack(c(previous, text), con)
}
```

With this definition, we can insert a second task on `stdin()`:

```
insertInput("options(check=T)", stdin(), 1)
```

In `insertInput`, it was important not to give `readLines` more than the number of lines of pushed-back text; otherwise, the function might try to do an actual read on the connection.

As the examples with `stdin()` point out, push-backs can be applied to connections that are open for read only. In particular, they are the only way to insert data into a text connection.

By default, each element of the character vector supplied to `pushBack` is followed by a newline character. Pushing back the result of a call to `readLine` is a standard way to read some lines and then push them back to be reread. Supplying the optional argument `newLines = F` suppresses the new lines.

10.5.2 Connection Modes; File Positions

File, fifo, and pipe connections are, by default, opened for both read and write; that is,

```
con = open("myfile")
```

creates a connection to the file that can be given to functions that expect to read or to write on the argument. If it turns out that the user does not have write permission on the file, the connection is opened anyway to permit reading. (An attempt to write on the connection will then cause an error.) Text connections are defined to be open for read only.

The default situation is suitable for most applications. Occasionally, however, you may want to specify a different relationship to the connection:

- to protect a file connection from accidentally being overwritten, even though the user does have permission to write on the file;
- to open a pipe so that the S process only writes to it (this will cause the output of the command to be printed to the standard output of the S process);
- to open a file so that it is truncated before writing on it, rather than appending to the current contents (the standard S `open` preserves existing data and appends new output).

These and other conditions can be obtained by providing a *mode* string when opening the file.

The modes discussed here are defined by the Posix standard operating system interface, and are not part of S (although S adds a few extra modes). The commonly occurring modes are listed in Table 10.1, but a string corresponding to any legal Posix mode can be supplied when opening a connection (see, for example, Chapter 3 of reference [7]).

For files (only), reading and writing can take place at any position of the file, assuming the mode allows it. S maintains for connections the concept of a read and a write position; reading only moves the read position and

"rw"	Open for read and write, truncating the file if it exists.
"ra"	Open for read and write, do not truncate, and allow writing *only* at the end of the file.
"r"	Open for read only.
"w"	Open for write only, truncating the file if it exists.
"a"	Open for write only, allow writing *only* at the end of the file.
"*"	The default. Open for read and write (or read only if write not allowed). Do not truncate, but allow writing anywhere. The initial write position is at the end of the file.
""	Don't open the connection.

Table 10.1: *The commonly occurring connection modes and their interpretation. Not all modes make sense for connections other than files.*

writing only the write position. For many applications, no explicit control over the file position is required. In the example on page 378, we defined a function that copied input onto a temporary file, then parsed that file when it encountered a special mark on the connection. Writing to the temporary file moved the write position on the file but not the read position, so that input still began at the start of the file, without any special action on our part.

If explicit control of file position is needed, the function `seek` provides a simple mechanism. The function takes as arguments a file connection, a position (measured in bytes from the start of the file), and `rw`, a choice of `"read"` or `"write"` position. The function returns the current read or write position, and as a side effect sets the corresponding position to the argument. If the position argument is omitted, the current position is unchanged. If the read/write choice is omitted, the last action determines the choice: read if the last action was input, write if output. To parse an expression from file `f` and then leave the file ready to reread the same expression:

```
pos = seek(f,rw = "read")
expr = parse(f, n=1)
seek(f, pos, "read")
```

The third argument in the second call to `seek` makes the intention clearer, but is not really needed since we know that input was the last operation.

For pipes and fifos, writing is only meaningful at the "end" of the connection. There is no concept of position for these connections. Data is read in the same order it is written. For text connections, only input is meaningful: (Pushbacks can simulate writing, however: see section 10.5.1.)

10.5.3 Blocking and Non-Blocking Connections

In addition to the mode string, the behavior of a connection can be specified to be *blocking* or *non-blocking* when it is opened. The concept is relevant mainly to connections accessible to more than one process. For example, suppose we are trying to read from a `fifo` connection. If another process is currently writing to this connection, the connection will not be available. Then the S process can either block, that is, wait until the connection becomes available, or else continue as if there was no input on the fifo. A similar distinction applies in the opposite direction, if the S session wants to write to a fifo or other connection when another process is currently reading. When a connection is opened, we can opt for either approach. A non-blocking connection will return immediately, a blocking connection will wait. The evaluation manager notes that the connection was not available and marks the last operation on the connection as *incomplete.*

A connection can be specified as blocking or non-blocking by supplying the argument `block` as `TRUE` or `FALSE` in the call to `open`, `file`, or `fifo`.

A connection to be used to synchronize output and input between processes generally needs to be non-blocking; otherwise, the entire S process will stop until the other process is finished. (This is probably not what you want, and in any case can lead to "deadlock" situations.) For other uses of connections, blocking is not typically an issue. The heuristic used by S is to make `file` connections blocking and `fifo` connections non-blocking, by default.

Non-blocking does not always come for free. Some software managing the connection has to be prepared for incomplete reading or writing; that is, for the possibility that not all the data written will make it out to the connection or that a read will not get all the data it expects to be available. The default effect is to treat an incomplete line as no data available. For a reader (section 10.6), this is fine as it is.

An S function can check for incomplete read or write on a non-blocking connection. If the last operation attempted on connection `con` was blocked,

```
isIncomplete(con)
```

returns `TRUE`; in all other cases it returns `FALSE`. On writing, a `TRUE` value can only be returned for a connection that is non-blocking and for which the connection is not available for the requested write (because some other process is currently reading it, usually). The flag is also set if an attempt to read a line got an empty or incomplete line on the non-blocking connection.

With fifo's and non-blocking connections generally, reading is done with the assumption that more data may be written to the connection eventually than is currently available. The "end-of-file" is not necessarily the last data that will appear on this connection. The line-oriented code supporting connections takes this into account. When an apparent end-of-file is reached, the final partial line, if any, is pushed back and a flag is set that causes `isIncomplete` to return `TRUE`. S functions can synchronize such connections by checking the incomplete flag and taking whatever action is appropriate when all the currently available lines of input have been read in.

Character input in S uses underlying code that reads a line at a time. This does not mean that S functions have to work only on whole lines. Many functions, such as `scan` and `parse`, are designed to be relatively independent of line boundaries. New input functions can be also; if they use functions such as `readLines` that are strictly line-oriented to do the actual reading, they should push back any leftover text by calling `pushBack`.

If an incomplete last line is encountered on a *blocking* connection, S will implicitly append a newline and generate a warning. Otherwise the absence of the final newline is irrelevant. For a *non-blocking* connection, the incomplete last line is pushed back (silently) and S behaves as if there were no more data currently available. Once a newline is written to the connection, the next read will pick up the entire line, including the pushed-back text.

The following little example shows the different behavior. First, a blocking connection is opened (and truncated) by a call to `file`. We then write some text to it with no terminating newline and read the connection with `readLines`. The text is read as if the newline had been there, but a warning is printed.

```
> Tfile = file("tmp/con1", "rw")
> cat("abc", file=Tfile)
> readLines(Tfile)
[1] "abc"
Warning messages:
  file "tmp/con1": incomplete last line in: readLines(Tfile)
```

If we change the situation only to reopen the connection non-blocking (`reopen` closes the connection, then calls `open` with the arguments given):

```
> Tfile = reopen("tmp/con1", "rw", block=F)
> cat("abc", file=Tfile)
> readLines(Tfile)
character(0)
```

More text can be appended to the connection to complete the line:

```
> cat(" def\n", file=Tfile)
> readLines(Tfile)
[1] "abc def"
```

This behavior is designed for connections that are expected to be updated as the S session goes along. Largely for this reason, `fifo` connections are non-blocking by default, since they work well for such updating applications.

10.5.4 Raw (Binary) Data on Connections

So far, all the data being read and written to connections has been text. Internal S data such as integers or numeric values have been converted from their raw (binary) form into text before being written, and from text into raw form when read. This is usually a good idea: the text on the connection can be read by a human and the external data can be saved, moved between one computer and another, and handled by the wide range of editors and other tools designed to handle text.

Occasionally, however, we would like to read or write data directly in its internal binary form. It may be that some other process generates or needs data in binary form, and we would like to communicate with that process through files or other connections. For example, some recording process may write observed numbers directly on a `fifo` in binary form, and we would like an S reader to pick up this data and process it. Perhaps, also, we just want to save the computing time needed to format the text data for input or output. Chances are the time saved is not very important in most applications; however, for very large amounts of data, binary files can provide an efficient communication mechanism.

The functions `readRaw` and `writeRaw` read and write data in raw form. For example, suppose `x` is a numeric vector of length 20:

```
writeRaw(x, "raw1")
```

This will write 20 numeric (double-precision) binary data values to the file `"raw1"`. Notice that *only* the data values are written. The contents of the file are not self-describing; any software in S or outside of S will need to know the kind of data on the file in order to read it in.

The natural correspondence to `writeRaw` is the function `readRaw`:

```
readRaw(con, what, length)
```

This reads binary data from connection `con`, expecting the data on the connection to be the same as the data in object `what`; e.g., if `what` is a numeric vector, binary values corresponding to C type `double` are expected. An object like `what` is returned, containing the data read.

If argument `length` is supplied, `readRaw` will try to read that many data values from the connection. If `length` is missing or negative, the function will read as many values as the length of `what`, if this length is positive, or the rest of the data on the file, if `what` has length 0.

```
readRaw("raw1", numeric(), 3) # read 3 numbers
readRaw("raw1", numeric(3))   # read 3 numbers
readRaw("raw1", numeric())    # read all the data on "raw1"
```

In contrast to many S functions, it is the `mode` of the data contained in `what` that matters, not the class of `what`. Classes other than `numeric` might have a numeric prototype. For all of them, `readRaw` works the same, reading numeric values. Section 5.5.2 discusses the various modes for data in S.

Although `readRaw` is a natural variant of other functions that read from connections, it behaves very differently than these when asked to read a large amount of data from a regular file. Instead of actually reading the data, `readRaw` maps the file to the memory of the S process. For large amounts of data this is more efficient, especially if you expect to only be using part of the data. In situations where another process writes data in binary form to a file, you can access that data efficiently in S through `readRaw`. The data does not need to be at the beginning of the file; calling `seek` before `readRaw` will allow the mapping to other positions on the file. A restriction requires the position to be an even multiple of the size of the data being mapped (e.g., if we're mapping numeric data and the corresponding C type takes k bytes, the position has to be a multiple of k); otherwise, memory mapping cannot be used.

10.6 Connections and Events

S provides a few functions that control how the S session responds to events, specifically to input waiting on some connection or to a timeout after some specified clock time. These provide useful and relatively simple tools for managing events in S. You can use them to specialize the user interface to your software and in general to customize the way S works for you and your users.

The facilities are not intended for detailed, small-grained control of events. A much more general and elaborate treatment of event management has been developed for use with S (see, e.g., the Ph. D. thesis of Duncan Temple Lang, reference [9]), but the discussion in this book is restricted to two simple control mechanisms: a reader on a connection and a monitor invoked after a timeout. Monitors are not directly related to connections, but we will discuss them in this section as well, on page 403.

10.6.1 Reader Connections

The function `setReader` arranges for the S evaluator manager to take tasks, such as parsing and evaluating expressions, from an arbitrary connection. The call to `setReader` provides two arguments, a connection and a reader.

```
tag = setReader(connection, reader)
```

Instead of reading immediately from the connection, `setReader` arranges that a new task will be started when input is waiting on the connection. The reader is a function, which takes a connection object as its argument. The S evaluator manager will call the reader function with the argument being the same connection given to `setReader`. The value returned by `setReader` is an identifying tag (actually the tag of the connection), which can be given as an argument to `dropReader` to stop looking for events on this connection.

By default, `reader` is `standardReader`. This behaves like the standard S session: expressions are parsed; when an expression is complete, it is evaluated as a task; the value of the task will be displayed via a call to `show` unless the task is an assignment or `.Auto.print` is set to `FALSE` in frame 1.

Initializing a reader via

```
setReader("recipes.S")
```

causes all the expressions on the file `"recipes.S"` to be parsed and evaluated, one at a time, from the file connection. As noted on page 389, the functions source and sink could also be used to evaluate the expressions on a file. The distinctive features of setReader are that the expressions are all done as separate tasks, asynchronously from tasks done by the current reader.

The standard S session begins with a reader defined on the connection stdin(), the standard input, with action defaulting to standardReader. Any number of additional readers can be specified for other connections by calls to setReader. Whenever a task is complete, the evaluation manager examines the current readers and monitors to see if any of them is ready to process a task. One of these readers or monitors will be started for the next task. For more details of how a task is selected, see page 404, but try to avoid relying on such details, since timing and event management in this form are subject to unpredictable behavior.

If the connection given to setReader is a file, all the tasks will be done and the connection will be closed, assuming it was not opened before the setReader call. If the connection is a fifo, the behavior is a little different and often more useful. The reader will not disappear when no further input is available, but will quietly wait. If input becomes available again, the evaluator manager will wake up the reader by calling the reader function. A fifo is therefore a good way to pick up tasks generated by another process, perhaps a graphical user interface.

If you want to stop reading from the fifo, close the connection or call dropReader with either the connection or the tag returned by setReader as its argument. If you don't have either of these handy, call showConnections to see the current open connections (see page 385).

The standard reader parses and evaluates S expressions, but many other actions can be useful as well. For example, suppose some test results are being appended to the end of a fifo "test.out". We want these results to also be appended automatically to an S object, results, as they appear. One way to do this is to set up a reader on the fifo:

```
> assign("testResults", numeric(), where=1)
> setReader(fifo("./test.out"),
    function(con) {
      data = get("testResults", where = 1)
      assign("testResults", where = 1,
          c(data, scan(con, -1)))
  })
```

Whenever data is available on the fifo, the reader function is called, reads all the available data (as many complete lines as have been written), and appends the items read to the object named `"testResults"` in the working database.

Let's look at another example, from testing S software. When writing any extensive software, it's a good idea to develop as extensive a set of *regression tests* as your energy and time permit. These tests are assertions about how the software should work; the term "regression" refers to the notion that new and improved versions of the software just might regress by failing to perform as previously specified.

S provides tools for regression testing of S software, one of them being a function `do.test`. This function takes as its argument the name of a file containing the text for S expressions. The expressions are supposed to evaluate to the value `TRUE`. If they don't, the implication is that something is wrong. When `do.test` is called, it silently parses and evaluates the expressions in the source file, except when an expression fails to evaluate to `TRUE`, in which case the expression and its value are printed. Source files of this form are provided to test the functions in S libraries, and are a useful (and recommended!) way to keep track of your own programming efforts. Suppose we have several such files (testing different kinds of functions, perhaps) and the names of the individual files are, say, in the object `testFiles`. We want to call `do.test` for each file, which could be done in the obvious way by:

```
for(what in testFiles) do.test(what)
```

The disadvantage is that the entire regression test is now one single task. If we have a number of test files and the files involve substantial computations, we may prefer to have each test file run as a separate task, to minimize interactions or simply for efficiency. An added property of separate tasks is that, should errors occur in one test file, we can arrange for testing to proceed with the next. (Whether this is an advantage depends perhaps on your philosophy of debugging, but it may be convenient at least at an early stage.)

To convert the loop to an iteration of tasks, we want to read one file name at a time. We could actually write the contents of `testFiles` to a file and read that back in, but the `textConnection` class is a simple way to do the same thing. The strings in `testFiles` become, conceptually, lines in the `textConnection`, and our reader function can just read those back in.

```
setReader(textConnection(testFiles),
    function(con)do.test(readLines(con, n = 1)))
```

Each call to the reader function reads a single string (i.e., line) from `con` and calls `do.test` with that string as its argument.

When the connection used by a reader is being both written to and read by the S process, some effort is needed to synchronize the input and output. The recommended approach involves three steps.

1. Open a fifo connection for both reading and writing.
2. Set up a reader on the same connection. The action function for the reader should, on each call, read *all* the currently available input on the connection, and do whatever the application wants done with that input.
3. Arrange for this fifo object to be supplied to whatever function(s) will be doing the output. Each burst of output should be sufficiently complete to make sense to the reader in the next step.

We will see shortly why the suggestions in steps 2 and 3 are helpful. First, here is an example. It doesn't accomplish anything useful other than reformatting all the standard S output, but the structure is typical.

```
> reprint
function(con)
{
    x = readLines(con, n = -1)
    cat(file = stderr(), paste("**", x, "**"), sep = "\n")
}
> my.fifo = fifo("junk.f", "rw", block=F)
> setReader(my.fifo, reprint)
> sink(my.fifo)
> cat ( 1:3, sep="\n")
> ** 1 **
** 2 **
** 3 **
"Hello, World"
> ** [1] "Hello, World" **
```

The call to `fifo` opens a two-way connection to the special fifo file `"junk.f"`, creating it if necessary. The call to `setReader` arranges for the S event-handler to invoke `reprint` if input is ready on this fifo. Precisely, at the end of each task, the event-handler examines the input file descriptor for `my.fifo`, via the `select` system call. If input is ready for that file descriptor, the S function is called with the S connection object as its argument. The details of the event-handler result in the prompt coming in the "wrong" place: on completion of the task of evaluating my typed input, the system decides it's time to emit another prompt. The `reprint` task is done after this, whereas standard printing is part of the evaluation task. (For more realistic uses of this style of synchronization the issue is unlikely to arise, since the reprinting would not be in the same place as the prompt.)

As required in step 3, `reprint` reads all available lines from the connection. For this reason, also, it is essential that the connection be non-blocking—otherwise, the whole S process would freeze while `readLines` waited for more input. It would be tempting to read just one line from the connection at a time, but this destroys synchronization because the stream-oriented input routines buffer the input. Suppose we modify the definition of `reprint` to read one line:

```
x = readLines(con, n = 1)
```

Then the behavior on the example above changes:

```
> cat(1:3, sep="\n")
> ** 1 **
"Hello, World"
> ** 2 **
** 3 **
** [1] "Hello, World" **
```

The second and third output lines were read in with the first to a buffer; unfortunately, after this the evaluator manager no longer believes there is input waiting on the connection. Only when further data is written to the fifo are those lines picked up.

If the reader function does not want to process all the input waiting, it can read everything and then call `pushBack` to return any unwanted lines to the connection. Pushed back lines will appear as input waiting for the next task.

Readers are associated with connections by the evaluation manager. A connection can have only one corresponding reader. Therefore, setting a reader on a connection automatically *replaces* any previous reader on the same connection. This applies most importantly to readers on the standard input connection. Unless your S session started with a nonstandard initialization script, `stdin()` is associated with the standard S reader, parsing and evaluating tasks. You can start another reader on `stdin()`, but be aware that it will replace the standard one. In most respects, this is a feature, since otherwise there would be confusion about which reader owned a particular piece of input. But if you want to return to the standard reader, the last step for the new reader must be to reset `stdin()`:

```
setReader(stdin())
```

You can also drop the reader on `stdin()` if you have initialized some other reader or monitor system for tasks, but if you have not, then S will decide there is no further source for tasks and will end the session.

10.6.2 Monitors; Timeout Events

The S evaluator manager will execute an action as a new task after a specified time.

```
tag = setMonitor(timeout, action)
```

The `timeout` argument is a time (in units of seconds). The evaluator manager will wait this long after the `setMonitor` call. When a task completes after this time, the manager may call `action`, which should be a function with no arguments. If you want the monitor to be reset (and then called again), the `action` should call

```
resetMonitor()
```

with an optional `timeout` argument, if you want to change the timeout. For example, the following monitor calls a function every five minutes to report the current number of data items read and appended to `testResults` in the example on page 399.

```
setMonitor(300,
  function() {
    message(length(testResults), " test results read")
    resetMonitor()
})
```

The value of the call to `setMonitor` is a tag that identifies the monitor. This tag should be saved if you plan to explicitly drop the monitor later on.

```
dropMonitor(tag)
```

More typically, though, the monitor action itself will test some condition and only conditionally call `resetMonitor`.

Monitors (but not readers) can also be started up within the current task, as *sub-events*. The call

```
setSubEvents(flag)
```

enables or disables sub-event checking according to the logical `flag` (initially sub-events are turned off). Sub-events allow reporting on, or even browsing in, ongoing computations.

10.6.3 Choosing a Task

When the evaluator completes a task, it examines the readers and monitors currently defined. If monitors are defined, and the monitor with the earliest wakeup time is ready, its action is evaluated as the next task. If no monitors are defined or ready, the evaluator examines the status of any reader connections defined. If any of these have input waiting, the evaluator selects one and calls its action function with the corresponding connection object as argument.

The available readers are selected in a roughly cyclical way; that is, after a particular reader completes a task, it will not get the next task unless no other reader is ready. This "round robin" mechanism can be upset by events that themselves alter the set of readers. Also, monitors get priority over readers. In principle, a monitor could lock out all readers by restarting itself with zero waiting time, but it's hard to see this as a useful programming technique.

Signals can also be processed between tasks, assuming no task is waiting. An error action specific to the signal received will be called between tasks, but not a general error action. See page 270 for the various error actions and how to specify them.

Chapter 11

Interfaces to C and Fortran

S provides *interface* functions to other languages: to the command shell and to subroutines written in C, C++, or Fortran. The shell interface is simple and provided by functions `shell` or `pipe`; calling subroutines is a little harder, and is described in this chapter. The functions `.C` and `.Fortran` provide the interface. You will need to write S functions that call particular subroutines through the interface. The source and object code for the subroutines is maintained in a `CHAPTER` directory and automatically linked with the S evaluator when the chapter is attached. C or Fortran programming is usually only needed to make the arguments match suitable S classes. Another interface is provided for C code that manipulates S objects directly (see Appendix A), but you are strongly encouraged to stick to the simpler interfaces described in this chapter, if possible.

A running S session has a *search list* of attached chapters, a mixture of S libraries and other chapter directories, plus potentially additional databases of various kinds. The chapters, whether S libraries or not, can dynamically link object code. Each chapter has an optional file, `S.so`, containing the result of compiling and linking an essentially arbitrary collection of software.

In particular, the file can contain the compiled version of subroutines in C or Fortran that are designed to be called from an S function. The *interface* from S is through one of a set of special S functions, each of which takes the name of a subroutine, some S objects designed to map into the arguments to that subroutine, and possibly some additional control arguments. The interface function returns an S object containing the results of the computations generated by the subroutine.

This chapter describes the two simplest of the interface functions, `.C` and `.Fortran`. To program an interface to subroutines using these functions, you need to do three things:

1. set up the subroutine so that its arguments correspond to data types in C or Fortran that match the allowed S classes;

2. write an S function that provides the arguments to the interface, making sure each has the right class, and then constructs from the value returned by `.C` or `.Fortran` a suitable S object to return to the user;

3. whenever the subroutine changes, use the `make` shell command to regenerate the `S.so` file.

For step 3 to work, you also need to set up your chapter to tell S what files need to be compiled and linked to make `S.so`. This only needs to be done when you change the list of such files.

Section 11.1 describes how to set up the chapter. Section 11.2 describes the use of `.C` and `.Fortran`. The remaining sections describe the control arguments and the treatment of missing values and `raw` data.

Throughout this chapter we will deal with "shell commands". They will be shown in examples preceded by a nominal shell prompt, `"$"`. The S shell commands reside in a directory, `cmd`, under the S or S-Plus home directory (defined by the shell variable `SHOME`). The `make` command will need to use the same shell variable to locate some critical files. Use either of the following approaches to running the shell commands:

- Make the command an argument to the shell command you use to run S itself, for example, `"Splus"`. The command `make`, for example, becomes

  ```
  Splus make
  ```

  (This works because the `Splus` command defines the needed shell variable and searches the `cmd` directory.)

- Before running the commands, define the `SHOME` shell variable, and then add `$SHOME/cmd` to the path that your command shell searches when looking for commands. The value of the `SHOME` shell variable can be printed by the `SHOME` command; for example,

```
$ Splus SHOME
/usr/local/splus
```

The second approach is a little more convenient if you use S shell commands frequently, and you probably *will*, if you're working with the techniques of this chapter frequently.

11.1 The S Chapter

C or Fortran routines to be used with S should be dynamically linked from an S *chapter* directory. You will likely already have built up an S chapter containing S functions and other objects related to a particular project. Chapter directories can also contain source and object files to be compiled and linked with the S session; the files specified for the chapter will be linked automatically when the chapter is attached. The mechanisms used are essentially identical to those S uses for its own libraries. Developing new interfaces to compiled code is another step in turning yourself from an S user into an S system developer. Enjoy!

11.1.1 Initializing the Chapter

If you haven't done so, run the shell command

```
CHAPTER
```

in the directory. Called without further arguments, it will look for any C or Fortran source files in the directory and initialize a file `makefile` to include all of these in the dynamic linking. It assumes any file names ending in `".c"`, `".cc"`, or `".f"` are the source files to be included. If you only want some specific files to be included, supply the names of the source files as command-line arguments to `CHAPTER`. You can run the command in an existing chapter as well, if you decide to add some compiled code to that chapter. The command will leave the existing S objects alone, but will still initialize the files needed for linking. If you already have a chapter with compiled code and want to change which files are compiled, or to change compile or load flags, you can instead edit the control files described in section 11.1.3.

Suppose we start with a clean directory, and we only want to link one source file, `scan_dollars.c`.

```
$ CHAPTER scan_dollars.c
Creating data directory for chapter
"make" will link: scan_dollars.c
Creating "makefile"

S chapter initialized.
```

The first output line tells us that there was no S database directory here; if there had been, it would have been left alone by the command. The second line tells us what files will be compiled and linked whenever we run the `make` command here. The third line says that a `makefile` was created with the rules necessary to do this compiling and linking, as well as rules for dumping and rebooting the contents of the chapter (see section 5.4.2).

To compile any source files and update the object file for linking, use the shell command `make`. Running the command would produce something like the following:

```
$ make
    gcc -I${SHOME}/include  -g  -c scan_dollars.c -o scan_dollars.o
    S LIBRARY S.so  scan_dollars.o
```

The `LIBRARY` command binds together the object code to be linked and creates a shared object file, `S.so`, which will be linked automatically when the chapter is attached to an S session. For the `make` command to work, it needs the shell variable `SHOME`, either set in your shell or by calling make indirectly, as described on page 406.

You can make an S chapter in a directory that already has a `makefile` used for other purposes, but unless you have a strong reason to do so, there is a danger of confusion between the different purposes of `make` in the directory. The `CHAPTER` command will try to accommodate such a makefile, however. If it encounters a directory with a `makefile` but not the special files used to control `make`, it will add the rules needed for S to the end of the existing `makefile`, and tell you so. If you are operating in this mode, you must explicitly state the target `S.so` when you want to relink the S code:

```
$ make S.so
```

If the `makefile` is only used for S, this rule is the default.

11.1.2 Attaching and Detaching the Chapter

An S chapter, database, or library is attached by a call to the `attach` or `library` function. The call is the same whether or not the chapter contains compiled code. The action of the S evaluator manager is also the same, except that the presence of the shared object file `S.so` triggers the manager to dynamically link that file. Global entry points in the compiled code are then accessible from S functions.

As an owner of a chapter including some C or Fortran code, you may want to do some initialization in C or Fortran. For example, there may be some global data that needs to be initialized *before* your subroutines are called. The S interface can make such initialization easier than it would be ordinarily. When the `attach` function is called, it looks in the chapter for an object named `.on.attach`, which should be a function definition or an S language object. If this exists in the chapter database, the last step in the `attach` will be to call the function or evaluate the object.

You can use this convention to initialize your C or Fortran code. Suppose the routine `waferChapterInit`, with no arguments, does the desired C initialization. Then the `.C` interface can invoke the routine, from a function `.on.attach` in the chapter database.

```
.on.attach = function() {
    .C("waferChapterInit")
}
```

The same kind of initialization can be done for Fortran routines (e.g., to initialize some common storage needed by the subroutines). Define an initialization subroutine and call that via `.Fortran` from your `.on.attach` function.

You can also arrange to do some computations just before your chapter is detached; for example, to free any large amounts of memory that you allocated for use of the code in this chapter. S has the same convention, this time looking for an object named `.on.detach` in the chapter about to be detached. The call to such a wrapup routine would be made from the S function `.on.detach` in your chapter.

C dynamic linking facilities also provide some techniques, independent of S, designed to initialize a dynamically linked object. But, given that you are using the S interface, the initialization described in this section is easier to program, more portable, and less “hidden”.

11.1.3 Modifying the Chapter

As you work on your C code, you will naturally need to re-make the S.so whenever the source code changes. So long as the set of files and the flags you want to use in compiling do *not* change, you only need to rerun make.

If you do need to add another file to the compile, or otherwise change what gets linked into S.so, you can edit a file S uses to control the make command. The makefile in your chapter is standard and will look the same for any S chapter. It reads in the file S_LOCAL_FILES to define the specifics needed for this particular chapter. Following the initialization on page 408, the contents would be as follows.

```
$ cat S_LOCAL_FILES
SRC= scan_dollars.c
OBJ= scan_dollars.o
```

The two lines of this file name the object files that will be targets for the make step, and the source files that produce those targets. To add the Fortran code in file dmatp.f to the chapter, add the source file to the SRC= line and the corresponding object file to the OBJ= line.

```
$ cat S_LOCAL_FILES
SRC= scan_dollars.c dmatp.f
OBJ= scan_dollars.o dmatp.o
```

Alternatively, you can just run CHAPTER again with the appropriate arguments for the new set of files:

```
$ CHAPTER scan_dollars.c dmatp.f
Data directory exists: leaving it alone
"make" will link: scan_dollars.c dmatp.f
Looks like a current S makefile exists: leaving it alone
```

As the messages suggest, the command leaves everything in the directory alone, *except* the S_LOCAL_FILES.

Another fairly common requirement is to link some additional object files or libraries, or to change some options to the linking that produces the S.so file. The actual linking is done by invoking the S utility LIBRARY. An additional make variable, LOCAL_LIBS, is passed to this utility. The LOCAL_LIBS variable is initially empty but can be set to contain any arguments that are legal in the compile–linking command. For example, if you needed the "-lF77" library, and also wanted to include the object file utils.o, you could add the line

```
LOCAL_LIBS=-lF77 utils.o
```

to the `S_LOCAL_FILES` file. The key distinction between `utils.o` here and the files in the `OBJ=` line is that the `LIBRARY` command does not try to `make` `utils.o`. Object files should be put in `LOCAL_LIBS` if there is no corresponding source to make them. The library flags might be needed if your routines require system libraries not normally used by S. To see the libraries linked by default, view the file

```
$SHOME/cmd/LIBRARY_FLAGS
```

and look for the line starting `DLLIBS=`. This file is a shell file, not an include file for `make`, so the format will be slightly different from `S_LOCAL_FILES`.

The three variables `SRC`, `OBJ`, and `LOCAL_LIBS` define the C and Fortran contents of your chapter: `SRC` defines the files to be dumped when the chapter is archived or moved; the files in `OBJ` will be targets for the `make` command and will be linked into `S.so`; the files or libraries in `LOCAL_LIBS` will also be linked into `S.so` but will *not* be targets in the make step.

You can, in fact, control the `LIBRARY` step further. The `LIBRARY` command looks for a shell file `LIBRARY_FLAGS` in the chapter directory, in addition to the one in `$SHOME/cmd`. The local file is empty by default, but you can edit in shell commands to override the settings in `$SHOME/cmd/LIBRARY_FLAGS`.

11.2 The Interface Functions from S to C and Fortran

The functions `.C` and `.Fortran` provide interfaces to routines written in C or Fortran. This section describes the general form of the call to the interfaces, which have identical argument lists and return the same structure in their result. The arguments come in three groups:

1. argument `NAME`, a character string supplying the name of the C or Fortran subroutine;

2. arguments `"..."` that must always correspond in number and type to the arguments of the subroutine;

3. optional control arguments, `NAOK`, `CLASSES`, and `COPY`.

We will look at the control arguments in sections 11.3 and 11.4. There are some special uses for `raw` data, discussed in section 11.5.

The value returned by `.C` and `.Fortran` is a list or named list, with as many elements as there were arguments to the subroutine. If any of the "..." arguments were named, the value is a named list with the same names, a convention used to extract elements easily from the value. The elements of the list will correspond to the objects given in the call, except that the subroutine is allowed to overwrite any of its arguments, and the new, overwritten data values are returned. This mechanism allows S to get computations back from C or Fortran routines that have no knowledge of S structures. In effect, the S function using the interface provides all the structure, and the subroutine just writes in some new data values.

Class	**C type**	**Fortran type**
`numeric`	`double *`	`double precision`
`logical`	`long *`	`integer`
`character`	`char **`	*Avoid!*
`integer`	`long *`	`integer`
`single`	`float *`	`real`
`complex`	`complex *`	`complex`
`raw`	`char *`	*Not allowed*
`list` and all recursive objects	`s_object **`	*Not allowed*

Table 11.1: *Mapping S objects to C and Fortran data types in* `.C` *and* `.Fortran` *interface calls. List-like objects should rarely be used; see Appendix A. The* `complex` *type in C is defined in the S header file,* `"S.h"`.

To use the interfaces, you need to understand how data in the S objects corresponds to type declarations in the subroutine. You may need to modify the subroutine to make the arguments have the appropriate data type. Each of the arguments to `.C` or `.Fortran` corresponding to an argument of the subroutine will be interpreted as a pointer in C or an array in Fortran, of some data type. The C data types corresponding to S objects are defined in Table 5.1 on page 201. The table is repeated here as Table 11.1, but with a column added for the corresponding Fortran data types.

11.2.1 The Interface to C

Let's proceed to an example of the `.C` interface. Suppose we have a C routine that takes an array of character strings and scans them for currency fields. It expects the strings to have a "$" sign, followed by a numerical digit amount, maybe with a decimal point and a fractional part; the corresponding double-precision converted amount is stored in the corresponding element of an array supplied as an argument. Any string that doesn't have this form is not converted. The routine `scan_dollars` implements this:

```
void
scan_dollars(char **fields, long *n_ptr, double *value)
{
  long i, n; double dollars;
  n = *n_ptr;
  for(i=0; i<n; i++)
    if(sscanf(fields[i], "$%lf", &dollars) == 1)
      value[i] = dollars;
}
```

The standard C library routine `sscanf` will return the number of fields found in this call; the C format string looks for a literal "$", followed by a floating-point numeric format (a little too forgiving for a currency field, but not bad).

The routine was modified to work with S by making sure the arguments all had one of the type declarations in Table 11.1. Otherwise, the second argument would likely have been `n`, the number of character strings, rather than a pointer to this value.

Now we need an S function to call this routine. The function's only argument is a vector of the currency items as character strings. The function creates a numeric vector of the same length, initialized with an implausible value, set here to `Inf`, a very large numeric value. Page 424 shows how to use missing values instead. The argument, its length, and the numeric vector are passed to `scan_dollars`. The S function only wants to return the numeric vector; this is the third argument to `scan_dollars` and so the third element in the list returned by `.C`.

```
"scanDollars" =
# convert the strings in 'text' to currency values, assuming the
# strings contain '"$"' followed by an amount in dollars.
# Anything else is returned as 'Inf'.
function(text) {
```

```
    text = as(text, "character")
    n = as(length(text), "integer")
    values = as(rep(Inf, n), "numeric")
    .C("scan_dollars",
       text,
       n,
       values)[[3]]
}
```

Both the C routine and the calling function are extremely simple, but they still illustrate some typical points about calls to C and Fortran.

- The S function must allocate all the data needed, for both input and results, and make absolutely sure that the arguments agree in class with the subroutine. The paranoid style in our example (actually, `Inf` is already `numeric`, and `length` always returns an integer) is much better than making a mistake, but section 11.3 describes a more systematic approach.

- The value returned from the S function comes from extracting some of the elements in the list returned by `.C` or `.Fortran`. You can extract one or more elements, or use names, as we'll show below.

- The arguments to the C routine usually include the lengths or dimensions of the S objects in addition to the objects themselves. The most common errors in programming the interface come from getting the lengths wrong.

- It's generally best to let the S function handle exceptions and special cases, with the C or Fortran code just setting values.

Here are a few details and alternatives.

For extracting results, list elements can be addressed in any way that is convenient. We used the `"[["` operator in the example; to return one or a few arguments from many arguments, you may prefer to generate a named list and extract the elements using the `"$"` operator. Arguments in the call to `.C` or `.Fortran` can have any name, other than `NAME`, `NAOK`, `CLASSES`, and `COPY`. The list returned will then be named, with the same names. Name the arguments you need in the call and extract the corresponding elements from the list. The `.C` call in our example could have used a name as follows:

```
.C("scan_dollars", text, n,
  x = values)$x
```

The name x (italicized in the example) is arbitrary, so long as it does not conflict with one of the special arguments; just stay away from capital letters, and there will be no problem.

Since the C or Fortran routine has no means of checking the lengths or classes of its arguments, getting the correct information from the S function is essential, and failing to do so is the most likely general source of errors. Even in as simple an example as ours, one could forget to initialize `n`, or mis-specify `n_ptr` as `long`. The usual symptom, a memory fault, is rather abrupt. For simple examples, some guesses or some *ad hoc* printing from the C routine may help. For more ambitious or difficult applications, debugging the S session via an interactive debugger such as `gdb` or `dbx` is feasible. By using the shell command `ps` while S is running, you can determine the executable program for S, typically `$SHOME/cmd/Sqpe`; this can be run directly in the debugger provided the shell variable `$SHOME` is set. Your version of S may not have internal C symbols, but the code in your chapter can, and that code is likely to be where you need to debug anyway.

Catching problems with the *use* of your routines will also eventually be an issue. The S error management can be invoked from C code, as is described on page 443 in Appendix A. For relatively simple situations, it's usually preferable to just indicate the problem via the returned values. This leaves the S function to handle the problem, perhaps more flexibly than can easily be done in C. Your C code then remains less dependent on the details of S, making it easier to adapt to other applications. In our example, the C code left untouched the values corresponding to erroneous elements in the text. The S function initialized these to an impossible value; after the `.C` call, the function could have checked for problems and issued a warning or error message.

Calls to C++

S is implemented in C, rather than C++, and therefore it provides no special support for C++features. However, the `.C` interface can be applied to C++routines, assuming that the chapter links the necessary support libraries for those routines. The available data types are the same as for C. Because S does not understand the "name mangling" of C++, the actual entry point called from S must be declared inside an

```
extern "C" {
}
```

declaration, within the C++ code. You don't want to use the mangled names, in any case, since they depend on the particular compiler.

Here is a little example, with thanks to Duncan Temple Lang. The routine intended for S is here in a separate file—not a requirement but convenient, since then the C++ method itself can be in a file that is pure C++. The file name should end in ".cc".

```
#include "classBond.h"
extern "C" {

void
S_bondYield(char **names, double *vals, long *len, double *yield)
{
  Bond f(names[0], vals, *len);

   yield[0] = f.yield();

return;
}

}
```

The file "classBond.h", to define the C++ class, might be as follows:

```
#ifndef CLASSBOND_H
#define CLASSBOND_H

class Bond {
  protected:
    char *name;
    double *data;
    long length;
  public:
    Bond(char *n, double *d, long l);
    double yield();
};

#endif
```

The remaining code, to define the `Bond` and `yield` methods, isn't shown, but need not depend on the S interface. As in the C and Fortran cases, the actual subroutine called has been modified to ensure that its arguments correspond to allowed data types. In this case the `name` and `length` are naturally a single string and a single `long`, but have to be extracted from pointers passed down in the `.C` call.

Symbolic Dump Format in C

The symbolic dump formats for writing data in the atomic modes are accessible in the "S.h" include file. They provide a portable way to write S data values from C to *full* precision. You must put the line

```
#include "S.h"
```

in the declarations of your source file. You then have access to the following declaration:

```
extern char *Integer_format, *Single_format,
     *Double_format, *Complex_format;
```

These are each suitable format strings for writing out one data item of mode "integer" or "logical" (format "Integer_format"), "numeric" (format "Double_format"), "single" (format "Single_format"), or "complex" (format "Complex_format"). These format strings print a single data value, with no following newline. You must use them from a running S session, since the formats are generated dynamically. While the formats are useful, it's better in general to create the output from the S dataPut function if possible, to handle various special issues such as NA and Inf values.

The complex Data Type

C does not have a datatype for numbers in the complex field. To use complex vectors with C, you need to include a suitable typedef, by including the S header file:

```
#include "S.h"
```

This declares a complex type, with two fields of type double, re, and im, for the real and imaginary parts of the number.

11.2.2 The Interface to Fortran

The .Fortran interface behaves much like the .C interface. As an example, let's look at dmatp, a Fortran subroutine that computes the matrix product of two double-precision matrices.

```
      SUBROUTINE DMATP(X, DX, Y, DY, Z)
      INTEGER DX(2), DY(2)
      DOUBLE PRECISION X(1), Y(1), Z(1)
```

```
      INTEGER I, J, N, P, Q, JJ
      INTEGER IJ
      DOUBLE PRECISION DDOT
      N = DX(1)
      P = DX(2)
      Q = DY(2)
      DO  3 I = 1, N
         JJ = 1
         IJ = I
         DO  2 J = 1, Q
            Z(IJ) = DDOT(P, X(I), N, Y(JJ), 1)
            IF (J .GE. Q) GOTO 1
               JJ = JJ+P
               IJ = IJ+N
 1          CONTINUE
 2          CONTINUE
 3       CONTINUE
      RETURN
      END
```

A function `matp` can call this subroutine through the `.Fortran` interface.

```
"matp" =
## the matrix product of the numeric matrices 'x' and 'y'.
function(x, y)
{
    dx = dim(x)
    dy = dim(y)
    if(length(dx) != 2 || length(dy) != 2 || dx[2] != dy[1])
        stop("invalid matrix arguments")
    z = .Fortran("dmatp",
        as(x, "numeric"),
        dx,
        as(y, "numeric"),
        dy,
        z = double(dx[1] * dy[2]))$z
    dim(z) = c(dx[1], dy[2])
    z
}
```

The general points mentioned for C interfaces on page 414 apply here too: the S function is responsible for all the data allocation; it must coerce the arguments to exactly the correct classes; it must pass down all lengths and dimensions to the subroutine; the value of the call to `.Fortran` is a list of all the arguments, possibly with data overwritten, and the relevant results are extracted from this list.

The example just wants to take one element from the subroutine, the numeric values to go into the matrix product. The name `z` (italicized in the example) is used to extract this element conveniently. The vector is then made into a matrix of the desired dimensionality. A more refined version of the `matp` function would also preserve row and column labels from the `dimnames` of the arguments.

We've left a dangerous bit of code in this example to illustrate the trickiness of getting the classes right. There is no check for the class of the arguments `dx` and `dy`. Presumably we "know" that `dim` always returns class `integer` and have checked this on some examples. But because of the dynamic nature of S, we can't *really* know any such thing. Nothing in principle prevents a library from containing a version of `dim` that delivers the expected numbers, but as `numeric` objects instead of `integer`. Ordinary S computations would proceed fine, but the `matp` interface would collapse by handing Fortran a `double precision` array where it expected `integer`. A systematic approach to handling argument classes is shown in section 11.3, and is recommended to avoid such traps.

Besides the general comments on the interfaces, two additional points apply specifically to Fortran.

- Many Fortran subroutines used with S do matrix computations. S matrix objects can be passed to Fortran as vectors, `"numeric"` for matrices with double-precision data. S stores matrix and other multiway array data in the conventional Fortran order (emphatically *not* a coincidence), so no reordering is needed.

- S is implemented in C, so the `.Fortran` interface depends on an interface between C and Fortran. To give your S functions that use Fortran the best chance to be portable (not to mention to run correctly), it's a good idea to restrict Fortran subroutines to numerical tasks, where the arguments can be `double precision`, `integer`, or `real`. Try not to use character arguments to Fortran subroutines; the correspondence to the internal S code will depend on the choice of Fortran compiler. Raw and non-atomic data are meaningless to Fortran.

11.3 Matching Argument Classes; Controlling Copying

The C and Fortran subroutines get their arguments as pointers or arrays, not as S objects. Therefore, the S side of the interface is responsible for making sure that the subroutine receives arguments of the right type. There are several approaches. The examples so far dealt with each argument separately. We ensured that an argument declared `double precision` in Fortran had the corresponding class by calling `as(x, "numeric")` and relied on an assertion that, for our example, `dim` always returns an integer vector.

A more systematic approach, however, is recommended instead. The *ad hoc* approach often involves extra computations, when the argument is already of the correct class. Much more serious is to omit the "extra" calls when they *are* needed. Matching arguments to classes must be done in S, since errors can't be detected in the subroutines themselves, and the consequences are likely to be wrong answers or memory faults.

The suggested technique is to provide S with an explicit list of the classes corresponding to the subroutine's arguments, either in the interface call itself or by creating an object in the chapter through a call to `setInterface`. The `.C` and `.Fortran` functions have an optional named argument, `CLASSES=`. The argument should be a character vector, with one element for each argument to the subroutine, containing the S classes that the corresponding arguments should have.

This approach can also treat another detail about the interfaces: making copies of the arguments. Because both C and Fortran can overwrite the data supplied to them, S will protect any argument by copying the data if there is a reference to it in some S frame. If you know that some arguments to the routine are "input-only" you can avoid this extra copy by declaring to S that this argument need not be copied. The special argument `COPY=` to `.C` or `.Fortran` controls copying. It should be a logical vector as long as the number of arguments, with `TRUE` for each argument that might be overwritten by the subroutine, and `FALSE` elsewhere. The `TRUE` elements tell the S evaluator to copy this object if it is shared data (i.e., part of an object assigned in some frame; section 4.9.7 discusses how S shares data among function calls). The `FALSE` elements assert there is *never* a need to copy this argument (naturally, if you tell S this and you're wrong, very subtle bugs can result).

Notice that this is a statement about the C or Fortran routine, *not* about the S objects. S knows whether or not it *needs* to copy the object; an object that was just created in the call to the interface, for example, by

```
integer(n)
```

won't be copied in any case. What S can't know is the behavior of the code inside the C or Fortran routine.

As examples, we modify the `scanDollars` and `matp` functions to use the special arguments. The C routine `scan_dollars` only overwrites the data pointed to by its third argument. The modified S function is then:

```
"scanDollars" =
# convert the strings in 'text' to currency values, assuming the
# strings contain '"$"' followed by an amount in dollars.  Anything
# else is returned as 'Inf'.
function(text)
.C("scan_dollars", text, length(text),
  x = rep(Inf, length(text)),
  CLASSES = c("character", "integer", "numeric"),
  COPY = c(F, F, T))$x
```

There is one slight danger: we did not coerce `text` to `character` before computing its length. All coerce methods on vectors are asserted to keep the length unchanged or produce an error. So long as no future coerce methods to `character` violate the assertion, the use of `CLASSES=` to do coercion is the right approach. On balance, the many advantages of a systematic approach to argument coercion outweigh the hypothetical risks. If we wanted to be truly paranoid, we could retain the line

```
text = as(text, "character")
```

from the previous version.

The Fortran example works similarly.

```
"matp" =
## the matrix product of the numeric matrices 'x' and 'y'.
function(x, y)
{
    dx = dim(x)
    dy = dim(y)
    if(length(dx) != 2 || length(dy) != 2 || dx[2] != dy[1])
        stop("invalid matrix arguments")
    z = .Fortran("dmatp", x, dx, y, dy,
            z = numeric(dx[1] * dy[2]),
```

```
            CLASSES = c("numeric", "integer", "numeric",
                        "integer", "numeric"),
            COPY = c(F,F,F,F,T))$z
    dim(z) = c(dx[1], dy[2])
    z
}
```

The last T value in the COPY= argument is actually irrelevant in both examples, since we generated this data in the call. It has no other references and the evaluator would not have copied it anyway. Still, it's safer not to lie, in case the function were later changed.

As an alternative to specifying the CLASSES and COPY information in each call to the interface, the same information can be stored as metadata in the chapter by a call to setInterface. The equivalent to the Fortran interface call would be:

```
setInterface("dmatp", "Fortran", where = ".",
  classes = c("numeric", "integer", "numeric", "integer",
              "numeric"),
  copy = c(F, F, F, F, T))
```

The first argument is the name of the subroutine, the second the language, either "Fortran" or "C"; the where argument optionally specifies the chapter. As with the special arguments to .C and .Fortran, classes tells S what class corresponds to the type of each of the subroutine's arguments and copy tells S whether it needs to protect this argument by copying it.

The effect of setInterface is to create an object in the specified S chapter describing the interface, essentially the classes= and copy= information. The object is part of the meta database for the chapter, and its name identifies it as an interface description for Fortran routine dmatp. When the S evaluator first calls the interface to the corresponding subroutine, it will search for such an object in the same way it searches for any object by name. If the interface description object is found, the information is stored in an internal table and used to coerce the arguments to the appropriate data type. Arguments that need copying are copied; others are not. The interface information is optional; if none is found, S assumes that all arguments must be copied and that the call to .C or .Fortran has done whatever coercing is needed.

Since setInterface creates a permanent object as a side effect it only needs to be called once, unless some of the information needs to be updated. The copy information can be supplied to the .C or .Fortran interface function explicitly, by using the special named argument COPY=, even after calling `setInterface`. The dynamic specification overrides the information in the metadata; useful, for example, if the need for copying depended on the particular arguments being supplied. Otherwise it is simpler and slightly more efficient to use `setInterface` once and rely on S to read in the information created.

The purpose of the `COPY=` argument is solely to save memory. The function `unset` provides a related trick. Often, you will assemble some object in the local frame, just to get the exact information needed by the routine. After the routine is called, you may have no further need for the local object. You would like to tell S that this object, though in the frame, can now be thrown away, once the current call is through with it. A call to `unset` does this: it takes the name of a local dataset as its argument, deletes reference to it from the the local frame, and returns the object as its value. In the two examples of this section, the only output argument was constructed in the call to the interface, so there were no wasted copies. If the output arguments included objects assigned in the function (including arguments), then `unset` can potentially save some copying. For example, here is an interface to a Fortran subroutine, `chol`, which writes its results back on top of the matrix supplied as its first argument.

```
"chol" =
function(x) {
  x = as(x, "matrix")
  p = nrow(x)
  if(!identical(p, ncol(x)))
    stop("only square matrices can be used")
  .Fortran("chol", unset(x), p, numeric(p),
          0, 0,
       CLASSES= c("numeric", "integer", "numeric",
          "integer", "integer"),
          COPY=c(T,F,T,F,F))[[1]]
}
```

Suppose the function `chol` was called in the form:

```
yy = chol(crossprod(y))
```

The actual argument corresponding to `x` is computed by the call `crossprod(y)`. Therefore, the only reference to it is in the `chol` function itself. After the call to `.Fortran`, `x` is not used again, so the Fortran routine could write on top of the data supplied in this case; the highlighted call to `unset(x)` tells the evaluator this. Its value will have no references and will not be copied.

Warning: the argument to `unset` will be gone from the frame; there must be *no* references to it in or after the `.Fortran` call.

As always, the recommended philosophy is not to worry about such efficiency issues until you are sure that this particular computation is important enough to justify the effort.

11.4 Dealing with NA's in C

Atomic vectors in S include the notion of `NA` or missing values. Other classes (such as `string`) can then pick up the notion of missing values from atomic slots.

By default, the interface to C or Fortran assumes that the subroutine doesn't know how to handle missing values. Missing values in the arguments will cause an S error. The optional argument `NAOK=T` to `.C` or `.Fortran` overrides the assumption and allows `NA`'s to be passed in. The option is useful even in cases where the subroutine is *not* dealing explicitly with missing values. In our `scanDollars` function, a numeric vector is passed in. The C routine fills in only those elements corresponding to legal elements in the `text` argument. We initialized the elements to `Inf`, a large numeric value, but a more natural initialization would have been to `NA`. Then illegal elements in `text` would produce missing values in the result. Using `NAOK=T` is required to get the missing values into the C subroutine.

```
"scanDollars" =
# convert the strings in 'text' to currency values, assuming the
# strings contain '"$"' followed by an amount in dollars.
# Anything else is returned as 'NA'.
function(text)
.C("scan_dollars", text, length(text),
  rep(NA, length(text)),
  NAOK=T,
  CLASSES = c("character", "integer", "numeric"),
  COPY = c(F, F, T))[[3]]
```

The `CLASSES=` argument was crucial here, to coerce the `NA` values to be numeric; `Inf` was already numeric so we could get away without coercing it. As with the `CLASSES` and `COPY` arguments, you have the option to specify the `NAOK` behavior of the routine by registering it. See page 420.

The internal C code manipulating S objects handles missing values by some tests that depend on the type of data representing the class. You can make use of the tests in your own C code, but for most purposes it's better to handle the NA issues in S. You can often select out the subsets of the data corresponding to missing values, pass down only the rest of the data, and then insert any recomputed values in the corresponding place. An alternative is to compute and pass down a logical vector corresponding to `is.na(x)`, or whatever test applies in the example. The C code can then know to skip or treat specially these cases.

Only if you believe neither of these solutions applies should you read on. To test for NA values in C, you will need to include the S declarations and macro definitions by the line:

```
#include "S.h"
```

in the declarations section of your C source file. Then the following C preprocessor macros allow you to test for NA. There is one macro for each relevant atomic class, and a final macro for the case that the type has to be passed in as an argument.

```
is_na_DOUBLE(d) /* class numeric; d is type (double *) */
is_na_INT(l) /* class integer; l is type (long *) */
is_na_LGL(l) /* class logical; l is of type (long *) */
is_na_REAL(l) /* class single; l is of type (float *) */
is_na_COMPLEX(c) /* class complex; c is of type (complex *) */

is_na(p, mode)  /* mode computed at run time */
na_set(p, mode) /* set contents of p to NA */
```

In `is_na` and `na_set`, the computed `mode` must be supplied set to one of the symbolic values `DOUBLE`, `INT`, `LGL`, `REAL`, or `COMPLEX`. The concept of NA is not defined for character or raw data, or for any list-like data.

The arguments to these macros are *always* pointers, not individual values. C is unforgiving about misusing pointer arguments, and the nature of these macros means that the errors are not found at compilation.

11.5 Raw Data in C

In addition to its direct uses, `raw` data is a useful way to communicate data to and from C code, for essentially arbitrary purposes. Because the contents of the bytes are not interpreted in S (beyond the basic vector operations), there is no particular problem in communicating arbitrary C structures. S provides a few tools for this.

You do not need to write S-dependent C code to use `raw` data; the `.C` interface interprets `raw` data as C type `char *`. S will allocate a block of n bytes for a raw vector of length n. The C routine can cast this block to the desired type, fill it, and pass the result back to S. Now the C information is available in S, to be passed around, saved, and later provided to any other C routine that needs it.

To keep the C code reasonably clean, you will likely want to treat such applications in terms of a C `struct` or `typedef`, when the application corresponds to `raw` data of a fixed length. Particularly if the C definition depends on types that could be non-portable in size, you will want to use some tools in S to help here.

As an example, suppose we have a `typedef` of interest, say `my_data`. We want to store this type of data in S objects, perhaps as a slot in some S class. The first step is to allocate a `raw` object of the right size for the data type `my_data`. While we could figure this out and simply call for the necessary number of bytes, the resulting S code would be potentially non-portable.

Instead, S allows the programmer to "register" any C `typedef`, so that the size is stored internally. After some C routine executes the call:

```
store_structure_size("my_data", sizeof(my_data));
```

then, for the rest of the session, data of the appropriate size can be generated by the S function call

```
raw("my_data")
```

The natural place to call `store_structure_size` is in an initialization routine for the S chapter: see page 409.

Appendix A

Programming in C with S Objects

> This appendix discusses the basics of writing C code that manipulates S objects. C routines called from the `.C` interface deal only with the data values in the objects—numbers and character strings, typically. This chapter describes the `.Call` interface; subroutines called from it deal with the structure of the S objects as well. This is an advanced topic, and liable to present challenges in writing and, especially, debugging. If the structure of the data can easily be set up in S, the `.C` interface of Chapter 11 should be used instead.

S objects can be created, used, and modified in C. Don't do this unless you feel comfortable with C programming *and* you believe there are serious advantages for working directly with S objects in C *and* you are prepared to spend substantial effort in writing and debugging. You will need to learn some specialized C coding techniques. Because these techniques deal directly with pointers in C to S objects, it's much easier to create bugs, some of which may prove very obscure. The level of security offered by the functional style of computing in S is compromised because you now have access to S objects in C. Many things can go wrong, with consequences from memory faults to corrupted data in other data frames.

The compensation is that you can operate directly with the S objects and the S evaluator. At the least, there will likely be efficiency improvements. A better justification comes from the need for some of the extra capabilities provided. Examples include:

- a convenient mechanism for evaluating S expressions from C;
- the ability to deal with different S classes of objects at the C level, coercing the objects as needed.
- the ability to allocate new objects of variable size and class, in response to computations in C;
- some mechanisms to map data from C pointers directly to S objects.

While nearly all these operations can be simulated without the `.Call` interface, the techniques described here are usually more natural.

A.1 The `.Call` Interface

The standard interface from S to C routines that manipulate S objects is via the function `.Call`. Its arguments are similar to those of the `.C` and `.Fortran` interfaces described on page 411 in Chapter 11.

- argument `NAME`, the name of the C routine being called;
- arguments `"..."` corresponding to each of the arguments to the C routine;
- optional control arguments, in this case mainly `COPY=`, saying which of the arguments to the routine should be protected by copying.

The `.Call` interface viewed from the S side is similar to the `.C` interface, except that unlike `.C`, `.Call` returns an arbitrary S object, whatever the C routine returns. All the arguments to the C routine and the value returned by that routine have the same C type, `"s_object *"`; that is, a pointer to a predefined C type representing S objects. The `COPY=` argument, if provided, should be a logical vector of length equal to the number of arguments to the C routine. As with `.C`, `TRUE` elements in the argument mean that the data in the corresponding argument should be copied if it needs protecting; `FALSE` elements mean that the C routine is asserted not to overwrite this argument. The interpretation is the same as for `.C`; see section 11.3. The

CLASSES= argument is interpreted similarly but is perhaps not as important, since we will show C facilities for coercing arguments. It can be provided, however, and in the case of .Call there is no restriction on the classes to which you want the arguments coerced.

We will show an example of the .Call interface on page 431, after introducing the C routine it calls.

A.2 C Routines Returning S Objects

C routines that are to be used via the .Call interface must follow three rules:

1. All the arguments and the return value are of C type "s_object *".

2. Each C source file must include the S standard header:

```
#include "S.h"
```

3. Each routine that deals with S objects and therefore with the S evaluator must declare the evaluator by a standard C preprocessor macro, S_EVALUATOR, appearing in the declaration part of the routine and *not* followed by a semicolon.

The include file contains the various macro definitions and prototypes for the C subroutines described in this section. The file is located in the include subdirectory of the S home directory; when CHAPTER initialized the makefile in the chapter directory, it arranged for this directory to be searched for C header files during compilation.

Here's an example, which we will examine throughout our discussion.

```
#include "S.h"

s_object *lapply_in_C(s_object *x,
              s_object *expr,
              s_object *name_obj,
              s_object *frame_obj)
{
  S_EVALUATOR
  long frame, n, i; char *name;
  s_object **els;

  x = AS_LIST(x);
  els = LIST_POINTER(x); n = LENGTH(x);
```

```
    frame = INTEGER_VALUE(frame_obj);
    name = CHARACTER_VALUE(name_obj);
    for(i=0; i<n; i++) {
      ASSIGN_IN_FRAME(name, els[i], frame);
      SET_ELEMENT(x, i, EVAL_IN_FRAME(expr, frame));
    }
    return(x);
}
```

The routine `lapply_in_C` implements in C a simplified version of the S function `lapply`. Its four arguments are a list, an S expression, the character string name of an object, and an evaluation frame. The C routine "applies" the S expression `expr` to each element of the list or list-like object `x`. The expression is written in terms of an object name, also passed in as an argument. What actually happens is that the calling S function takes the argument name and the function body from a user-supplied function of one argument, and passes these both to the C routine, along with the numeric index, `frame`, of an evaluation frame. In a loop, the routine assigns each element of the list in the frame, and then evaluates the expression in the same frame. The successive values are stored back in the original list.

The C computations simulate a loop in S that calls the function each time. By extracting the body and using a previously generated frame, some overhead in the S evaluator is eliminated. Frankly, the savings are unlikely to be worth the programming effort for most applications, but the example helps illustrate C programming for S. A related, slightly more complicated version of the same idea *does* provide a serious example. The function `rapply` recursively applies a function to all the elements and sub-elements of a recursive object, to any depth, when the elements match chosen classes. The S and C computations for `rapply` follow the style of the example here.

The computations in `lapply_in_C` use a number of C macros to manipulate S objects and to invoke the S evaluator. We will return to those in section A.3. The overall style is the immediate point of the example: a line to `include` the S header file; a declaration of the S evaluator; and all the arguments as well as the return value having type `s_object *`.

One detail, though, needs emphasizing. Any computation that replaces elements in a list or other recursive object *must* use the `SET_ELEMENT` macro, and must never set the element directly in a C assignment. Otherwise, storage requirements for valid objects may be violated. See page 434.

An S function to make use of this routine could be defined as follows:

```
"qapply" =
## a (slightly) quicker version of 'lapply'. The S function 'f'
## will be applied to each element of list 'x'; 'f' must be
## a function of one argument.
function(x, f)
{
  name = formalArgs(f)
  if(length(name) != 1)
    stop("Need a function of 1 argument, not ", length(name))
  if(!is.recursive(x))
    x = as(x, "list")
  frame = new.frame()
  .Call("lapply_in_C", x, functionBody(f), name, frame)
}
```

The `COPY=` argument was omitted from the `.Call`, so by default the S evaluator will protect all arguments by copying. Section A.4 will discuss the issues involved, but copying is the safe default.

Since nearly any C computation on S objects can also be done in S, you have a choice of language to use for such computations as checking the class of arguments or supplying default values. When in doubt, do the computation in S, and particularly so when you find the C computation getting complicated. In the `qapply` function, we require the user to supply a function of one argument; since the name of that argument is used in the C routine, we have a simple way to check, and to provide the user an error message. We also arrange that the argument `x` is a recursive object; but only in the broad sense implied by `is.recursive`. This ensures that we can extract the elements and pass them to the chosen function. (See page 199 in Chapter 5.) Objects from classes with slots will qualify, in general, causing `qapply` to iterate on the slots—not necessarily the best idea for all applications. Since we're storing the computed results back in `x`, leaving the class of the object arbitrary could produce an invalid object from that class. The recommended style, as done on page 429, is to coerce `x` explicitly to class `list` via the `AS_LIST` macro before dealing with its elements.

A.3 S Objects from Basic S Classes

All S objects in C start out as a pointer to an object of type `s_object`. The header file `"S.h"` defines this type. For C computations with S objects, you should always use a set of C preprocessor macros that manipulate such objects; the macros hide much detail that is easy to get wrong. The macros in this section deal with basic S classes. Section A.6 describes how to adapt your computations to other, arbitrary classes.

Class	**C type**	**Macro** *type*
`numeric`	`double *`	`NUMERIC`
`logical`	`long *`	`INTEGER`
`character`	`char **`	`CHARACTER`
`integer`	`long *`	`LOGICAL`
`single`	`float *`	`SINGLE`
`raw`	`char *`	`RAW`
`list` and all recursive objects	`s_object **`	`LIST`

Table A.1: *Basic S objects in C. The "type" name will be combined with a macro name in Table A.2 to give the name of macros to allocate, coerce, and otherwise operate on the corresponding objects.*

When the S objects belong to the "basic" classes, the C macros allow computations with corresponding C data types, as defined in Table A.1. The C code you write will be specialized to one or more of the types in the second column of the table. The third column of the table gives symbolic names for the available types. C macros specialized to a particular type will have names containing the corresponding symbol; for example, `NEW_NUMERIC` for the `NEW` macro specialized to `NUMERIC` data. Macro call `NEW_NUMERIC(n)` creates an S numeric vector of length `n` and returns a value of type pointer-to-`s_object`, pointing to the corresponding object.

```
obj = NEW_NUMERIC(n);
```

This is the equivalent in C of the S expression `numeric(n)`. By replacing `NUMERIC` in the name of the macro with any of the other symbolic names in the third column of the table, you get an S object of the corresponding type.

Similar macros exist for other operations on the objects. Table A.2 summarizes them; the rest of this section gives examples of their use. Macros from the first two rows of the table return a pointer to type `s_object`; the *type*`_POINTER` macros return one of the C types in the second column of Table

A.1; and the *type*_VALUE macros have a type of the contents of the types in that column. So, for example, CHARACTER_POINTER returns type "char **", and CHARACTER_VALUE returns type "char *".

Macro	**Meaning**
NEW_*type*(n)	A pointer to an S object of this class and length n
AS_*type*(obj)	A pointer to an S object of this class, from coercing obj
IS_*type*(obj)	A C boolean, TRUE or FALSE, according to whether obj is (in the S sense) an object of the corresponding class
*type*_POINTER(obj)	A pointer to the data part of obj, with the type as shown in table A.1
*type*_VALUE(obj)	The single value of obj, which should have length 1

Table A.2: *C preprocessor macros for S objects. The actual macro names are obtained by substituting* NUMERIC, *etc. for "type" in the first column, using any of the symbolic types in the third column of Table A.1. The C variables* obj *and* n *are of type* s_object * *and* long.

The AS macros work like the as function in S. They return a pointer to an S object of the corresponding class.

```
obj = AS_NUMERIC(obj)
```

If the coercion is not possible, the macro returns the C NULL pointer value—it does *not* generate an error . The IS macros correspond to the is function in S, testing whether the object extends the corresponding basic class. These macros return a single value in the C type boolean, defined in "S.h" to enumerate the values TRUE and FALSE, consistently with usual C logical computations. It's important either to use an IS macro before applying the corresponding AS macro or to test the pointer returned from the AS macro for being NULL, since the AS macros do not generate an error. Also, you need to use the AS macro even after the IS test if you plan to work with the data in the object. An S object can extend a class but have a different representation from that class. A typical use of the macros would be something like the following:

```
if(IS_LIST(x))
    x = AS_LIST(x)
else
    PROBLEM "expected x to be a list object" ERROR;
```

Section A.7 discusses the `PROBLEM` macro, used to generate S error handling from C.

The `POINTER` macros return C pointers to the data part of S objects.

```
double *d;
  ....
d = NUMERIC_POINTER(obj);
```

Notice that all recursive classes, not just class `LIST`, use macros `NEW_LIST`, `LIST_POINTER`, etc. The C type of `LIST_POINTER` is `s_object **`, a pointer to an array of pointers to S objects. To deal with an S class having explicit slots, see section A.6.

Important: Dealing with recursive objects in C can be dangerous and requires care. The evaluator maintains a system of data-sharing, so that elements of a recursive object may be shared with other objects. Any computation that changes any element of a recursive object must be done so as to keep the data-sharing valid; for example, before changing any element, the code must protect shared references from being changed at the same time. Two steps must be taken in C code that plans to alter any recursive object.

1. The object must be protected by copying the top level. This is done automatically by the `.Call` interface at the top level, unless you suppress copying through the `COPY` argument or by setting an interface explicitly, as described on page 422 in section 11.3. If you plan to replace elements at lower levels, you need to protect these explicitly via the `COPY` or `COPY_ALL` macros, and assign the protected elements back into the higher-level object. See section A.4.

2. *All* assignments of objects to elements of lists, at any level, must be done by the `SET_ELEMENT` macro.

   ```
   SET_ELEMENT(x, i, value);
   ```

   This replaces element `i` of the S object `x` with the S object `value`. The index goes from 0 in the C style, not from 1 in the S style. Never use a `LIST_POINTER` variable on the left of a C assignment; instead, always apply the `SET_ELEMENT` macro to the whole object.

The example on page 430 shows a simple application. The object `x` is an argument to the interface and so is protected by the evaluator. The value of the expression is an S object, assigned to the `i`-th element of `x`.

You need to be sure that `obj` really is a pointer to the right type of object, before applying one of the `POINTER` macros to it. The `POINTER` macros themselves do not check. So before applying `NUMERIC_POINTER` to `obj`, it should either be the result of `NEW_NUMERIC`, or `AS_NUMERIC`, or of some other computation you've done that assures you really do have numeric data. Don't spare the calls to the `AS` macros. They take little time if the object already belongs to the relevant class and if it does not, failure to coerce it will likely prove fatal to the computation.

The `VALUE` macros are shortcuts to computing a single value from an object of length 1. The computations will produce error messages if the object is not suitable to compute the value. The value returned is of type `*(`*ptr*`)`, where *ptr* is the type in the second column of Table A.1: `NUMERIC_VALUE(obj)` will be of type `double`.

Besides the macros in Table A.2, there are some related computations that don't fall into the paradigm of the table. Three special `IS` tests are defined:

```
IS_RECURSIVE(obj)
IS_ATOMIC(obj)
IS_NUMERIC_COMPATIBLE(obj)
```

The first tests whether that object has any recursive type of data; in contrast, `IS_LIST` tests specifically for an extension of the `list` class. The `IS_ATOMIC` macro tests for any of the non-recursive basic types; `IS_NUMERIC_COMPATIBLE` excludes from these the `CHARACTER` data.

So far, the actual data in the S objects has all come from S. Either the data was passed in as an argument or the object was created by a `NEW` macro. This is an important safeguard: S handles the data part of objects by some careful special code. Never set the data pointer of an S object to some arbitrary block of storage! Very occasionally, however, it may be important to tell S to handle some block of storage as if it were data for an S object. Various mechanisms of interprocess communication or other computation may leave you, as a C programmer, holding a pointer to a block of storage containing atomic data, for example, `double`'s. If the block is large, you would rather S took the whole block as the data part of a `numeric` object than have it create another object and copy the data.

When the data is on a file, this is just what `readRaw` does. If the data is available as a pointer, the macros MAP_*type* do the analogous operation, with *type* again being one of the types in Table A.1, but *not* LIST or CHARACTER.

```
obj = MAP_type(pointer, length, notify)
```

These macros expect pointer to be a C pointer of the corresponding type, and length to be the number of elements in the data being supplied. The notify argument is a pointer to a C routine to be called when S no longer has any reference to the object constructed.

The macros return an object of the corresponding S class. The object is very special; S regards the data as having been mapped from somewhere else. It makes no effort to manage the data directly; in particular, the evaluator will never free the data. As an example, suppose our C code has a pointer, big_num_block, to len_block values of type double.

```
s_object *obj; double *big_num_block;
long len_block;
  ....
obj = MAP_NUMERIC(big_num_block, len_block, NULL);
```

After this, obj is a legitimate S object; in particular, it can be returned as the value of a .Call.

Each mapped object is given a unique tag when it is created. The notify routine is a pointer to a C routine that will be called with this tag as its only argument. You can then take any action in the notify routine; S has told you that the memory previously used for mapping this object is no longer needed. Since the S manager did not allocate the memory in the first place, it will never try to free it. That is the responsibility of the calling routine. If no notification is needed, supply the notify argument as NULL. The tags are unique positive values of type s_index; this is some unsigned integer type chosen to be large enough on the local machine.

Since the tags are unique, you can use the same notify routine for multiple objects, so long as you keep some sort of table by tag. The tag for a memory-mapped object is the value of the C expression

```
MAP_TAG(obj)
```

and is zero if obj is not mapped. The notify routine can also be used with objects that are mapped to a file, although it's not usually needed in this case, since S will handle closing the file when there is no further need for it. If you do want to be notified when an object mapped by a call to readRaw, for example, is no longer active, use the C macro

```
SET_MAP_NOTIFY(obj, notify)
```

with `notify` again being a pointer to a routine of type `void` taking an `s_index` argument.

A.4 Protecting S Objects in C

The most dangerous aspect of operating on S objects in C comes when data is modified in any way. You *must* protect any such data; otherwise, the data may be shared between S objects and your modifications can corrupt other S objects. Bugs produced this way can be extremely nasty and hard to find. By default, S will copy all the arguments to the subroutine if there are other references to them. You may want to have greater control, however, to save memory. Also, for list-like objects, you may need to be more careful in any case. If either situation applies, read on.

If an object, say `obj`, is to be modified, you should protect that object by a `TRUE` element in the `COPY=` argument to `.Call`, or just omit the `COPY=` argument so that all objects are protected. If you assert that the C code can never overwrite the elements of an argument, use a `FALSE` element to prevent copying. To be specific, "copy" means the following. The object will be examined to see whether it has any references other than that in the `.Call` itself. If so, the array of elements will be copied.

You can also do the same copy protection in the C code itself, by invoking the macro

```
COPY(obj);
```

before doing any modification to `obj`. You may need to use `COPY` if the question of whether the object gets modified depends on other computations.

For recursive, list-like objects the situation is a little more subtle. The copy protection will copy the pointers to the elements of `obj` but not these elements themselves. Usually that's what you want: if you plan only to replace some of the elements with newly computed values, it would be wasteful to copy the entire object. However, keep in mind that you then may need to protect those elements if, instead of replacing them with something new, you are going to modify the element itself. In a few cases, you may not be able to control the level at which the modification takes place. Such exceptions need to be handled by executing

```
COPY_ALL(obj);
```

which will copy all the data, unconditionally. For atomic data, the elements are not pointers to objects, so you can always get by with the `COPY` argument or macro.

If it is not obvious whether copy protection is needed for an object, here are two rules of thumb that cover most cases:

1. If you explicitly overwrite any of the elements of `obj`, you must always protect it.

2. If you pass the elements to the S evaluator for an evaluation or assignment, using the macros of section A.5, the evaluator should handle the protection for you.

Remember always that any actual assignment of elements of a list-like object must be done via the `SET_ELEMENT` macro (page 434).

If what you are doing does not seem to fall in either of the two cases, you will need to consider the evaluation and storage model very carefully. In fact, it's a good idea to keep to either simple manipulation of the object or else one of the standard evaluator operations, just for the reason that other manipulations can be tricky. For example, calling the C routines underlying various S functions directly from C is legal, but you must be careful to protect any objects that might be needed later. It's safer to evaluate a call to the S function, in spite of the slight additional overhead. And inserting objects into evaluation frames, other than by the `ASSIGN` macro (see page 439), is potentially very bad because it can break reference counting. Fortunately, it's also not easy to do.

Before working with S data, make sure also that the object has been coerced to the corresponding type. S objects are fully dynamic; any S object in any function or on any database can in principle contain any kind of data. C is at the other extreme, totally unforgiving and unintelligent about data types. If you treat `d` as if it points to numeric data, C will believe you, and compute garbage if you have lied. And if you treat numeric data as if it contained pointers to character strings or to S objects, you are likely to get memory faults. If the testing or coercing gets any more complicated than is handled by the `IS` and `AS` macros, it's safer to let the S function do the work, either by special tests as in our `qapply` example, or by using the `CLASSES` argument to `.Call`.

A.5 C Evaluation Utilities for S Objects

To carry out an S evaluation from C, we need an S object representing the expression to be evaluated, say `expr`. Then, to evaluate this in the frame of the function that invoked `.Call`:

```
obj = EVAL(expr);
```

A more general version, `EVAL_IN_FRAME(expr, n)`, evaluates the expression in evaluation frame `n`. Nearly always, you want to compute the expression in S and include it as an argument in the `.Call`, as in the `qapply` example on page 430. The `EVAL_IN_FRAME` version of calling the evaluator is the one you usually need if the expression has been passed in to the function invoking `.Call`. Like the expression, the numeric index of the frame will likely be computed in S and passed to the C code.

S objects can be retrieved and assigned by name in C routines. The macros `GET` and `ASSIGN` behave like the corresponding `get` and `assign` functions: `GET(name)` looks for the object of the given `name`; `ASSIGN(name, obj)` assigns the object as `name` in the local evaluation frame. In both cases, `name` is a C pointer of type `char *`.

As with `EVAL`, both `GET` and `ASSIGN` have versions that include the evaluation frame, and these versions are more likely what you will need.

```
GET_FROM_FRAME(name, n)
ASSIGN_IN_FRAME(name, obj, n)
```

The integer value `n` is again the index of an evaluation frame.

Some specialized computations on objects have analogues in C. In addition to the `NEW`, `IS`, and `AS` operations, user code can extract some of the basic information in the object:

```
long n; s_class *cl;
 ....
n = GET_LENGTH(obj);
cl = GET_CLASS(obj);
```

The `GET_CLASS` macro returns a pointer to an object of datatype `s_class`, as the example shows. For printing messages and other applications, it may be more useful to have a `"char *"` pointer to the name of the class. Use the expression

```
GET_CLASS_NAME(obj)
```

for this purpose. See page 443 for an example.

Parallel operations allow setting the class and length.

```
SET_LENGTH(obj, n);
SET_CLASS(obj, cl);
```

Before using these or any other computations to modify an S object, make sure that the object can be modified without danger, according to the rules on page 437. The `SET_CLASS` computation is essentially identical to `AS(obj, cl)`, but explicitly done in place.

Special operations exist also for some classes; for example, `named` and `array`, to get or set their slots:

```
s_object *namedObj, *matrixObj, *dim, *names, *dimnames;
 ....
names = GET_NAMES(namedObj);
 ....
SET_NAMES(namedObj, names);
 ....
dim = GET_DIM(matrixObj);
dimnames = GET_DIM(matrixObj, dimnames);
 ....
SET_DIM(matrixObj, dim);
SET_DIMNAMES(matrixObj, dimnames);
```

The `names`, `dim`, and `dimnames` objects used in setting the attributes should have the correct information, just as they would if the assignment were done in S, but they do not need to be explicitly coerced to the class of the corresponding slot.

These specialized C computations are *not* generic function calls; they operate on the specific classes noted and on a few extensions, but do not generally dispatch methods for the corresponding functions. So, for example, if you wanted to get the value of `dim(x)` for an argument `x`, and have this work regardless of the class of `x`, you need to supply `dim(x)` as a separate argument in the `.Call` interface, rather than counting on `GET_DIM` in the C code.

A.6 Dealing with Arbitrary S Classes in C

Occasionally, you may need to deal explicitly with classes and their representation at the level of C code. The C typedef `s_class` is a structure that contains versions in C of the key information about an S class. To begin dealing with the class, construct a pointer to the class, and ensure that S has looked up the class definition in the metadata for classes, by calling the `MAKE_CLASS` macro in C.

```
s_class *track_class;
 ...
track_class = MAKE_CLASS("track");
```

As part of defining the class, S will look up the representation. Therefore, this code must be executed when the representation is available. If the class is defined on a particular library or chapter, you need to ensure that the `library` or `attach` call has been evaluated. A good general strategy is to call an initialization routine from an S function, `.on.attach`, in the same chapter; see page 409.

An object can be created in C from any S class. For example, given the pointer to the `"track"` class above:

```
obj = NEW(track_class);
```

This returns a copy of the prototype object from the class. If the class has slots, the object will be recursive, and its data elements will be the slots. Otherwise, the object will just be whatever was supplied as the prototype in the definition of the class. Executing `NEW` in C corresponds to evaluating a call to `new` in S. In C, the argument is a pointer to the `s_class` type, not a pointer to a character string, whereas in S the two need not be distinguished.

The C code can also use the `IS` macro to test an object's `is` relation to a class and the `AS` macro to coerce an object to a class, by analogy to the S functions `is` and `as`.

```
if(!IS(x, track_class))
    PROBLEM
    "class to be or to extend class \"track\""
    ERROR;
x = AS(x, track_class);
```

(See section A.7 for the `PROBLEM` paradigm.) As with `NEW`, the class must be specified by a pointer to type `s_class`. The coercion done by `AS` will protect non-local objects from being overwritten.

Once a class is defined you can find out the positions in the class representation of various slots. There are two basic ways to do this: `SLOT_POSITION` takes pointers to two S objects as arguments. The first is an object from the relevant class, the second is the name of the slot (as an S character or name object). This is a good way to find out slot positions when you don't necessarily know the exact class of the object.

More commonly, you may want to set up indices for the slots to make repeated access to them faster and more convenient. To do this, you want the offset in the object's data for a particular slot: `GET_SLOT_OFFSET` gives that information. In the `track` example:

```
int xoffset, yoffset;
 ...
xoffset =  GET_SLOT_OFFSET(track_class, "x");
yoffset =  GET_SLOT_OFFSET(track_class, "y");
```

It is an error to try to get a slot that does not exist in the given class.

For convenience, you can store the offsets in some global place, arrange to initialize them when your library is attached, and use them whenever the C code needs to get the corresponding slot:

```
s_object *obj, *x, *y;
  /* suppose obj points to a 'track' object */
 ...
x = LIST_POINTER(obj)[xoffset];
y = LIST_POINTER(obj)[yoffset];
```

An S object with slots is always recursive (sometimes a list, sometimes a structure): `LIST_POINTER(obj)` gives a pointer, of C type `s_object **`, to the vector of slot pointers.

The initialization computations for the class should be in a C routine that is called from the `.on.attach` function in your chapter. Here is a `track_init` routine incorporating the steps outlined so far in this section to initialize the class `track`.

```
static s_class *track_class;
static int xoffset, yoffset;

void track_init(void)
{
  S_EVALUATOR

  track_class = MAKE_CLASS("track");
  xoffset =  GET_SLOT_OFFSET(track_class, "x");
  yoffset =  GET_SLOT_OFFSET(track_class, "y");
}
```

This version assumes that access to the information is only needed in this C source file, so that track_class, xoffset, and yoffset can be static (and we don't need to worry about name conflicts). Whenever you need to manipulate slots in C, explicit initialization is recommended. By no means should you build into the code fixed constants for the slot offsets: if the class definition should change, truly obscure bugs will likely spring up.

Of course, if the slot *names* should change, the C code needs revision in any case. Since GET_SLOT_OFFSET generates an S error if it cannot find the slot, you will get quick notice if the C and S views of the class get out of sync.

A.7 Handling Errors in C

Whether or not your C code actually manipulates S objects, it may well encounter situations that prevent completing the current computations. Since S provides extensive debugging facilities, the C routine should invoke these, providing as useful a message as possible to describe the situation. The message is constructed in the style of C's printf routine:

```
PROBLEM format,a1, a2, ..., an ERROR;
```

where format is a quoted format string and a1, etc. are arguments matching the printf-style conversion specifications in the format string.

```
if(d[i] < 0.)
  PROBLEM
  "expected positive data, item %d was %lg",
   i+1, d[i] ERROR;
```

The format string should not contain new lines; S will handle printing.

The sequence PROBLEM .... ERROR; is the C equivalent of a call to the S function stop. Three other S functions, message, warning, and terminate deal with user messages also (see section 6.4.2 on page 274). The first just prints the message on standard error, the others generate a warning and terminate the S session. Replacing macro ERROR with PRINT_IT, WARN, or TERMINATE, respectively, will give the equivalent effect in C.

```
PROBLEM
  "had to convert object from class \"%s\" to class \"track\"",
  GET_CLASS_NAME(obj) WARN;
```

In addition, replacing ERROR with END_MESSAGE just completes the message, after which you can get back a copy, of type char *, by invoking the macro GET_MESSAGE:

```
PROBLEM
 "name of object %ld was \"%s\"",
 i+1, names[i]
END_MESSAGE;
msg[i] = GET_MESSAGE;
```

The character string returned by GET_MESSAGE is dynamically allocated in the current S evaluation frame. It is suitable for use there, or for returning as an element of a character vector, but *not* for a permanent C pointer—for that apply a C routine such as strdup to the pointer.

Appendix B

Compatibility of S with Older Versions

This appendix discusses steps needed to use functions and other objects created in older versions of S and S-Plus, specifically, the version described in references [1] and [3]. It should be read mainly by those who own libraries or databases of such objects, and want to make them available with current S. While the current evaluator can use old-style objects, the recommended approach is to create a new S chapter to which the whole contents of the old database will be copied and converted. Some functions may need conversion, object code needs to use dynamic linking, and documentation files need to be converted to documentation objects. The function `convertOldLibrary` does most of this automatically. For a variety of data objects, conversion to more modern equivalents is an option, provided by the function `modernize`.

The general compatibility goal for the current version of S is to support the techniques described in the two books, [1] and [3]. Versions of S-Plus through Version 4, or releases through 1997, all correspond to this earlier version of S. Data objects created by these versions can be used with the newer version; the evaluator turns the objects into valid current S objects when reading them. Functions written for the older versions may need conversion: a few functions have changed their names, old-style classes should be detected and specified, and the approach to loading compiled object code needs to be updated. Documentation files in the old library should be converted to documentation objects. This conversion is handled, essentially automatically, by the function `convertOldLibrary`, described in section B.1.

The owner of an old-style library creates a new S chapter; a call to `convertOldLibrary` copies and converts the contents of the old library into the new chapter. From then on, all further programming can be done in the new chapter. The old library hasn't been touched, but should not be needed for any further work.

The approach to classes and methods described in Appendix A of reference [3] remains available as well. Here too, the approach described in the reference is supported, but by different computations embedded in the current evaluator. Techniques that took advantage of the earlier implementation may need revising.

Occasionally, old classes cannot be detected automatically, because there are no objects with the class in the library and no function sets the class explicitly. These classes should be specified by calling `setOldClass` as shown in section B.2. Handling of methods and extensions will be smoother if all old-style classes are explicitly specified.

If the old library has C or Fortran code, the source should be copied to the new chapter directory before doing the conversion. A `make` process then creates a shared object for dynamic linking with S; see section B.4 for details.

While conversion of old objects is otherwise not usually required, it may be a good idea. The function `modernize` will convert objects into the recommended modern form, and can make it easier to work with many classes of objects. Section B.2 discusses modernizing old objects and other class-compatibility issues.

Documentation in the earlier version of S did not use documentation objects. The old-style documentation files in an old library will be converted to documentation objects in the new chapter, as part of the general conversion. The underlying tools can, if necessary, be used independently as well. Section B.5 discusses how.

The tools described provide a fairly mechanical conversion of old libraries into new chapters. The hard question is: Will the old computations still work? Compatibility with older versions of S is defined by the two references mentioned; they are as close as we can come to a "standard" for the earlier version of S. The code that implements S has changed radically, however. As a result it can not be true that all the functions written for the earlier version will work now. Many functions quite naturally used techniques that were not explicitly sanctioned in the references, but that worked. Some of these will still work, some will not. And the books were user guides, not standards documents, so they sometimes gave a vague or inconsistent definition. Making the converted version of an old library work *fully* may be a challenge. The web site for the book will accumulate experience with incompatibilites; meanwhile, a few general points may be helpful.

1. Computations that don't work may be using some property of the implementation not asserted in the books; looking back at the references may help to pinpoint the extra assumption. Particular candidates are computations that make assumptions about *how* old-style methods were evaluated; e.g., without creating a new evaluation frame. Look for calls to `substitute` or `sys.parent` as a clue.

2. Several deliberate changes enforce aspects of the S functional programming model: objects have to be assigned locally before being altered by replacements; generic functions, their methods and their group generic must have identical argument lists; classes must have consistent inheritance. Some restrictions are relaxed partially for old-style classes and methods.

3. There is a fundamental choice between fixing the old-style code and converting to a library using current classes and methods. If the computational goals of the library seem to fit naturally with current concepts in S, writing a new library that reuses most of the computational tools is likely to be more enjoyable and to produce a better result.

The tools in this appendix carry out a mechanical conversion that should provide a start in any case.

B.1 Converting Old Libraries and Databases

The S evaluator can read the binary files storing old-style S objects, so you can attach an old database of S objects. Be *very* careful, however, not to assign into such a database, if you ever plan to compute on it with an old version of S. Newly assigned objects have information not in the old-style files and the old version, naturally enough, can not read these files.

The recommended strategy is not to attach old-style databases, but instead to convert the database by copying the objects into a new chapter, making some conversions in the process. The function `convertOldLibrary` automatically makes changes in functions and specifies old-style classes, while copying all objects to the new database. Suppose my home directory has an old-style library in subdirectory `"tools"`. The objective is to create a valid library for current S, say in subdirectory `"Ntools"` of my home directory. The essential steps to convert the old library are as follows.

1. Create the new directory and initialize it as an S chapter by invoking the `S CHAPTER` command. If the old library has compiled code to be used in the new library, the source files for that code should be copied to the new directory before invoking the `CHAPTER` command, so that the `make` procedure will compile them.

2. In a running S session, invoke the `convertOldLibrary` function to copy and update the objects in the old library. In our example, this would have the form:

   ```
   convertOldLibrary("$HOME/tools", "$HOME/Ntools")
   ```

   Neither the old nor the new library needs to be attached before running the conversion.

3. If there is compiled code associated with the library, run `S make` to create the linkable object code (see section B.4 for details).

At this point the new library should be ready for testing.

Running `convertOldLibrary` on objects already converted should cause no problems. In particular, if you find you need to make some editing changes on the objects copied to the new database that might introduce more old-style code to convert, you can run `convertOldLibrary` on `"Ntools"` only, in which case it will only copy modified objects. An additional argument to the function is the list of objects to be examined, so individual functions can be updated if they need to be added to the new library. By default, `convertOldLibrary` also converts old-style documentation files in the library to documentation objects; see section B.5.

The main effects of `convertOldLibrary` are to convert some function calls and to create metadata for old-style classes. Some old-style functions were superseded because their old definitions were incompatible with current S, and their names were too natural to "retire" them. The most important example is the `class` function. Any calls to this function will be changed by `convertOldLibrary` to call a newly-named function, `oldClass`. The class extracted by `oldClass` will be consistent with the definition in reference [3]. Assignments of old-style classes will also be converted, by using `oldClass` on the left of the assignment, so that the object will simulate the class "attribute" used in the earlier version of S.

Some additional functions have had their definitions changed for various reasons. For example, calls to `log` with a second argument for the base were legal in previous versions of S. The current `log` function has only one argument, in order to be a legitimate member of the `Math` group; The function `logb` takes a base as its second argument. Processing by `convertOldLibrary` will convert two-argument calls to `log` to call `logb` instead.

Old-style classes should be declared by a call to `setOldClass`, with the vector of character-string names in the old-style class as the argument. The class names are stored in the new library's metadata. The evaluator can then find old-style inheritance reliably and can treat current and old-style classes differently when necessary. The call to `convertOldLibrary` will generate the calls to `setOldClass` whenever it detects the presence or use of a known old-style class. Section B.2 discusses the changes in classes further.

Aside from any changes, there are moderate efficiency advantages to converting the objects. The S evaluator has to do the conversion of old-style binary files into current S objects each time the object is used. Assigning the objects on the new library performs the conversion permanently. For this reason, and for possible modernizing of some objects, you should generally convert old databases even if they do not contain functions.

B.2 Classes

Earlier versions of S introduced the concept of classes and methods [3, Appendix A], but without most of the explicit tools of current classes. Objects only had classes optionally, and a class was only defined per-object, by an attribute. There was no central information about either classes or methods—no class representation, no explicit interclass relations, and no metadata. Methods were dispatched by looking for functions whose names were formed from concatenating the generic name and the class name. To provide inherited methods, class attributes were allowed to contain more than one string. The first string was the class of the object and the following strings indicated other classes which, in the current terminology, that class extended.

The current class/method model offers substantial advantages over the earlier model. There is more consistency in the contents of objects, providing greater security against creating invalid objects. A wider range of both classes and methods can be specified. S can do more for the user, because knowledge about classes is kept internally and used to make decisions during evaluation.

The earlier model is, however, still largely supported by the current implementation. Principally, this is to maintain compatibility with the large body of existing code that uses the previous model. One step is strongly recommended: declare old-style classes to current S by calling `setOldClass` as described below. This may have been done for you automatically by `convertOldLibrary`; if not, or if you are in doubt, you should call `setOldClass` explicitly. Computations with old-style objects *may* work anyway, but the S evaluator will be able to handle the classes more consistently if it has explicit knowledge of them.

Old-style classes will be picked up automatically in the conversion if there is an object in the database that has the class attribute or if one of the functions explicitly assigns this class. The evaluator warns on encountering undeclared inheritance. A call to `setOldClass`, such as:

```
setOldClass(c("chron", "times"), where = "$HOME/Ntools")
```

will declare the class. The `where` argument should be a library that provides computations to be used with the class. Then any users attaching the library will get the correct definition of the class as well. The right place for the declaration is with the functions that *generate* objects from the class, rather than on a database only containing the objects themselves. The first argument to `setOldClass` should be the character vector that old-style functions would use in assigning the class: the name of the class followed by any inheritance.

Metadata generated by `setOldClass` is used to dispatch old-style methods and to distinguish old-style classes from similar-looking modern classes. Old-style inheritance is *not* considered in dispatching current methods, mainly because there is no way to guarantee the necessary coercion of objects to the target class. If you want to apply new-style methods to old-style classes, you must make all the relevant signatures explicit. However, mixing old and new ingredients this way should usually be avoided.

One potential inconsistency in old-style classes needs to be noted, but let's hope it does not exist in practice. Nothing in the old-style approach required the classes to be used consistently. In principle, two objects could agree in the first string of their class attribute and disagree in that class's inheritance. As a strictly hypothetical example, in addition to objects with the inheritance from "chron" to "times" above, other objects could have had:

```
> oldClass(weird)
[1] "chron" "dates"
```

In current S, each class has to have a consistent inheritance. No consistent call to `setOldClass` can cover this case. Such inconsistency would have been a bad idea, but since there was no central definition of the class, earlier software could not check for the situation. If there are actual examples, the suggested fix is to include all the classes in the defined inheritance, merging the classes in what seems the most plausible order. For example,

```
setOldClass(c("chron", "dates", "times"))
```

To get full consistency with the old behavior, however, you would need to define two new classes, say "chronDates" and "chronTimes" in our example, convert the individual objects appropriately and duplicate the old-style methods for the two new classes.

B.3 Modernizing Old Data

Only occasionally are you *required* to convert an old-style data object to something new. For a number of old-style classes, though, you may *prefer* to modernize the data. For example, categories, factors, and ordered factors are old-style objects designed to deal with repeated values from a set of levels. These objects run into various confusions between the character strings defining the levels and numeric indices used internally in the objects. Depending on how the objects are handled, they can look like integer data, character data or neither. The newer class `string` and its two extensions `stringFactor` and `stringOrdered` provide similar functionality but with advantages of consistency, explicit control, and efficiency for large objects.

You may want to modernize data from old classes by converting to a new class or otherwise modifying the object. S provides a centralized approach to such conversions via the function `modernize`. If by now you have read much about classes and methods, you will not be surprised to learn that `modernize` is a generic function, with methods for old-style classes that convert the objects to the recommended new-style object. The function takes an object and returns a possibly modified object as its value. By defining methods for classes that need updating, the behavior of modernize can be extended and modified as you wish. Here is the method for factor objects:

```
> showMethod("modernize", "factor")
# turns the factor object into an object of class '"stringFactor"'
# ('"string"', but required to have levels defined).
function(object, very = 0)
as(object, "stringFactor")
```

The default method does nothing for most classes of objects, but invokes modernize recursively on each element of lists and other recursive objects. In writing methods for modernize, keep in mind that current S does *not* use old-style inheritance in dispatching current methods. Therefore, separate method declarations have to be generated for each old class. For example, the inheritance of class "ordered" is class "factor", but would need methods, even if the methods did the same thing (they don't), and even if we assumed all old-style "ordered" objects inherited from "factor".

The original version of classes and methods in S was presented along with software for statistical models, in reference [3]. The largest quantity of related software deals with the models library and, in particular, *data frame* objects (see section 4.7.3 for an overview of this software in current S).

The class data.frame benefits from being modernized, though it does not require it, and the object returned will still have the old-style class "data.frame". The main reason for modernizing data frames is to turn objects with old-style classes "factor" and "ordered" into objects with the analogous true classes "stringFactor" and "stringOrdered". In addition, variables in data frames can have class "string". The main distinction between string and stringFactor is that the levels for string objects are just those needed for the actual data; therefore, subsetting the data changes the levels. String factor objects are treated as having fixed levels, changed only by explicit action by the user. Although the automatic conversion is as above, you may want to set the class of some variables in data frames to "string".

The advantages of the class string and its extensions include explicit control over conversion, better behavior for string matching, and the ability to have labelling information for data frames in string objects, without levels. The function levelsIndex makes explicit the old idea of getting the numeric index values from a factor or category. The call levelsIndex(x) returns an integer object with 1 wherever the element of x matched the first element of levels(x), and so on. The coercion of string objects to either character or integer is explicitly defined by as methods.

B.4 Old-Style Interface to C and Fortran

The current version of S includes the interfaces to C and Fortran subroutines from earlier versions. In older versions, the object code for these routines had to be loaded with S, either statically or dynamically. Static loading meant generating a separate executable program for S by running a C-language loader. Dynamic loading used various techniques (specialized for the operating system and hardware) to bind individual compiled object files (".o" files) to the running S process. In the current version, both static and dynamic loading are replaced by *dynamic linking*. If an S chapter has been set up to include some C or Fortran routines, then a shared object file including the compiled version of these routines will be automatically linked with the running S process, when the chapter is attached. So long as the owner of the library re-makes the shared object file when the C or Fortran code changes, no other action is needed. In Chapter 11, this mechanism is described; read that chapter if you want to understand the general concepts.

If the old-style library to be converted has C or Fortran code, copy the files needed to the new directory before running the `CHAPTER` command in that directory as described on page 448. By default, all such files in the directory will be included in the file, `S_LOCAL_FILES`, used to make the shared object library. If for some reason you don't want to include all the files, provide arguments to the `CHAPTER` command. See page 408 for details.

After creating the chapter, running the `make` command in the library will generate the linked object. You should remove any calls to `dyn.load` in the S functions of the library (if the library used old-style dynamic loading).

Old-style chapters generated an initialization C file, named `tools_i.c` in library `"tools"`, for example. You don't need or want this file in current S; *don't* copy it to the new chapter. You can call for any initialization you do need from an attach action; see page 409.

B.5 Old-Style Documentation

Earlier versions of S did not have documentation objects; instead, documentation was kept on text files, using a variant of `nroff`-style documentation. As part of the general conversion process, these files will be converted into documentation objects in the new chapter. The tools for viewing and editing the documentation are then available for the converted help files. Chapter 9 discusses documentation objects.

As with the rest of the old library, the recommended approach is to convert the entire library, using `convertOldLibrary`, and then work with the new chapter for further refinements. Sometimes devotees of `nroff/troff` may have inserted some code into the help files that does not fit well with the structure of current S documentation objects, or that simply exceeds the scope of the conversion utility. You will need to do some editing to replace this material, but less confusion will result if the editing is done on the documentation object, as opposed to editing the old-style file and redoing the conversion.

The documentation conversion is available as a separate computation, if converting a complete library is not sufficient for your needs. The function `convertOldDoc` converts old-style documentation for specified library and topics, and assigns a documentation object for each topic. Like the function `convertOldLibrary`, it takes as argument the old library from which to get the documentation, the new chapter to which objects should be assigned, and optionally a list of what to convert:

```
convertOldDoc(from, to, what)
```

In this case, `what` is a list of *files*, not objects. Old-style libraries stored documentation in directory `.Data/.Help` in the library directory; the argument `what` is expected to be a vector of the names of files within that directory.

The conversion process is heuristic, but usually works fairly well. Material that can't be interpreted or that does not conform to the structure of documentation objects will be copied essentially literally in the conversion. The displayed documentation will be ugly but no information will be lost. A later call to `dumpDoc` will allow you to edit the material (assuming you can figure out what it was meant to do). As a general rule of thumb, if the `nroff` command had to do with detailed control of spacing or other formatting, it's best just to leave it out. Otherwise, there may be some SGML equivalent to obtain a similar effect. The web site for the book has more pointers on this and other documentation issues.

Bibliography

[1] R. A. Becker, J. M. Chambers, and A. R. Wilks. *The New S Language.* Chapman and Hall, 1988.

[2] John Chambers, Mark Hansen, and Jonathan Mattingly. S and Java: Experimenting with Data Analysis on the Web *Bell Labs Memorandum, 1997.* (`http://cm.bell-labs.com/stat/doc/SJava.ps`)

[3] John M. Chambers and Trevor Hastie (eds). *Statistical Models in S.* Chapman and Hall, 1993.

[4] Mark Hansen and David A. James. A computing environment for spatial data analysis in the microelectronics industry. *BLTJ*, vol. 2 (1997), 114-129. (`http://www.lucent.com/bltj/winter_97/paper09`)

[5] Donald E. Knuth. *The Art of Computer Programming*, Volume 3: Sorting and Searching. Addison-Wesley, 1973.

[6] Andreas Krause and Melvin Olson. *The Basics of S and S-PLUS.* Springer-Verlag, 1997.

[7] Donald Lewine. *POSIX Programmer's Guide.* O'Reilly & Associates, Sebastopol, CA, 1991.

[8] Phil Spector. *An Introduction to S and S-Plus.* Duxbury Press, 1994.

[9] Duncan Temple Lang. *A Multi-Threaded Extension to a High Level Interactive Statistical Computing Environment.* PhD thesis, Univ. Cal. at Berkeley, 1997.
(`http://cm.bell-labs.com/stat/doc/multi-threaded-S.ps`)

[10] W. N. Venables and B. D. Ripley. *Modern Applied Statistics with S-Plus (2nd ed.).* Springer-Verlag, 1997.

Index